AF389470

Biocontrol of Plant Diseases: Antagonist Microorganisms, Biostimulants, Induced Resistance (IR)

Biocontrol of Plant Diseases: Antagonist Microorganisms, Biostimulants, Induced Resistance (IR)

Editors

Carlos Agustí-Brisach
Eugenio Llorens

Basel • Beijing • Wuhan • Barcelona • Belgrade • Novi Sad • Cluj • Manchester

Editors
Carlos Agustí-Brisach
University of Cordoba (UCO)
Cordoba
Spain

Eugenio Llorens
Universitat Jaume I de
Castellón (UJI)
Castellon
Spain

Editorial Office
MDPI
St. Alban-Anlage 66
4052 Basel, Switzerland

This is a reprint of articles from the Special Issue published online in the open access journal *Plants* (ISSN 2223-7747) (available at: https://www.mdpi.com/journal/plants/special_issues/plant_diseases_biocontrol).

For citation purposes, cite each article independently as indicated on the article page online and as indicated below:

Lastname, A.A.; Lastname, B.B. Article Title. *Journal Name* **Year**, *Volume Number*, Page Range.

ISBN 978-3-0365-8690-8 (Hbk)
ISBN 978-3-0365-8691-5 (PDF)
doi.org/10.3390/books978-3-0365-8691-5

Cover image courtesy of Carlos Agustí-Brisach

Contents

About the Editors

Carlos Agustí-Brisach

Carlos Agustí-Brisach is an Assistant Professor in Plant Pathology in the Department of Agronomy (Unit of Excellence 'María de Maeztu' 2020-24) at the University of Cordoba, Spain. He graduated with a BSc in Agricultural Engineering (2008), an MSc in Plant Protection (2010), and a PhD in Plant Pathology (2013) from the Polytechnic University of Valencia (Valencia, Spain). He served as a Postdoctoral Researcher at the University of Angers, France (2013-14); at Kearny Agricultural Research and Extension Centre, UC Davis, Fresno, USA (4 months; 2017); and at the University of Cordoba, Spain (2016-2021). He also served as a visiting Professor at the University of Catania (Sicily, Italy) in July 2022. In addition, in 2015, he was recruited by IDAI Nature S.L, a private company on Biostimulation for plant protection, where he coordinated the I+D+i Department until he joined the University of Cordoba in 2016. Currently, his main research lines are 'Etiology, epidemiology and control of wood diseases in Mediterranean woody crops'; and 'Biocontrol of diseases in woody plants by means of antagonistic microorganisms, bio-stimulants and resistance host inducers'. The research activity of the Dr. Agustí-Brisach is focused on applied plant pathology. He is the author of 70 papers in JCR scientific journals (Hi = 17), 2 book chapters, 36 outreach publications in national journals, and 95 contributions to national or international congresses.

Eugenio Llorens

Eugenio Llorens is a researcher in Plant Protection in the Department of Biology, Biochemistry and Natural Sciences at the University Jaume I, Spain. He graduated with a BSc in Agricultural Engineering (2007), an MSc in Molecular Biology (2009), and a PhD in Plant Science (2014). He has expanded his training with stays at the Valencian Institute of Agrarian Research (Valencia, Spain), at the Lake Alfred Citrus Research and Education Center (Florida, USA), at the National Institute of Agricultural and Food Research and Technology (Madrid, Spain); and in the department of Molecular Biology and Plant Ecology at the University of Tel Aviv (Israel). His research has focused on the search for new methods for pest control in order to reduce treatments with chemical products, looking for more sustainable alternatives. He is currently focused on the search for and characterization of beneficial microorganisms. His research has been published in more than 30 scientific articles and has been presented at more than 100 national and international congresses.

 *plants*

Editorial

Biocontrol of Plant Diseases by Means of Antagonist Microorganisms, Biostimulants and Induced Resistance as Alternatives to Chemicals

Eugenio Llorens [1,*] and Carlos Agustí-Brisach [2,*]

[1]   Departament of Biology, Biochemistry and Natural Sciences, Universitat Jaume I de Castellón (UJI), 12006 Castellón de la Plana, Spain

[2]   Department of Agronomy (DAUCO, Unit of Excellence María de Maeztu 2020-23), ETSIAM, University of Cordoba (UCO), 14071 Córdoba, Spain

*   Correspondence: ellorens@camn.uji.es (E.L.); cagusti@uco.es (C.A.-B.)

Citation: Llorens, E.; Agustí-Brisach, C. Biocontrol of Plant Diseases by Means of Antagonist Microorganisms, Biostimulants and Induced Resistance as Alternatives to Chemicals. *Plants* **2022**, *11*, 3521. https://doi.org/10.3390/plants11243521

Received: 16 November 2022
Accepted: 1 December 2022
Published: 14 December 2022

Publisher's Note: MDPI stays neutral with regard to jurisdictional claims in published maps and institutional affiliations.

Plant diseases are one of the biggest problems in conventional agriculture as they reduce both yield and crop value. Therefore, it is necessary to establish systems that protect against pests to maintain or improve yields while ensuring high food quality. The chemical pesticides used in conventional agriculture over the last 50 years contaminate the environment, including soil and water; may leave residues in food; and affect other beneficial organisms that are not targeted by treatment. Therefore, researchers are searching for natural and eco-friendly alternatives for crop protection that avoid the problems associated with chemical pesticides.

In recent decades, the natural compounds and beneficial microorganisms for use as biological control agents (BCAs) or plant biostimulators have attracted interest. These treatments have several advantages compared to classic pesticides; for example, they exhibit low toxicity, zero residue in foods, and their mode of action often allows their use to be preventive or curative. These advantages make them suitable for sustainable agriculture, for fulfilling new demands from the agro-food sector of the Euroregion, as well as for society in the frame of the European Green Deal.

The term biocontrol comprises a group of treatments that are based on natural compounds, extracts or microorganisms. These treatments have demonstrated effectiveness in protecting plants against both biotic and abiotic stresses. This protection can be achieved through several systems, depending on the BCA used, among which we highlight the use of antagonistic microorganisms and host resistance induction.

Antagonistic microorganisms can colonize the rhizosphere or the aerial parts of the plant without causing any damage to the plant, preventing colonization by other pathogenic microorganisms. These microorganisms, in addition to competing for the ecological niche, can release compounds into their environment with diverse functions. It has been demonstrated that some exudates are able to prevent the development of pathogens, providing partial protection to plants. On the other hand, some microorganisms also excrete molecules that improve plant growth and performance under stressful conditions, such as hormones and siderophores.

Plants also possess a response system that allows them to fight infection by pathogens. It has been shown that these responses can be enhanced through the application of certain beneficial compounds and microorganisms, resulting in a faster and stronger response. This enhanced response is often enough to reduce the infection below the economic threshold of damage. The activation of the plant's immune system against subsequent stress can be also divided in two types. On the one hand, there is direct activation of the defensive mechanisms, in which the hormonal signaling pathways (mainly jasmonic acid and salicylic acid) activate the expression of genes and other defense mechanisms involved in the response. On the other hand, the enhancement of resistance can be a "priming" response.

In this case, plants do not show a strong defensive response when the resistance inducer is applied, but primed plants show a faster and stronger defensive response upon exposure to stress.

This Special Issue highlights a selection of cutting-edge research on beneficial microorganisms, including cyanobacteria, actinomycetes, bacteria and fungi. Abdelaziz et al. [1] showed that foliar spraying applications of cyanobacteria were effective in relieving the toxic influences of *Fusarium oxysporum* on infected pepper plants. Elshafie and Camele [2], Carlucci et al. [3] and Diaz-Diaz et al. [4] demonstrated the effectiveness of actinobacterial strains (mainly *Streptomyces* spp.) in controlling soil-borne pathogens of tomato, fennel and bean crops, respectively. The effect of bacterial strains of *Bacillus velezensis* was evaluated, with successful results obtained against the nematode *Meloidogyne incognita* in cucumber plants by Asaturova et al. [5], and against several fungal diseases of grapevine by Hamaoka et al. [6]. López-Moral et al. [7] demonstrated the influence of several BCAs and biofertilizers on the effect of olive stem extracts on the viability of *Verticillium dahliae* conidia. The effectiveness of *Trichoderma* spp. strains as BCAs was also demonstrated against *Macrophomina phaseolina* in peanut crop by Martínez-Salgado et al. [8]. These articles reflect the potential of a broad diversity of microorganisms against a range of plant diseases.

Moreover, the selection of articles on biostimulation based on fertilizers, soil amendments or waste extracts provides a vision of how plants can be protected without the use of synthetic chemical pesticides. Likewise, Stavridou et al. [9] demonstrated that the use of biosolids for soil amendment had positive effects not only on plant health and protection, but also on the growth of non-pathogenic antagonistic microorganisms against *F. oxysporum* f. sp. *radicis-lycopersici* in the tomato rhizosphere. El Boumlasy et al. [10] showed the effectiveness of new natural substances, obtained from shrimp wastes using minimal processing, against fungi and oomycetes belonging to the genera *Alternaria*, *Colletotrichum*, *Fusarium*, *Penicillium*, *Plenodomus* and *Phytophthora*. A site-targeted copper-based biofertilizer was evaluated in combination with a commercial *Trichoderma atroviride* by Reis et al. [11], who reported that this combination may constitute a promising long-term approach to mitigating the impact of Botryosphaeria dieback in grapevine.

Finally, Liu et al. [12] reviewed the literature on the role of Cruciferae glucosinolates in resistance against diseases and pests. This review explores the mechanisms via which glucosinolates act as a defensive substance, participate in responses to biotic stress, and enhance plant tolerance to stress.

We are grateful to the co-authors who have contributed to this Special Issue and hope that the knowledge provided in the published papers will encourage the use of eco-friendly management strategies for protection against plant diseases.

Author Contributions: Conceptualization, E.L. and C.A.-B.; writing—original draft preparation, E.L. and C.A.-B.; writing—review and editing, E.L. and C.A.-B. All authors have read and agreed to the published version of the manuscript.

Funding: We acknowledge financial support of MICINN, the Spanish State Research Agency, through the Severo Ochoa and María de Maeztu Program for Centres and Units of Excellence in R&D (Ref. CEX2019-000968-M).

Data Availability Statement: Not applicable.

References

1. Abdelaziz, A.M.; Attia, M.S.; Salem, M.S.; Refaay, D.A.; Alhoqail, W.A.; Senousy, H.H. Cyanobacteria-Mediated Immune Responses in Pepper Plants against *Fusarium* Wilt. *Plants* **2022**, *11*, 2049. [CrossRef] [PubMed]
2. Elshafie, H.S.; Camele, I. Rhizospheric *Actinomycetes* Revealed Antifungal and Plant-Growth-Promoting Activities under Controlled Environment. *Plants* **2022**, *11*, 1872. [CrossRef] [PubMed]
3. Carlucci, A.; Raimondo, M.L.; Colucci, D.; Lops, F. *Streptomyces albidoflavus* Strain CARA17 as a Biocontrol Agent against Fungal Soil-Borne Pathogens of Fennel Plants. *Plants* **2022**, *11*, 1420. [CrossRef] [PubMed]

4. Díaz-Díaz, M.; Bernal-Cabrera, A.; Trapero, A.; Medina-Marrero, R.; Sifontes-Rodríguez, S.; Cupull-Santana, R.D.; García-Bernal, M.; Agustí-Brisach, C. Characterization of Actinobacterial Strains as Potential Biocontrol Agents against *Macrophomina phaseolina* and *Rhizoctonia solani*, the Main Soil-Borne Pathogens of *Phaseolus vulgaris* in Cuba. *Plants* **2022**, *11*, 645. [CrossRef] [PubMed]

5. Asaturova, A.M.; Bugaeva, L.N.; Homyak, A.I.; Slobodyanyuk, G.A.; Kashutina, E.V.; Yasyuk, L.V.; Sidorov, N.M.; Nadykta, V.D.; Garkovenko, A.V. *Bacillusvelezensis* Strains for Protecting Cucumber Plants from Root-Knot Nematode *Meloidogyne incognita* in a Greenhouse. *Plants* **2022**, *11*, 275. [CrossRef] [PubMed]

6. Hamaoka, K.; Aoki, Y.; Suzuki, S. Isolation and Characterization of Endophyte *Bacillus velezensis* KOF112 from Grapevine Shoot Xylem as Biological Control Agent for Fungal Diseases. *Plants* **2021**, *10*, 1815. [CrossRef] [PubMed]

7. López-Moral, A.; Agustí-Brisach, C.; Leiva-Egea, F.M.; Trapero, A. Influence of Cultivar and Biocontrol Treatments on the Effect of Olive Stem Extracts on the Viability of *Verticillium dahlia* Conidia. *Plants* **2022**, *11*, 554. [CrossRef] [PubMed]

8. Martínez-Salgado, S.J.; Andrade-Hoyos, P.; Lezama, C.P.; Rivera-Tapia, A.; Luna-Cruz, A.; Romero-Arenas, O. Biological Control of Charcoal Rot in Peanut Crop through Strains of *Trichoderma* spp., in Puebla, Mexico. *Plants* **2021**, *10*, 2630. [CrossRef] [PubMed]

9. Stavridou, E.; Giannakis, I.; Karamichali, I.; Kamou, N.N.; Lagiotis, G.; Madesis, P.; Emmanouil, C.; Kungolos, A.; Nianiou-Obeidat, I.; Lagopodi, A.L. Biosolid-Amended Soil Enhances Defense Responses in Tomato Based on Metagenomic Profile and Expression of Pathogenesis-Related Genes. *Plants* **2021**, *10*, 2789. [CrossRef] [PubMed]

10. El boumlasy, S.; La Spada, F.; Tuccitto, N.; Marletta, G.; Mínguez, C.L.; Meca, G.; Rovetto, E.I.; Pane, A.; Debdoubi, A.; Cacciola, S.O. Inhibitory Activity of Shrimp Waste Extracts on Fungal and Oomycete Plant Pathogens. *Plants* **2021**, *10*, 2452. [CrossRef] [PubMed]

11. Reis, P.; Gaspar, A.; Alves, A.; Fontaine, F.; Rego, C. Combining an HA + Cu (II) Site-Targeted Copper-Based Product with a Pruning Wound Protection Program to Prevent Infection with *Lasiodiplodia* spp. in Grapevine. *Plants* **2021**, *10*, 2376. [CrossRef] [PubMed]

12. Liu, Z.; Wang, H.; Xie, J.; Lv, J.; Zhang, G.; Hu, L.; Luo, S.; Li, L.; Yu, J. The Roles of Cruciferae Glucosinolates in Disease and Pest Resistance. *Plants* **2021**, *10*, 1097. [CrossRef]

Review

The Roles of Cruciferae Glucosinolates in Disease and Pest Resistance

Zeci Liu [1,2], Huiping Wang [2], Jianming Xie [2], Jian Lv [2], Guobin Zhang [2], Linli Hu [2], Shilei Luo [2], Lushan Li [2,3] and Jihua Yu [1,2,*]

1   Gansu Provincial Key Laboratory of Aridland Crop Science, Gansu Agricultural University, Lanzhou 730070, China; liuzc@gsau.edu.cn
2   College of Horticulture, Gansu Agriculture University, Lanzhou 730070, China; wanghp@st.gsau.edu.cn (H.W.); xiejianming@gsau.edu.cn (J.X.); lvjian@gsau.edu.cn (J.L.); zhanggb@gsau.edu.cn (G.Z.); hull@gsau.edu.cn (L.H.); luosl@gsau.edu.cn (S.L.); lils@st.gsau.edu.cn (L.L.)
3   Panzhihua Academy of Agricultural and Forestry Sciences, Panzhihua 617000, China
*   Correspondence: yujihua@gsau.edu.cn; Tel.: +86-931-763-2188

Abstract: With the expansion of the area under Cruciferae vegetable cultivation, and an increase in the incidence of natural threats such as pests and diseases globally, Cruciferae vegetable losses caused by pathogens, insects, and pests are on the rise. As one of the key metabolites produced by Cruciferae vegetables, glucosinolate (GLS) is not only an indicator of their quality but also controls infestation by numerous fungi, bacteria, aphids, and worms. Today, the safe and pollution-free production of vegetables is advocated globally, and environmentally friendly pest and disease control strategies, such as biological control, to minimize the adverse impacts of pathogen and insect pest stress on Cruciferae vegetables, have attracted the attention of researchers. This review explores the mechanisms via which GLS acts as a defensive substance, participates in responses to biotic stress, and enhances plant tolerance to the various stress factors. According to the current research status, future research directions are also proposed.

Keywords: *Brassicaceae*; glucosinolates; hydrolytic products; pathogen; insect resistance; secondary metabolites

Citation: Liu, Z.; Wang, H.; Xie, J.; Lv, J.; Zhang, G.; Hu, L.; Luo, S.; Li, L.; Yu, J. The Roles of Cruciferae Glucosinolates in Disease and Pest Resistance. *Plants* **2021**, *10*, 1097. https://doi.org/10.3390/plants10061097

Academic Editors: Carlos Agustí-Brisach and Eugenio Llorens

Received: 15 April 2021
Accepted: 21 May 2021
Published: 30 May 2021

Publisher's Note: MDPI stays neutral with regard to jurisdictional claims in published maps and institutional affiliations.

1. Introduction

Plants are exposed to complex and highly variable environmental conditions in the course of their growth and development, and are often at risk of death or even extinction under the influence of diverse biotic and abiotic stress factors [1–3]. To survive such challenges in their habitats and environments, plants have evolved numerous adaptive mechanisms, including the production of diverse metabolites, which exhibit obvious species specificity [4]. Depending on their structure and type, plant secondary metabolites are mainly divided into terpenoids, phenols, and nitrogen-containing compounds [5–7]. Numerous studies have shown that there are about 90,000–200,000 types of metabolites in plants, and they play essential roles in signal transduction, adaptive regulation, growth and development, and plant defense [8–10]. Since many of the secondary metabolites act as defenses, it is presumed that biological invasion played a primary role in the evolution of the compounds [11,12].

Among the secondary metabolites, glucosinolates (GLS) are a type of anion hydrophilic secondary metabolite containing nitrogen and sulfur; GLS are water-soluble and can easily be dissolved in ethanol, methanol, and acetone [13,14]. GLS are found in 16 species of dicotyledonous angiosperms, and their contents are relatively high in the Cruciferae, Cleomaceae, and Caricaceae, and especially in the genus *Brassica*, such as in *B. rapa* ssp. *pekinensis*, *B. oleracea*, *B. napus*, *B. juncea*, and *B. rapa*, as well as in *Arabidopsis thaliana* [13,15–18]. In addition, the GLS biosynthetic pathway has been extensively studied in the model plant *Arabidopsis*, and the regulatory genes have been comprehensively described. Such studies have sparked interest in the unconventional metabolites derived from amino acids, with a lot of research focusing on GLS in *Brassica* plants [19–21].

Since the first type of GLS was isolated from mustard seeds, the associated plant species and GLS degradation products have been gradually recognized. Currently, the structures of more than 200 types of GLS have been identified [13,17,18,22], with more than 15 detected in Cruciferae [23]. Naturally occurring GLS have a common chemical structure: the structures are generally composed of β-D-glucosinyl, a sulfide oxime group, and side-chain R groups (including alkyl, hydroxyalkyl, hydroxyalkenyl, alkenyl, methyl-sulfinylalkyl, methylsulfonylalkyl, methylthioalkyl, arylalkyl, and indolyl) derived from amino acids; furthermore, GLS are generally in the form of potassium or sodium salts [15]. Based on the amino-acid side chain R groups, GLS can be divided into three categories, including aliphatic GLS (side chains are mainly derived from methionine, alanine, valine, leucine, or isopropyl leucine), indole GLS (side chains mainly derived from tryptophan), and aromatic GLS (side chains mainly derived from phenylalanine or tyrosine) [19,24].

GLS are mainly found in plant seeds, roots, stems, and leaf vacuole cells, and are relatively stable in nature with no associated biological activity; conversely, glucosinase (also known as myrosinase), which is responsible for hydrolyzing glucose residues in the GLS core skeleton, is located in specific protein bodies [25–28]. In intact plants, the hydrolytic systems containing GLS and myrosinase are spatially isolated; however, when tissue is damaged, for example following infestation or mechanical injury, the two rapidly combine, which leads to the rapid formation of GLS hydrolytic products [29–33]. In addition, food processing techniques, such as chopping, juicing, chewing, cooking, high temperature treatment, and thawing, can also break down GLS [34]. The hydrolytic products of GLS breakdown include glucose and unstable sugar glycoside ligands, and the glycoside ligands are rearranged to form isothiocyanates, nitriles, oxazolidinethiones, thiocyanate, epithionitriles, and other products, which all exhibit a wide range of biological activity [35–38].

GLS and their degradation products influence the taste and flavor of cruciferous vegetables [39,40], and there were significant difference in GSL content among *Brassica* plants; the total GSL content in the freeze-dried samples ranged from 621.15–42,434.21 μmol kg^{-1}, with an average value of 14,050.97 μmol kg^{-1} [41]. The spicy taste in radish is caused primarily by volatile allyl, 3-butane, and 4-methyl thiocyanate (ITC). Furthermore, over the past few decades, it was established that some of the metabolite classes containing nitrogen and sulfur exhibit immunosuppressant and anticancer properties [16,35,42–45]. Sulforaphane, the degradation product of glucoraphanin, exhibits anticancer activity, can relieve neuropathic pain caused by chemotherapy, and has significant inhibitory effects against prostate, rectal, breast, pancreatic, and bladder cancers [46–52]. In contrast, a progoitrin degradation product, goitrin (5-vinyloxazolidine-2-thione), can cause goiter and abnormalities in the internal organs of animals [53].

In the field, the mustard oil bomb is a major defense mechanism deployed against insect herbivory [54–56], pathogen infection [57–61], and various abiotic stress factors (such as drought, low or high temperature, light, and salt stress) [62–70]. The findings of such studies have prompted research on the potential application of GLS extracts and metabolites in crop pest and disease control in recent years [8,71,72].

Cruciferous vegetables are the largest leafy vegetables in the world, and are widely cultivated globally, and the cultivated area is expanding year by year according to the statistics of Food and Agriculture Organization of the United Nations (FAO) (Figure 1). Cruciferous plants in cultivation are often affected by various fungi (*Plasuwdiophora brassicae*, *Fusarium oxysporum*, *Peronospora parasitica* (Pers), *Sclerotinia sclerotiorum*) [73–76], bacteria (*Xanthomonas campestris* pv. *campestris*, *Erwinia carotovora* pv. *carotovora* Dye, *Maculicola pseudomonas syringae*) [77–79], and viruses (*Turnip mosaic virus*) [80], which cause club root, Fusarium wilt, downy mildew, sclerotinose, black rot, soft rot, black spot, mosaic, etc. Furthermore, *Plutella xylostella*, aphids, and *Pieris rapae* seriously affect the growth and development of cruciferous plants, and greatly reduce the productivity of cruciferous vegetable farms [81–83].

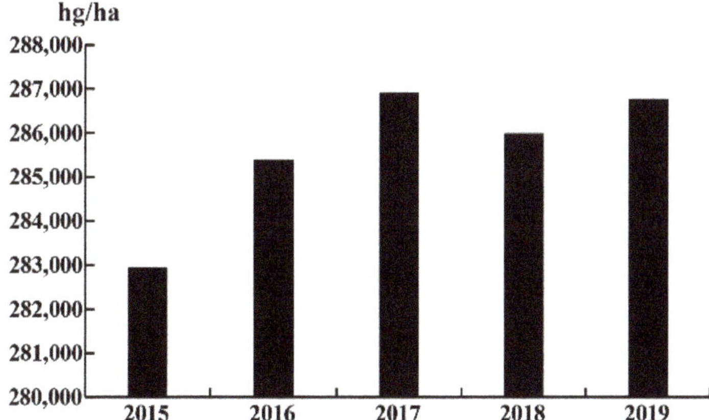

Figure 1. Changes in the cultivated area of cruciferous vegetables in recent years.

Currently, chemical control using pesticides is the primary method used to prevent and manage the diseases and insect pests that impair cruciferous vegetable cultivation and productivity. Despite the agricultural production industry currently advocating reducing pesticide application, the use of pesticides is still high according to the data from the FAO (Figure 2). Mass application of chemical pesticides not only increase production costs and deposit excessive pesticide residues on vegetables, but also pose threats to the environment and human health. Consequently, studies and comprehensive data on the potential of GLS derived from Cruciferae to control diseases are required. This review explores and summarizes the latest research on the disease and insect resistance function of GLS, in addition to the underlying resistance mechanisms, in cruciferous plants and in *Arabidopsis*. The present review could provide a theoretical basis for the application of GLS in disease and pest resistance, and the breeding of resistant cruciferous vegetables.

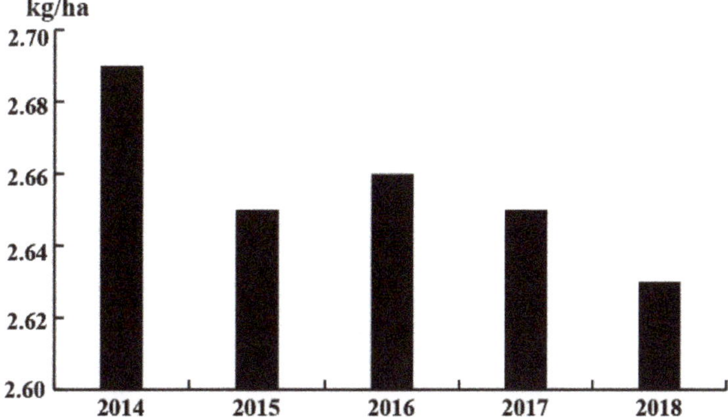

Figure 2. Pesticide application per hectare in recent years.

2. Defense Response of GLS to Fungal Diseases

The main diseases affecting agricultural production are fungal diseases, which have caused serious losses to the production of cruciferous vegetables. Consequently, investigating the potential effects of GLS extracts and enzymolysis products in resistance against fungal diseases, in addition to their underlying mechanisms of action, could facilitate efforts to improve agricultural productivity in cruciferous crops. Aqueous extracts containing ITC can inhibit the growth of *Alternaria brassicicola* in vitro by 50% [84]. Following exposure to allyl-ITC (Al-ITC), *A. brassicicola* exhibits a response similar to that observed during oxidative stress, based on the results of a study examining the transcriptomic responses of *Arabidopsis* challenged with *A. brassicicola*. In addition, ITCs play major roles in *Arabidopsis* resistance against *Plectosphaerella cucumerina*, *Botrytis cinerea*, *Fusarium oxysporum*, and *Peronospora parasitica* inoculation, demonstrated in a study using a GLS biosynthesis mutant gsm1-1 and wild-type *Arabidopsis* [85]. Humphry et al. (2010) investigated the accumulation of indole GLS in several insertion lines, and the results suggested that MYB51 participates in the regulation of genes critical for GLS metabolism, which also influences antifungal defense [86]. Meanwhile, S-deficiency in oilseed rape can reduce GLS biosynthesis, which negatively affects resistance against *Leptosphaeria maculans*, *B. cinerea*, and *Phytophthora brassicae* [57].

According to Giamoustaris and Mithen (2010), the levels of *Alternaria* infection are positively correlated with napus GLS contents, and there is no significant relationship between the GLS content and *Leptosphaeria maculans* resistance [87]. In addition, Robin et al. (2020) found that GLS biosynthetic genes were induced following a study carried out on two resistant and two susceptible cabbage in-bred lines after inoculation with two *Leptosphaeria maculans* isolates, and GLS (aliphatic and indolic GLS) accumulation was enhanced [88]. In a study investigating the indolyl-3-acetonitrile, 4-methoxyglucobrassicin, and indole GLS concentrations in *B. rapa* inoculated with *Albugo candida*, Pedras et al. (2008) observed increased levels of indole GLS in inoculated leaves when compared to the control leaves [89]. *B. rapa* indole GLS has also been reported to limit *Colletotrichum gloeosporioides* and *Colletotrichum orbiculare* infection [90], and tryptophan pathway genes involved in indole-GLS biosynthesis are upregulated in *F. oxysporum*-infected plants [91,92].

Based on dynamic transcriptomic analyses of *B. rapus* defense response to *S. sclerotiorum* post-inoculation, Zhao et al. (2004), Borge et al. (2015), and Wu et al. (2016) observed that not only the GLS content but also indolic GLS biosynthesis are associated with *S. sclerotiorum* resistance, and that *S. sclerotiorum* infection can induce GLS biosynthesis [8,93,94]. Unlike in the case of *S. sclerotiorum*, *B. cinerea* does not induce GLS biosynthesis [95]. A comparison of the disease symptoms of wild-type and transgenic *Arabidopsis* lines following inoculating with arbuscular mycorrhizal fungi (AMF), based on the production or enhancement of GLS levels, revealed a previously undocumented role of GLS biosynthesis in reducing AMF colonization [96].

After *Plasmodiophora brassicae* infection, the aliphatic, indolic, and aromatic GLS contents of susceptible *B. napus* exhibit increased accumulation; however, only aromatic GLS contents are significantly increased in resistant *Matthiola incana* L. [97]. The major aliphatic GLS, gluconapin, is significantly increased during secondary infection in *B. napus*, and exogenous jasmonic acid (JA) treatment induces aliphatic GLS in *B. napus* and aromatic GLS in *M. incana*. The expression of *BnMYB28.1*, which regulates the contents of aliphatic GLS in *B. napus*, is significantly increased following both treatment with exogenous JA and *P. brassicae* inoculation. Similarly, after *B. cinerea* infection, the genes involved in indole GLS biosynthesis are upregulated in the *Arabidopsis* UGT80A2 and UGT80B1 double mutant, and the upregulation was correlated with increased levels of JA and the upregulation of two marker genes (PDF1.2 and PR4) of the ERF branch of the JA signaling pathway [98].

3. Defense Responses of GLS to Bacterial Diseases

The bacteria that infect cruciferous plants are all rod-shaped bacteria, which can invade the host through stomata, hydathodes, and wounds, and then be retransmitted by running water, rain, insects, etc. Bacterial diseases in cruciferous have widespread occurrence, are highly destructive, and are challenging to control. Meanwhile, because the pathogens are different from the fungal diseases, the corresponding disease resistance mechanism of host and the GLS involved in resistance may be different. Several studies have demonstrated that GLS are involved in plant defense against a variety of bacterial diseases. Similar to the case in fungal disease infection, infection by *Burkholderia cepacia*, *Pseudomonas syringae*, and *Xanthomonas campestris* pv. *campestris* (*Xcc*) led to the upregulation of the GLS biosynthesis [99]. In addition, the introduction of CYP79 influenced *Arabidopsis* disease resistance by increasing the GLS synthesis, and overexpressing the CYP79D2 from cassava increased the accumulation of the aliphatic isopropyl and methylpropyl GLS, which also enhanced resistance against the soft-rot pathogen, *Erwinia carotovora*; however, overexpressing the sorghum CYP79A1 or CYP79A2 increased the accumulation of p-hydroxybenzyl and benzyl GLS, respectively [100].

Mishina et al. (2007) observed that the knockout of PAL1 increased leaf survival after *P. syringae* infection in an analysis conducted on *Arabidopsis* mutants and wild type plants, while Truman et al. (2007) and Aires et al. (2011) observed that indole GLS biosynthesis decreased after *P. syringae* infection [101–103]. Following the transcriptional and metabolic profiling of *A. thaliana* mutants, Clay et al. (2009) reported that the PEN2 and PEN3 genes are necessary for resistance to PtoDC3000 pathogens [104]. Furthermore, Geng et al. (2012) demonstrated that coronatine, a toxin produced by *P. syringae*, suppresses the salicylic acid (SA)-independent pathway, facilitating callose deposition by reducing the accumulation of an indole GLS upstream of the PEN2 myrosinase activity [105]. In addition, a positive correlation has been reported between total GLS content and *Xcc* disease severity, and *Xcc* infection enhanced GLS biosynthesis during the early infection period [106,107]. *Pectobacterium carotovorum* ssp. *carotovorum* infection in *B. rapa* can trigger the upregulation of the JA and ethylene (ET) biosynthesis genes in sr gene mutants and increase resistance capacity via GLS accumulation [108].

4. Defense Response of GLS to Pests

With global warming, the loss caused by pests is increasing. Meanwhile, because pests have migration ability, once the control is not effective, it will cause serious damage [109]. Hence, pest control has been a hot spot in agriculture. At present, pest control is mainly focused on chemical agents, but how *Brassicaceae* plants perceive and defend themselves from such threats remain poorly understood. Investigating the mechanisms via which *Brassicaceae* resist insect pests could facilitate efforts to improve crop productivity. Brown and Morra (1997) were the first to report that GLS-containing plants could control soil-borne plant pests [110]. Since then, numerous studies have demonstrated that GLS contents in tissues are positively correlated with damage caused by *Pieris rapae* and *Spodoptera littoralis* [111,112], but negatively correlated with the damage caused by slugs [86]. Furthermore, GLS accumulation induced by *Spodoptera exigua* required functional NPR1 and ETR genes [113].

In another study, the weights of *Trichoplusiani* and *Manduca sexta* on the TGG1 and TGG2 double myrosinase mutants were significantly higher than in wild-type *Arabidopsis* [27]. Similarly, *Mamestra brassicae* larvae gained less weight and exhibited stunted growth when fed on MINELESS (lacking myrosin cells) plants compared to when fed on wild-type plants, with the myrosinase activity in the wild-type seedlings reducing; however, the levels of indol-3-yl-methyl, 1-methoxy-indol-3-yl-methyl, and total GLS in both the wild-type and MINELESS seedlings increased [114]. Conversely, *M. brassicae* and *P. rapae* weighed more on the high-sinigrin concentration plants than in low-sinigrin concentration plants; however, their weights decreased in the high-sinigrin, high-glucoiberin, and high-glucobrassicin genotypes; furthermore, development time increased under high glucobrassicin concentrations [115].

By testing the GLS and phenolic concentrations trends in *Brassica nigra* (L.) Koch before and after herbivory by *Pratylenchus penetrans* Cobb and the larvae *Delia radicum* L., Van et al. (2005) observed that the total GLS levels were affected by herbivory by the two root feeders [116]. Besides, *Spodoptera litura Fabricius* was more affected by induced GLS responses than *Plutella xylostella* L. [117]. In addition, following a comparison of GLS levels and the expression profiles of GLS biosynthesis genes before and after *Plutella xylostella* infestation, Liu et al. (2016) observed a difference in the proportions of stereoisomers of hydroxylated aromatic GLS between G-type (pest-resistant) and P-type (pest-susceptible) *Barbara vulgaris* [56]. Using m/z 60 as a marker of Al-ITC formation from the sinigrin GLS, Van et al. (2012) analyzed the GLS profiles and volatile organic compound emissions in five *Brassicaceae* species before and after artificial injury or infestation by cabbage root fly larvae (*D. radicum*). According to the results, m/z 60 in *B. nigra*, *B. juncea*, and *B. napus* was primarily emitted directly after artificial injury or root fly infestation, sulfide and methanethiol emissions from *B. nigra* and *B. juncea* increased after infestation, and *B. oleracea* and *Brassica carinata* exhibited increases in fig m/z 60 emissions following larval damage [118].

Long-term feeding on GLS-free *Brassicaceae* diets hardly affects *P. xylostella* oviposition preference and larvae survival; thus, high GLS content varieties are likely to be more susceptible to damage by *P. xylostella* than lower GLS content varieties [119]. Similarly, Chen et al. (2020) generated single or double mutant gss1 and gss2 lines using the CRISPR/Cas9 system and analyzed their resistance to *P. xylostella* [120]. According to the results of the bioassays, when fed on their usual artificial diet, there were significant reductions in egg hatching rates and final larval survival rate of the single mutant gss2 lines when compared with the original strain or mutant gss1 lines, and the absence of GSS1 or GSS2 reduced the survival rate of *P. xylostella* and prolonged the duration of the larval stage. In addition, feeding by *Spodoptera littoralis*, *Pieris brassicae*, and *P. rapae* led to upregulation of the aliphatic GLS pathway [121–123], and the GLS contents were negatively correlated with *P. brassicae* damage. Furthermore, methyl jasmonate (MeJA) can enhance resistance to *P. brassicae* by inducing GLS accumulation [124,125].

5. Defense Response of GLS to Insects and Aphids

Insects and aphid not only have a wide range of species and rapid reproduction, but also can cause wounds to the plant when feeding, leading to the invasion of pathogenic bacteria, and then cause secondary damage. Moreover, the GLS synthesis and response mechanisms following insect and aphid herbivory are qualitatively and quantitatively different [126]. Agerbirk et al. (2001) observed no correlation between *B. vulgaris* ssp. *arcuata* GLS content and resistance against *Phyllotreta nemorum* [127], while Kroymann et al. (2003) observed a positive correlation between GLS content and damage caused by *Psylliodes chrysocephala* [128]. According to Ulmer et al (2006), total GLS levels did not influence *Ceutorhynchus obstrictus* larval growth or development; however, high levels of specific GLS, such as p-hydroxybenzyl and 3-butenyl GLS, were associated with increased development time or reduced weight [129]. After *Brevicoryne brassicae* herbivory, Myrosinase binding protein (MBP), myrosinase associated protein (MyAP), and myrosinase transcripts, and the synthesis of indolyl and aliphatic GLS, particularly 3-hydroxypropyl and ITC, are induced [103,130,131].

By comparing the larval instar weights and mortality of cabbage stem flea beetle (*P. chrysocephala*) larvae, after feeding on different species, Döring et al. (2020) observed that aliphatic GLS contents increased in the infested turnip rape, and aliphatic and benzenic GLS decreased in infested Indian rape [132]. Although larval weight was not correlated with total GLS, it was positively correlated with progoitrin and 4-hydroxyglucobrassicin contents. Furthermore, decreasing the side chain length of aliphatic GLS and the degree of hydroxylation of butenyl GLS could increase the extent of feeding by adult flea beetles [87].

Numerous intermediate synthetic genes participate in GLS resistance to insects and aphids. For instance, Mewis et al. (2006) observed that GLS accumulation caused by *B. brassicae* and *Myzus persicae* required functional NPR1 and ETR1 genes [113]. After *Myzus persicae* feeding and aphid saliva treatment, a set of O-methyltransferases involved in the synthesis of aphid-repellent GLS were significantly up-regulated based on qRT-PCR analyses of 78 genes. However, ITC production was not correlated with these gene expression level, suggesting that aphid salivary components trigger a defense response in *Arabidopsis* that is independent of the aphid-deterrent GLS [133]. In addition, aphid attack could increase indolyl GLS concentrations three-fold [134]. Using a combination of QTL fine-mapping and microarray-based transcript profiling methods, CYP81F2 was revealed to facilitate defense against *B. brassicae* but not resistance against herbivory by larvae from four lepidopteran species [135]. By comparing the survival of the *Bemisia tabaci* MEAM1 and *B. tabaci* MED following exposure to sinigrin and myrosinase, Hu et al. (2020) reported that exposure to the toxic hydrolysates of GLS hydrolysates and myrosinase is greater for MED than for MEAM1 [136].

6. Conclusions and Future Research Outlook

Cruciferous vegetables are the most important leafy vegetables; however, the cultivation of cruciferous plants is affected by various fungi, bacteria, aphids, and other pest insects. At present, the prevention and control of these diseases and insect pests mainly focus on chemical agents, and the dosage of the chemical pesticides is also increasing, which not only leads to excessive pesticide residues on cruciferous plants, causing great damage to the environment, but also threatens people's health. Understanding how host-plant characteristics influence the physiological and behavioral responses is essential for the development of resistant cruciferous germplasms. A large number of studies have shown that GLS, esters, and flavonoids are closely related to Cruciferae disease resistance [137,138]. As an important secondary metabolite in cruciferous vegetables, GLS are closely related to biotic and abiotic stresses. Numerous studies have demonstrated a positive relationship between GLS content and disease and insect resistance [57,88,99,106,107,111,112,116,128,134] (Table 1). Consequently, in future cruciferous vegetable breeding activities, varieties with high GLS contents can be selected appropriately to improve plant disease resistance and reduce pesticide use. The degradation products (isothiocyanate and thiocyanate) of GLS are involved in the resistance to a variety of fungi, bacteria, insects, and soil-borne pests [8,71,72]; the aqueous extracts of cruciferous leaves also contain ITC, which can restrict the growth of a variety of fungi, bacteria, and pests [84,85,110]. Moreover, the resistance of these degradation products to pests and diseases is a broad-spectrum resistance, and thus can be used to develop botanical pesticides.

Pest invasion and disease infestation can increase GLS, especially indole GLS, in cruciferous plants [8,9,89–95,105,117,121–123,133,134]. In the case of rapeseed and other species that have low GLS, molecular biology techniques can be used to increase indole GLS production, which could improve resistance to diseases and insect, without increasing total GLS synthesis. Similar to other secondary metabolites, GLS synthesis is regulated by plant hormones. By controlling the amount of sulfur fertilizer applied and exogenous plant hormone treatments, such as JA, ET, MeJA, and SA, GLS synthesis can be modulated, and, in turn, disease resistance [57,105,125,126,139–142]. In addition, pathogen infection and insect herbivory can trigger the upregulation of the JA and ET biosynthesis genes, and increase defensive capacity via GLS accumulation [108]. Therefore, in subsequent cruciferous vegetable production activities, appropriate plant hormones could be sprayed as a novel pest management strategy to improve their stress resistance and minimize pesticide use (Figure 3).

Table 1. Correlation of the GLS components and their metabolites in corresponding pathogen, pest, and insect resistance.

Component	Species	Names	Correlation	References
ITC; Allyl-ITC	Fungal	*Alternaria brassicicola*	positive	[84–86]
		Plectosphaerella cucumerina	positive	[86]
		Botrytis cinerea	positive	[86]
		Fusarium oxysporum	positive	[86]
		Peronospora parasitica	positive	[86]

Table 1. *Cont.*

Component	Species	Names	Correlation	References
Total GLS	Fungal	*Alternaria brassicicola*	positive	[92]
		Leptosphaeria maculans	No; positive	[92,93]
		Sclerotinia sclerotiorum	positive	[7,98,99]
		Arbuscular mycorrhizal fungi	positive	[101]
	Bacteria	*Burkholderia cepacia*	positive	[104]
		Pseudomonas syringae	positive	[104]
		Xanthomonas campestris	positive	[104,111,112]
		Pectobacterium carotovorum	positive	[113]
	Pest	*Pieris rapae*	positive	[114]
		Spodoptera littoralis	positive	[115]
		Slug	negative	[93]
		Spodoptera exigua	positive	[116]
		Trichoplusia ni	positive	[117]
		Manduca sexta	positive	[117]
		Mamestra brassicae	positive	[118]
		Pratylenchus penetrans	positive	[118]
		Delia radicum L.	positive	[118]
		Spodoptera litura Fabricius	positive	[121]
		Plutella xylostella L.	positive	[121,123,124]
		Pieris brassicae	positive	[128,129]
	Insect	*Phyllotreta nemorum*	No	[131]
		Psylliodes chrysocephala	positive	[132]
		Ceutorhynchus obstrictus	No	[133]
Indole GLS	Fungal	*Albugo candida*	positive	[94]
		Colletotrichum gloeosporioides	Positive	[95]
		Colletotrichum orbiculare	Positive	[95]
		Fusarium oxysporum	positive	[96,97]
		Plasmodiophora brassicae	positive	[102]
	Bacteria	*Pseudomonas syringae*	positive	[106–108]
Aliphatic GLS	Fungal	*Plasmodiophora brassicae*	positive	[102]
	Pest	*Spodoptera littoralis*	positive	[125]
		Pieris brassicae	positive	[125]
		Pieris rapae	positive	[126]
	Insect	*Psylliodes chrysocephala*	positive	[137]
Aromatic GLS	Fungal	*Plasmodiophora brassicae*	positive	[102]
	Pest	*Plutella xylostella* L.	positive	[60]
Benzenic GLS	Insect	*Psylliodes chrysocephala*	positive	[136]
Indolyl-3-acetonitrile, 4-methoxyglucobrassicin,	Fungal	*Albugo candida*	positive	[94]
Aliphatic isopropyl; methylpropyl GLS	Bacteria	*Erwinia carotovora*	positive	[105]

Table 1. *Cont.*

Component	Species	Names	Correlation	References
Indol-3-yl-methyl; 1-methoxy-indol-3-yl-methyl	Pest	*Mamestra brassicae*	positive	[118]
P-hydroxybenzyl; 3-butenyl	Insect	*Ceutorhynchus obstrictus*	positive	[133]
Sinigrin	Pest	*Pieris rapae*	negative	[119]
Glucobrassicin	Pest	*Pieris rapae*	positive	[119]

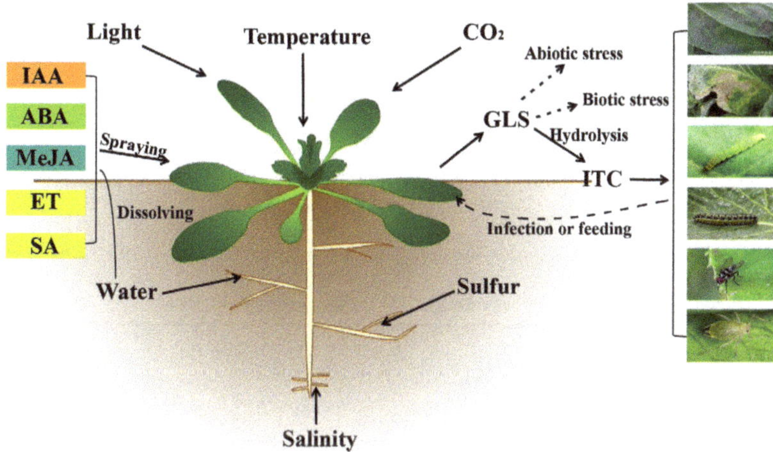

Figure 3. Factors affecting the synthesis of glucosinolate in cruciferous plants.

Some studies have demonstrated that the different stereoisomer structures of hydroxylated aromatic GLS is one of the important factors influencing the varying disease resistance levels between non-cultivars and resistant cultivars [56]. Consequently, by determining and analyzing the GLS responsible for resistance in tolerant materials, chemical synthesis or biotechnology tools can be used to mass-produce the corresponding GLS for widespread application. The present review on the GLS responsible for disease and pest resistance in cruciferous vegetables and their underlying mechanisms could not only offers insights on how cruciferous plants could respond to increased biotic stress in the future but could also facilitate the development of novel disease and pest-resistant plants and the development of safe and high-yield cruciferous vegetable germplasms globally.

Despite there being many research studies on cruciferous plant resistance, which indicated the invasion of the diseases, insects and pests will lead to the synthetic form of the hormones and GLS, and total GLS, especially the indole GLS content, has a positive correlation with cruciferous resistance, but only a few studies have identified the specific resistance due to GLS and other metabolites. This may be related to the determination methods of GLS composition. At present, the conventional determination methods of GLS are HPLC (high performance liquid chromatography) and HPLC-MS; these two methods are not only expensive and difficult, but also often lack some standard samples. So, efficient and accurate GLS determination methods need to be improved or developed in subsequent research. Moreover, the synthesis mechanism and corresponding intermediate pathway of GLS are mainly focused on the model plant *A. thaliana*; studies on other species are thus few. The genomes and cultivation of most cruciferous vegetable patterns are different from *A. thaliana*. Compared with *Arabidopsis*, the genomes and cultivation patterns of Cruciferous plants are more complex. Although the genomes of many cruciferous plants have been sequenced, the genomes of the majority of species are still unknown, which has limited the study on the anabolism of GLS and other disease-resistance-related substances. With the reduction of the cost of genome sequencing, transcriptome sequencing, and omics analysis, it is believed that people will have a new understanding of the mechanism of GLS against diseases and insects.

Author Contributions: Z.L., H.W., L.H. and J.Y. designed the research; S.L. and J.L. performed the data analysis and prepared the manuscript; J.X., G.Z. and L.L. revised the manuscript. All authors have read and agreed to the published version of the manuscript.

Funding: This research was funded by Gansu Provincial Key Laboratory of Aridland Crop Science, Gansu Agricultural University, grant number GSCS-2020-11, Natural Science Foundation of Gansu Province, grant number 20JR10RA537, Gansu Agricultural University doctoral Research Initiation fund, grant number GAU-KYQD-2019-30, and Special Project of Central Government Guiding Local Science and Technology Development, grant number ZCYD-2020-5.

Institutional Review Board Statement: Not applicable.

Informed Consent Statement: Not applicable.

Conflicts of Interest: The authors declare no conflict of interest.

References

1. Ahmad, M.; Ali, Q.; Hafeez, M.M.; Malik, A. Improvement for Biotic and Abiotic Stress Tolerance in Crop Plants. *Biol. Clin. Sci. Res. J.* **2021**, *2021*, e004.
2. Kliebenstein, D.J.; Osbourn, A. Making New Molecules-evolution of Pathways for Novel Metabolites in Plants. *Curr. Opin. Plant Biol.* **2012**, *15*, 415–423. [CrossRef] [PubMed]
3. Rejeb, I.B.; Pastor, V.; Mauch-Mani, B. Plant Responses to Simultaneous Biotic and Abiotic Stress: Molecular Mechanisms. *Plants* **2014**, *3*, 458–475. [CrossRef] [PubMed]
4. Pastorczyk, M.; Bednarek, P. The Function of Glucosinolates and Related Metabolites in Plant Innate Immunity. *Adv. Bot. Res.* **2016**, *80*, 171–198.
5. Moller, B.L. Functional Diversifications of Cyanogenic Glucosides. *Curr. Opin. Plant Biol.* **2010**, *13*, 337–346. [CrossRef] [PubMed]
6. Wu, J.; Baldwin, I.T. New Insights into Plant Responses to the Attack from Insect Herbivores. *Annu. Rev. Genet.* **2010**, *44*, 1–24. [CrossRef] [PubMed]
7. Bouranis, D.L.; Malagoli, M.; Avice, J.C.; Bloem, E. Advances in Plant Sulfur Research. *Plants* **2020**, *9*, 256. [CrossRef] [PubMed]
8. Borges, A.; Abreu, A.C.; Ferreira, C.; Saavedra, M.J.; Simoes, L.C.; Simoes, M. Antibacterial Activity and Mode of Action of Selected Glucosinolate Hydrolysis Products Against Bacterial Pathogens. *J. Food Sci. Technol.* **2015**, *52*, 4737–4748. [CrossRef]
9. Jeandroz, S.; Lamotte, O. Editorial: Plant Responses to Biotic and Abiotic Stresses: Lessons from Cell Signaling. *Front. Plant Sci.* **2017**, *8*, 1772. [CrossRef]
10. Gupta, A.; Hisano, H.; Hojo, Y.; Matsuura, T.; Ikeda, Y.; Mori, I.C.; Senthil-Kumar, M. Global Profiling of Phytohormone Dynamics during Combined Drought and Pathogen Stress in Arabidopsis Thaliana Reveals ABA and JA as Major Regulators. *Sci. Rep.* **2017**, *7*, 4017. [CrossRef] [PubMed]
11. Chisholm, S.T.; Coaker, G.; Day, B.; Staskawicz, B.J. Host-microbe Interactions: Shaping the Evolution of the Plant Immune Response. *Cell* **2006**, *124*, 803–814. [CrossRef] [PubMed]
12. Beran, F.; Kollner, T.G.; Gershenzon, J.; Tholl, D. Chemical Convergence between Plants and Insects: Biosynthetic Origins and Functions of Common Secondary Metabolites. *New Phytol.* **2019**, *223*, 52–67. [CrossRef]
13. Ishida, M.; Hara, M.; Fukino, N.; Kakizaki, T.; Morimitsu, Y. Glucosinolate Metabolism, Functionality and Breeding for the Improvement of Brassicaceae Vegetables. *Breed Sci.* **2014**, *64*, 48–59. [CrossRef] [PubMed]
14. Blazevic, I.; Montaut, S.; Burcul, F.; Olsen, C.E.; Burow, M.; Rollin, P.; Agerbirk, N. Glucosinolate Structural Diversity, Identification, Chemical Synthesis and Metabolism in Plants. *Phytochemistry* **2020**, *169*, 112100. [CrossRef] [PubMed]
15. Fahey, J.W.; Zalcmann, A.T.; Talalay, P. The Chemical Diversity and Distribution of Glucosinolates and Isothiocyanates among Plants. *Phytochemistry* **2001**, *56*, 5–51. [CrossRef]
16. Halkier, B.A.; Gershenzon, J. Biology and Biochemistry of Glucosinolates. *Annu. Rev. Plant Biol.* **2006**, *57*, 303–333. [CrossRef]
17. Clarke, D.B. Glucosinolates, Structures and Analysis in Food. *Anal. Methods* **2010**, *2*, 4. [CrossRef]
18. Agerbirk, N.; Olsen, C.E. Glucosinolate Structures in Evolution. *Phytochemistry* **2012**, *77*, 16–45. [CrossRef] [PubMed]
19. Sonderby, I.E.; Geu-Flores, F.; Halkier, B.A. Biosynthesis of Glucosinolates—Gene Discovery and Beyond. *Trends Plant Sci.* **2010**, *15*, 283–290. [CrossRef]
20. Li, Z.; Liu, Y.; Li, L.; Fang, Z.; Yang, L.; Zhuang, M.; Zhang, Y.; Lv, H. Transcriptome Reveals the Gene Expression Patterns of Sulforaphane Metabolism in Broccoli Florets. *PLoS ONE* **2019**, *14*, e0213902. [CrossRef]
21. Andini, S.; Dekker, P.; Gruppen, H.; Araya-Cloutier, C.; Vincken, J.P. Modulation of Glucosinolate Composition in Brassicaceae Seeds by Germination and Fungal Elicitation. *J. Agric. Food Chem.* **2019**, *67*, 12770–12779. [CrossRef] [PubMed]

22. Frerigmann, H.; Bottcher, C.; Baatout, D.; Gigolashvili, T. Glucosinolates are Produced in Trichomes of Arabidopsis Thaliana. *Front. Plant Sci.* **2012**, *3*, 242. [CrossRef] [PubMed]
23. Hwang, I.M.; Park, B.; Dang, Y.M.; Kim, S.Y.; Seo, H.Y. Simultaneous Direct Determination of 15 Glucosinolates in Eight Brassica Species by UHPLC-Q-Orbitrap-MS. *Food Chem.* **2019**, *282*, 127–133. [CrossRef]
24. Bekaert, M.; Edger, P.P.; Hudson, C.M.; Pires, J.C.; Conant, G.C. Metabolic and Evolutionary Costs of Herbivory Defense: Systems Biology of Glucosinolate Synthesis. *New Phytol.* **2012**, *196*, 596–605. [CrossRef] [PubMed]
25. Kliebenstein, D.J.; Kroymann, J.; Mitchell-Olds, T. The Glucosinolate-myrosinase System in an Ecological and Evolutionary Context. *Curr. Opin. Plant Biol.* **2005**, *8*, 264–271. [CrossRef] [PubMed]
26. Hopkins, R.J.; van Dam, N.M.; van Loon, J.J. Role of Glucosinolates in Insect-plant Relationships and Multitrophic Interactions. *Annu. Rev. Entomol.* **2009**, *54*, 57–83. [CrossRef]
27. Barth, C.; Jander, G. Arabidopsis Myrosinases TGG1 and TGG2 have Redundant Function in Glucosinolate Breakdown and Insect Defense. *Plant J.* **2006**, *46*, 549–562. [CrossRef]
28. Burow, M.; Halkier, B.A. How does a Plant Orchestrate Defense in Time and Space? Using Glucosinolates in Arabidopsis as Case Study. *Curr. Opin. Plant Biol.* **2017**, *38*, 142–147. [CrossRef]
29. Koroleva, O.A.; Davies, A.; Deeken, R.; Thorpe, M.R.; Tomos, A.D.; Hedrich, R. Identification of a New Glucosinolate-Rich Cell Type in Arabidopsis Flower Stalk. *Plant Physiol.* **2000**, *124*, 599–608. [CrossRef] [PubMed]
30. Husebye, H.; Chadchawan, S.; Winge, P.; Thangstad, O.P.; Bones, A.M. Guard Cell- and Phloem Idioblast-specific Expression of Thioglucoside Glucohydrolase 1 (myrosinase) in Arabidopsis. *Plant Physiol.* **2002**, *128*, 1180–1188. [CrossRef]
31. Thangstad, O.P.; Gilde, B.; Chadchawan, S.; Seem, M.; Bones, A.M. Cell Specific, Cross-species Expression of Myrosinases in Brassica Napus, Arabidopsis Thaliana and Nicotiana Tabacum. *Plant Mol. Biol.* **2004**, *54*, 597–611. [CrossRef] [PubMed]
32. van Dam, N.M.; Tytgat, T.O.G.; Kirkegaard, J.A. Root and Shoot Glucosinolates: A Comparison of Their Diversity, Function and Interactions in Natural and Managed Ecosystems. *Phytochem. Rev.* **2008**, *8*, 171–186. [CrossRef]
33. Bjorkman, M.; Klingen, I.; Birch, A.N.; Bones, A.M.; Bruce, T.J.; Johansen, T.J.; Meadow, R.; Molmann, J.; Seljasen, R.; Smart, L.E.; et al. Phytochemicals of Brassicaceae in Plant Protection and Human Health—Influences of Climate, Environment and Agronomic Practice. *Phytochemistry* **2011**, *72*, 538–556. [CrossRef]
34. Blažević, I.; Radonić, A.; Mastelić, J.; Zekić, M.; Skočibušić, M.; Maravić, A. Glucosinolates, Glycosidically Bound Volatiles and Antimicrobial Activity of Aurinia sinuata (Brassicaceae). *Food Chem.* **2010**, *121*, 1020–1028. [CrossRef]
35. Bones, A.M.; Rossiter, J.T. The Enzymic and Chemically Induced Decomposition of Glucosinolates. *Phytochemistry* **2006**, *67*, 1053–1067. [CrossRef] [PubMed]
36. Zhang, Z.; Ober, J.A.; Kliebenstein, D.J. The Gene Controlling the Quantitative Trait Locus EPITHIOSPECIFIER MODIFIER1 Alters Glucosinolate Hydrolysis and Insect Resistance in Arabidopsis. *Plant Cell* **2006**, *18*, 1524–1536. [CrossRef]
37. Burow, M.; Bergner, A.; Gershenzon, J.; Wittstock, U. Glucosinolate Hydrolysis in Lepidium sativum—Identification of the Thiocyanate-forming Protein. *Plant Mol. Biol.* **2007**, *63*, 49–61. [CrossRef] [PubMed]
38. Esteve, M. Mechanisms Underlying Biological Effects of Cruciferous Glucosinolate-Derived Isothiocyanates/Indoles: A Focus on Metabolic Syndrome. *Front. Nutr.* **2020**, *7*, 111. [CrossRef]
39. Bongoni, R.; Verkerk, R.; Steenbekkers, B.; Dekker, M.; Stieger, M. Evaluation of Different Cooking Conditions on Broccoli (*Brassica oleracea* var. *italica*) to Improve the Nutritional Value and Consumer Acceptance. *Plant Foods Hum. Nutr.* **2014**, *69*, 228–234. [CrossRef]
40. Novotny, C.; Schulzova, V.; Krmela, A.; Hajslova, J.; Svobodova, K.; Koudela, M. Ascorbic Acid and Glucosinolate Levels in New Czech Cabbage Cultivars: Effect of Production System and Fungal Infection. *Molecules* **2018**, *23*, 8. [CrossRef]
41. Rhee, J.H.; Choi, S.; Lee, J.E.; Hur, O.S.; Assefa, A.D. Glucosinolate Content in *Brassica* Genetic Resources and Their Distribution Pattern within and between Inner, Middle, and Outer Leaves. *Plants* **2020**, *9*, 1421. [CrossRef]
42. Keck, A.S.; Finley, J.W. Cruciferous Vegetables: Cancer Protective Mechanisms of Glucosinolate Hydrolysis Products and Selenium. *Integr. Cancer Ther.* **2004**, *3*, 5–12. [CrossRef]
43. Dinkova-Kostova, A.T.; Kostov, R.V. Glucosinolates and Isothiocyanates in Health and Disease. *Trends Mol. Med.* **2012**, *18*, 337–347. [CrossRef] [PubMed]
44. Lippmann, D.; Lehmann, C.; Florian, S.; Barknowitz, G.; Haack, M.; Mewis, I.; Wiesner, M.; Schreiner, M.; Glatt, H.; Brigelius-Flohe, R.; et al. Glucosinolates from Pak Choi and Broccoli Induce Enzymes and Inhibit Inflammation and Colon Cancer Differently. *Food Funct.* **2014**, *5*, 1073–1081. [CrossRef] [PubMed]
45. Lee, Y.R.; Chen, M.; Lee, J.D.; Zhang, J.; Lin, S.Y.; Fu, T.M.; Chen, H.; Ishikawa, T.; Chiang, S.Y.; Katon, J.; et al. Reactivation of PTEN Tumor Suppressor for Cancer Treatment through Inhibition of a MYC-WWP1 Inhibitory Pathway. *Science* **2019**, *364*, 6441. [CrossRef] [PubMed]
46. Liu, B.; Mao, Q.; Cao, M.; Xie, L. Cruciferous Vegetables Intake and Risk of Prostate Cancer: A Meta-analysis. *Int. J. Urol.* **2012**, *19*, 134–141. [CrossRef] [PubMed]
47. Wu, Q.J.; Yang, Y.; Vogtmann, E.; Wang, J.; Han, L.H.; Li, H.L.; Xiang, Y.B. Cruciferous Vegetables Intake and the Risk of Colorectal Cancer: A Meta-analysis of Observational Studies. *Ann. Oncol.* **2013**, *24*, 1079–1087. [CrossRef] [PubMed]
48. Liu, X.; Lv, K. Cruciferous Vegetables Intake is Inversely Associated with Risk of Breast Cancer: A Meta-analysis. *Breast* **2013**, *22*, 309–313. [CrossRef]

49. Lucarini, E.; Micheli, L.; Trallori, E.; Citi, V.; Martelli, A.; Testai, L.; de Nicola, G.R.; Iori, R.; Calderone, V.; Ghelardini, C.; et al. Effect of Glucoraphanin and Sulforaphane against Chemotherapy-induced Neuropathic Pain: Kv7 Potassium Channels Modulation by H2S Release *In Vivo*. *Phytother. Res.* **2018**, *32*, 2226–2234. [CrossRef] [PubMed]

50. Abbaoui, B.; Lucas, C.R.; Riedl, K.M.; Clinton, S.K.; Mortazavi, A. Cruciferous Vegetables, Isothiocyanates, and Bladder Cancer Prevention. *Mol. Nutr. Food Res.* **2018**, *62*, e1800079. [CrossRef] [PubMed]

51. Kellingray, L.; Le Gall, G.; Doleman, J.F.; Narbad, A.; Mithen, R.F. Effects of In Vitro Metabolism of a Broccoli Leachate, Glucosinolates and S-methylcysteine Sulphoxide on the Human Faecal Microbiome. *Eur. J. Nutr.* **2020**, 1–14.

52. Maina, S.; Misinzo, G.; Bakari, G.; Kim, H.Y. Human, Animal and Plant Health Benefits of Glucosinolates and Strategies for Enhanced Bioactivity: A Systematic Review. *Molecules* **2020**, *25*, 16. [CrossRef] [PubMed]

53. Dinkova-Kostova, A.T.; Talalay, P. Persuasive Evidence that Quinone Reductase Type 1 (DT diaphorase) Protects Cells against the Toxicity of Electrophiles and Reactive Forms of Oxygen. *Free Radic. Biol. Med.* **2000**, *29*, 231–240. [CrossRef]

54. Rask, L.; Andréasson, E.; Ekbom, B.; Eriksson, S.; Bo, P.; Meijer, J. Myrosinase: Gene Family Evolution and Herbivore Defense in Brassicaceae. *Plant Mol. Biol.* **2000**, *42*, 93. [CrossRef]

55. Kos, M.; Houshyani, B.; Wietsma, R.; Kabouw, P.; Vet, L.; Loon, J.; Dicke, M. Effects of Glucosinolates on a Generalist and Specialist Leaf-chewing Herbivore and an Associated Parasitoid. *Phytochemistry* **2012**, *77*, 162–170. [CrossRef] [PubMed]

56. Tongjin, L.; Xiaohui, Z.; Haohui, Y.; Niels, A.; Yang, Q. Aromatic Glucosinolate Biosynthesis Pathway in Barbarea Vulgaris and Its Response to Plutella Xylostella Infestation. *Front. Plant Sci.* **2016**, *7*, 83.

57. Dubuis, P.-H.; Marazzi, C.; Städler, E.; Mauch, F. Sulphur Deficiency Causes a Reduction in Antimicrobial Potential and Leads to Increased Disease Susceptibility of Oilseed Rape. *J. Phytopathol.* **2005**, *153*, 27–36. [CrossRef]

58. Rausch, T.; Wachter, A. Sulfur Metabolism: A Versatile Platform for Launching Defence Operations. *Trends Plant Sci.* **2005**, *10*, 503–509. [CrossRef] [PubMed]

59. Mocniak, L.E.; Elkin, K.; Bollinger, J.M., Jr. Lifetimes of the Aglycone Substrates of Specifier Proteins, the Autonomous Iron Enzymes That Dictate the Products of the Glucosinolate-Myrosinase Defense System in Brassica Plants. *Biochemistry* **2020**, *59*, 2432–2441. [CrossRef]

60. Santamaria, M.E.; Garcia, A.; Arnaiz, A.; Rosa-Diaz, I.; Romero-Hernandez, G.; Diaz, I.; Martinez, M. Comparative Transcriptomics Reveals Hidden Issues in the Plant Response to Arthropod Herbivores. *J. Integr. Plant Biol.* **2021**, *63*, 312–326. [CrossRef]

61. Calmes, B.; N'Guyen, G.; Dumur, J.; Brisach, C.A.; Campion, C.; Iacomi, B.; Pigné, S.; Dias, E.; Macherel, D.; Guillemette, T.; et al. Glucosinolate-derived Isothiocyanates Impact Mitochondrial Function in Fungal Cells and Elicit an Oxidative Stress Response Necessary for Growth Recovery. *Front. Plant Sci.* **2015**, *6*, 414. [CrossRef]

62. Pereira, F.; Rosa, E.; Fahey, J.W.; Stephenson, K.K.; Carvalho, R.; Aires, A. Influence of Temperature and Ontogeny on the Levels of Glucosinolates in Broccoli (*Brassica oleracea* Var. *italica*) Sprouts and Their Effect on the Induction of Mammalian Phase 2 Enzymes. *J. Agric. Food Chem.* **2002**, *50*, 6239–6244. [CrossRef]

63. LoPez-Berenguer, C.; MartiNez-Ballesta, M.D.; Moreno, D.; Carvajal, M.; Garcia-Viguera, C. Growing Hardier Crops for Better Health: Salinity Tolerance and the Nutritional Value of Broccoli. *J. Agric. Food Chem.* **2009**, *57*, 572–578. [CrossRef] [PubMed]

64. Yuan, G.; Wang, X.; Guo, R.; Wang, Q. Effect of Salt Stress on Phenolic Compounds, Glucosinolates, Myrosinase and Antioxidant Activity in Radish Sprouts. *Food Chem.* **2010**, *121*, 1014–1019. [CrossRef]

65. Safavi Fard, N.; Heidari Sharif Abad, H.; Shirani Rad, A.H.; Majidi Heravan, E.; Daneshian, J. Effect of Drought Stress on Qualitative Characteristics of Canola Cultivars in Winter Cultivation. *Ind. Crop. Prod.* **2018**, *114*, 87–92. [CrossRef]

66. Eom, S.H.; Baek, S.A.; Kim, J.K.; Hyun, T.K. Transcriptome Analysis in Chinese Cabbage (*Brassica rapa* ssp. pekinensis) Provides the Role of Glucosinolate Metabolism in Response to Drought Stress. *Molecules* **2018**, *23*, 5. [CrossRef] [PubMed]

67. Huseby, S.; Koprivova, A.; Lee, B.R.; Saha, S.; Mithen, R.; Wold, A.B.; Bengtsson, G.B.; Kopriva, S. Diurnal and Light Regulation of Sulphur Assimilation and Glucosinolate Biosynthesis in *Arabidopsis*. *J. Exp. Bot.* **2013**, *64*, 1039–1048. [CrossRef]

68. del Carmen Martinez-Ballesta, M.; Moreno, D.A.; Carvajal, M. The Physiological Importance of Glucosinolates on Plant Response to Abiotic Stress in *Brassica*. *Int. J. Mol. Sci.* **2013**, *14*, 11607–11625. [CrossRef] [PubMed]

69. Vale, A.P.; Santos, J.; Brito, N.V.; Fernandes, D.; Rosa, E.; Oliveira, M.B. Evaluating the Impact of Sprouting Conditions on the Glucosinolate Content of *Brassica Oleracea* Sprouts. *Phytochemistry* **2015**, *115*, 252–260. [CrossRef] [PubMed]

70. Xuan, W.; Beeckman, T.; Xu, G. Plant Nitrogen nutrition: Sensing and Signaling. *Curr. Opin. Plant Biol.* **2017**, *39*, 57–65. [CrossRef]

71. Sanchez-Pujante, P.J.; Borja-Martinez, M.; Pedreno, M.A.; Almagro, L. Biosynthesis and Bioactivity of Glucosinolates and Their Production in Plant In Vitro Cultures. *Planta* **2017**, *246*, 19–32. [CrossRef]

72. Park, S.; Rim, S.J.; Jo, M.; Lee, M.G.; Kim, C.E. Comorbidity of Alcohol Use and Other Psychiatric Disorders and Suicide Mortality: Data from the South Korean National Health Insurance Cohort, 2002 to 2013. *Alcohol. Clin. Exp. Res.* **2019**, *43*, 842–849. [CrossRef] [PubMed]

73. Korbas, M.; Jajor, E.; Budka, A. Clubroot (*Plasmodiophora Brassicae*)—A Threat for Oilseed Rape. *J. Plant Prot. Res.* **2009**, *49*, 4. [CrossRef]

74. Yu, F.; Zhang, W.; Wang, S.; Wang, H.; Yu, L.; Zeng, X.; Fei, Z.; Li, J. Genome Sequence of *Fusarium Oxysporum* f. sp. *Conglutinans*, the Etiological Agent of Cabbage Fusarium Wilt. *Mol. Plant Microbe Interact.* **2021**, *34*, 210–213. [CrossRef] [PubMed]

75. Jensen, B.; Hockenhull, J.; Munk, L. Seedling and Adult Plant Resistance to Downy Mildew (*Peronospora parasitica*) in Cauliflower (*Brassica oleracea* convar. *botrytis* var. *botrytis*). *Plant Pathol.* **1999**, *48*, 604–612. [CrossRef]

76. Mahalingam, T.; Chen, W.; Rajapakse, C.S.; Somachandra, K.P.; Attanayake, R.N. Genetic Diversity and Recombination in the Plant Pathogen *Sclerotinia Sclerotiorum* Detected in Sri Lanka. *Pathogens* **2020**, *9*, 4. [CrossRef] [PubMed]
77. Samal, B.; Chatterjee, S. New Insight into Bacterial Social Communication in Natural Host: Evidence for Interplay of Heterogeneous and Unison Quorum Response. *PLoS Genet.* **2019**, *15*, e1008395. [CrossRef]
78. Zhang, S.-H.; Yang, Q.; Ma, R.-C. *Erwinia Carotovora* ssp. *Carotovora* Infection Induced "Defense Lignin" Accumulation and Lignin Biosynthetic Gene Expression in Chinese Cabbage (*Brassica rapa* L. ssp. *pekinensis*). *J. Integr. Plant Biol.* **2007**, *49*, 993–1002. [CrossRef]
79. Takikawa, Y.; Takahashi, F. Bacterial Leaf Spot and Blight of Crucifer Plants (Brassicaceae) Caused by *Pseudomonas Syringae* pv. *Maculicola* and *P. cannabina* pv. *alisalensis*. *J. Gen. Plant Pathol.* **2014**, *80*, 466–474. [CrossRef]
80. Gong, J.; Ju, H.K.; Kim, I.H.; Seo, E.Y.; Cho, I.S.; Hu, W.X.; Han, J.Y.; Kim, J.K.; Choi, S.R.; Lim, Y.P.; et al. Sequence Variations Among 17 New Radish Isolates of *Turnip mosaic virus* Showing Differential Pathogenicity and Infectivity in *Nicotiana benthamiana*, *Brassica rapa*, and *Raphanus sativus*. *Phytopathology* **2019**, *109*, 904–912. [CrossRef] [PubMed]
81. Iamba, K.; Malapa, S. Efficacy of Selected Plant Extracts against Diamondback Moth (*Plutella xylostella* L.) on Round Cabbage In Situ. *J. Entomol. Zool. Stud.* **2020**, *8*, 1240–1247.
82. Gabryś, B.; Pawluk, M. Acceptability of Different Species of Brassicaceae as Hosts for the Cabbage Aphid. In *Proceedings of the 10th International Symposium on Insect-Plant Relationships*; Springer: Dordrecht, The Netherlands, 1999; pp. 105–109.
83. Griese, E.; Pineda, A.; Pashalidou, F.G.; Iradi, E.P.; Fatouros, N.E. Plant Responses to Butterfly Oviposition Partly Explain Preference–Performance Relationships on Different Brassicaceous Species. *Oecologia* **2020**, *192*, 293–309. [CrossRef] [PubMed]
84. Tierens, M.J.; Thomma, B.; Brouwer, M.; Schmidt, J.; Kistner, K.; Porzel, A.; Mauch-Mani, B.; Broekaert, C. Study of the Role of Antimicrobial Glucosinolate-Derived Isothiocyanates in Resistance of Arabidopsis to Microbial Pathogens. *Plant Physiol.* **2001**, *125*, 1688–1699. [CrossRef]
85. Sellam, A.; Dongo, A.; Guillemette, T.; Hudhomme, P.; Simoneau, P. Transcriptional Responses to Exposure to the Brassicaceous Defence Metabolites Camalexin and Allyl-isothiocyanate in the Necrotrophic Fungus *Alternaria Brassicicola*. *Mol. Plant Pathol.* **2007**, *8*, 195–208. [CrossRef] [PubMed]
86. Humphry, M.; Bednarek, P.; Kemmerling, B.; Koh, S.; Stein, M.; Gobel, U.; Stuber, K.; Pislewska-Bednarek, M.; Loraine, A.; Schulze-Lefert, P.; et al. A Regulon Conserved in Monocot and Dicot Plants Defines a Functional Module in Antifungal Plant Immunity. *Proc. Natl. Acad. Sci. USA* **2010**, *107*, 21896–21901. [CrossRef] [PubMed]
87. Giamoustaris, A.; Mithen, R.F. Glucosinolates and Disease Resistance in Oilseed Rape (*Brassica napus* ssp. *oleifera*). *Plant Pathol.* **2010**, *46*, 271–275. [CrossRef]
88. Robin, A.H.K.; Laila, R.; Abuyusuf, M.; Park, J.I.; Nou, I.S. Leptosphaeria Maculans Alters Glucosinolate Accumulation and Expression of Aliphatic and Indolic Glucosinolate Biosynthesis Genes in Blackleg Disease-Resistant and -Susceptible Cabbage Lines at the Seedling Stage. *Front. Plant Sci.* **2020**, *11*, 1134. [CrossRef]
89. Pedras, M.S.; Zheng, Q.A.; Gadagi, R.S.; Rimmer, S.R. Phytoalexins and Polar Metabolites from the Oilseeds Canola and Rapeseed: Differential Metabolic Responses to the Biotroph Albugo Candida and to Abiotic Stress. *Phytochemistry* **2008**, *69*, 894–910. [CrossRef]
90. Hiruma, K.; Onozawa-Komori, M.; Takahashi, F.; Asakura, M.; Bednarek, P.; Okuno, T.; Schulze-Lefert, P.; Takano, Y. Entry Mode-dependent Function of an Indole Glucosinolate Pathway in Arabidopsis for Nonhost Resistance against Anthracnose Pathogens. *Plant Cell* **2010**, *22*, 2429–2443. [CrossRef] [PubMed]
91. Kidd, B.N.; Kadoo, N.Y.; Dombrecht, B.; Tekeoglu, M.; Kazan, K. Auxin Signaling and Transport Promote Susceptibility to the Root-infecting Fungal Pathogen *Fusarium Oxysporum* in *Arabidopsis*. *Mol. Plant Microbe Interact. MPMI* **2011**, *24*, 733. [CrossRef] [PubMed]
92. Zhu, Q.H.; Stephen, S.; Kazan, K.; Jin, G.; Fan, L.; Taylor, J.; Dennis, E.S.; Helliwell, C.A.; Wang, M.B. Characterization of the Defense Transcriptome Responsive to *Fusarium Oxysporum*-infection in *Arabidopsis* Using RNA-seq. *Gene* **2013**, *512*, 259–266. [CrossRef] [PubMed]
93. Zhao, J.; Peltier, A.J.; Meng, J.; Osborn, T.C.; Grau, C.R. Evaluation of Sclerotinia Stem Rot Resistance in Oilseed *Brassica napus* Using a Petiole Inoculation Technique Under Greenhouse Conditions. *Plant Dis.* **2004**, *88*, 1033–1039. [CrossRef]
94. Wu, J.; Zhao, Q.; Yang, Q.; Liu, H.; Li, Q.; Yi, X.; Cheng, Y.; Guo, L.; Fan, C.; Zhou, Y. Comparative Transcriptomic Analysis Uncovers the Complex Genetic Network for Resistance to *Sclerotinia Sclerotiorum* in *Brassica Napus*. *Sci. Rep.* **2016**, *6*, 19007. [CrossRef] [PubMed]
95. Stotz, H.U.; Sawada, Y.; Shimada, Y.; Hirai, M.Y.; Sasaki, E.; Krischke, M.; Brown, P.D.; Saito, K.; Kamiya, Y. Role of Camalexin, Indole Glucosinolates, and Side Chain Modification of Glucosinolate-derived Isothiocyanates in Defense of Arabidopsis against *Sclerotinia Sclerotiorum*. *Plant J.* **2011**, *67*, 81–93. [CrossRef] [PubMed]
96. Anthony, M.A.; Celenza, J.L.; Armstrong, A.; Frey, S.D. Indolic Glucosinolate Pathway Provides Resistance to Mycorrhizal Fungal Colonization in a Non-host Brassicaceae. *Ecosphere* **2020**, *11*, 4. [CrossRef]
97. Xu, L.; Yang, H.; Ren, L.; Chen, W.; Liu, L.; Liu, F.; Zeng, L.; Yan, R.; Chen, K.; Fang, X. Jasmonic Acid-Mediated Aliphatic Glucosinolate Metabolism Is Involved in Clubroot Disease Development in *Brassica napus* L. *Front. Plant Sci.* **2018**, *9*, 750. [CrossRef]
98. Castillo, N.; Pastor, V.; Chavez, A.; Arro, M.; Boronat, A.; Flors, V.; Ferrer, A.; Altabella, T. Inactivation of UDP-Glucose Sterol Glucosyltransferases Enhances *Arabidopsis* Resistance to *Botrytis cinerea*. *Front. Plant Sci.* **2019**, *10*, 1162. [CrossRef]

99. Tinte, M.M.; Steenkamp, P.A.; Piater, L.A.; Dubery, I.A. Lipopolysaccharide Perception in Arabidopsis Thaliana: Diverse LPS Chemotypes from Burkholderia Cepacia, Pseudomonas Syringae and Xanthomonas Campestris Trigger Differential Defence-related Perturbations in the Metabolome. *Plant Physiol. Biochem.* **2020**, *156*, 267–277. [CrossRef]

100. Brader, G.; Mikkelsen, M.D.; Halkier, B.A.; Tapio Palva, E. Altering Glucosinolate Profiles Modulates Disease Resistance in Plants. *Plant J.* **2006**, *46*, 758–767. [CrossRef] [PubMed]

101. Mishina, T.E.; Zeier, J. Bacterial Non-host Resistance: Interactions of *Arabidopsis* with Non-adapted *Pseudomonas Syringae* Strains. *Physiol. Plant* **2007**, *131*, 448–461. [CrossRef]

102. Truman, W.; Bennettt, M.; Kubigsteltig, I.; Turnbull, C.; Grant, M. Arabidopsis Systemic Immunity Uses Conserved Defense Signaling Pathways and is Mediated by Jasmonates. *Proc. Natl. Acad. Sci. USA* **2007**, *104*, 1075–1080. [CrossRef] [PubMed]

103. Aires, A.; Dias, C.S.P.; Carvalho, R.; Oliveira, M.H.; Monteiro, A.A.; Simões, M.V.; Rosa, E.A.S.; Bennett, R.N.; Saavedra, M.J. Correlations between Disease Severity, Glucosinolate Profiles and Total Phenolics and *Xanthomonas Campestris* pv. *Campestris* Inoculation of Different Brassicaceae. *Sci. Hortic.* **2011**, *129*, 503–510. [CrossRef]

104. Clay, N.K.; Adio, A.M.; Denoux, C.; Jander, G.; Ausubel, F.M. Glucosinolate Metabolites Required for an *Arabidopsis* Innate Immune Response. *Science* **2009**, *323*, 95–101. [CrossRef] [PubMed]

105. Geng, X.; Cheng, J.; Gangadharan, A.; Mackey, D. The Coronatine Toxin of *Pseudomonas Syringae* is a Multifunctional Suppressor of *Arabidopsis* Defense. *Plant Cell* **2012**, *24*, 4763–4774. [CrossRef]

106. Barah, P.; Winge, P.; Kusnierczyk, A.; Tran, D.H.; Bones, A.M. Molecular Signatures in *Arabidopsis Thaliana* in Response to Insect Attack and Bacterial Infection. *PLoS ONE* **2013**, *8*, e58987. [CrossRef]

107. Sun, Q.; Zhang, E.; Liu, Y.; Xu, Z.; Hui, M.; Zhang, X.; Cai, M. Transcriptome Analysis of Two Lines of *Brassica Oleracea* in Response to Early Infection with *Xanthomonas Campestris* pv. *Campestris*. *Can. J. Plant Pathol.* **2020**, *43*, 127–139. [CrossRef]

108. Liu, M.; Wu, F.; Wang, S.; Lu, Y.; Chen, X.; Wang, Y.; Gu, A.; Zhao, J.; Shen, S. Comparative Transcriptome Analysis Reveals Defense Responses against Soft Rot in Chinese Cabbage. *Hortic. Res.* **2019**, *6*, 68. [CrossRef] [PubMed]

109. Deutsch, C.A.; Tewksbury, J.J.; Tigchelaar, M.; Battisti, D.S.; Merrill, S.C.; Huey, R.B.; Naylor, R.L. Increase in Crop Losses to Insect Pests in a Warming Climate. *Science* **2018**, *361*, 916–919. [CrossRef]

110. Brown, P.D.; Morra, M.J. Control of Soil-Borne Plant Pests Using Glucosinolate-Containing Plants. *Adv. Agron.* **1997**, *61*, 168–231.

111. Strauss, S.Y.; Lambrix, I. Optimal Defence Theory and Flower Petal Colour Predict Variation in the Secondary Chemistry of Wild Radish. *J. Ecol.* **2004**, *92*, 132–141. [CrossRef]

112. Vadassery, J.; Reichelt, M.; Hause, B.; Gershenzon, J.; Boland, W.; Mithofer, A. CML42-mediated Calcium Signaling Coordinates Responses to *Spodoptera* Herbivory and Abiotic Stresses in Arabidopsis. *Plant Physiol.* **2012**, *159*, 1159–1175. [CrossRef] [PubMed]

113. Mewis, I.; Appel, H.M.; Hom, A.; Raina, R.; Schultz, J.C. Major Signaling Pathways Modulate Arabidopsis Glucosinolate Accumulation and Response to Both Phloem-feeding and Chewing Insects. *Plant Physiol.* **2005**, *138*, 1149–1162. [CrossRef] [PubMed]

114. Ahuja, I.; van Dam, N.M.; Winge, P.; Traelnes, M.; Heydarova, A.; Rohloff, J.; Langaas, M.; Bones, A.M. Plant Defence Responses in Oilseed Rape *MINELESS* Plants after Attack by the Cabbage Moth *Mamestra Brassicae*. *J. Exp. Bot.* **2015**, *66*, 579–592. [CrossRef]

115. Santolamazza-Carbone, S.; Sotelo, T.; Velasco, P.; Cartea, M.E. Antibiotic Properties of the Glucosinolates of *Brassica oleracea* var. *acephala* similarly Affect Generalist and Specialist Larvae of Two Lepidopteran Pests. *J. Pest Sci.* **2016**, *89*, 195–206. [CrossRef]

116. Dam, N.; Raaijmakers, C.E.; Putten, W. Root Herbivory Reduces Growth and Survival of the Shoot Feeding Specialist *Pieris Rapae* on *Brassica Nigra*. *Entomol. Exp. Appl.* **2005**, *115*, 161–170.

117. Mathur, V.; Ganta, S.; Raaijmakers, C.E.; Reddy, A.S.; Vet, L.E.M.; van Dam, N.M. Temporal Dynamics of Herbivore-induced Responses in *Brassica Juncea* and Their Effect on Generalist and Specialist Herbivores. *Entomol. Exp. Appl.* **2011**, *139*, 215–225. [CrossRef]

118. van Dam, N.M.; Samudrala, D.; Harren, F.J.; Cristescu, S.M. Real-time Analysis of Sulfur-containing Volatiles in *Brassica* Plants Infested with Root-feeding *Delia Radicum* Larvae Using Proton-transfer Reaction Mass Spectrometry. *AoB Plants* **2012**, *2012*, pls021. [CrossRef]

119. Badenes-Pérez, F.R.; Gershenzon, J.; Heckel, D.G. Plant Glucosinolate Content Increases Susceptibility to Diamondback Moth (Lepidoptera: Plutellidae) Regardless of Its Diet. *J. Pest Sci.* **2019**, *93*, 491–506. [CrossRef]

120. Chen, W.; Dong, Y.; Saqib, H.S.A.; Vasseur, L.; Zhou, W.; Zheng, L.; Lai, Y.; Ma, X.; Lin, L.; Xu, X.; et al. Functions of Duplicated Glucosinolate Sulfatases in the Development and Host Adaptation of *Plutella Xylostella*. *Insect Biochem. Mol. Biol.* **2020**, *119*, 103316. [CrossRef]

121. Ogran, A.; Faigenboim, A.; Barazani, O. Transcriptome Responses to Different Herbivores Reveal Differences in Defense Strategies between Populations of *Eruca Sativa*. *BMC Genom.* **2019**, *20*, 843. [CrossRef] [PubMed]

122. Gols, R.; van Dam, N.M.; Reichelt, M.; Gershenzon, J.; Raaijmakers, C.E.; Bullock, J.M.; Harvey, J.A. Seasonal and Herbivore-induced Dynamics of Foliar Glucosinolates in Wild Cabbage (*Brassica oleracea*). *Chemoecology* **2018**, *28*, 77–89. [CrossRef]

123. Robert, C.A.M.; Pellissier, L.; Moreira, X.; Defossez, E.; Pfander, M.; Guyer, A.; van Dam, N.M.; Rasmann, S. Correlated Induction of Phytohormones and Glucosinolates Shapes Insect Herbivore Resistance of Cardamine Species Along Elevational Gradients. *J. Chem. Ecol.* **2019**, *45*, 638–648. [CrossRef]

124. Buckley, J.; Pashalidou, F.G.; Fischer, M.C.; Widmer, A.; Mescher, M.C.; De Moraes, C.M. Divergence in Glucosinolate Profiles between High- and Low-Elevation Populations of *Arabidopsis halleri* Correspond to Variation in Field Herbivory and Herbivore Behavioral Preferences. *Int. J. Mol. Sci.* **2019**, *20*, 174. [CrossRef] [PubMed]

125. Papazian, S.; Girdwood, T.; Wessels, B.A.; Poelman, E.H.; Dicke, M.; Moritz, T.; Albrectsen, B.R. Leaf Metabolic Signatures Induced by Real and Simulated Herbivory in Black Mustard (*Brassica nigra*). *Metabolomics* **2019**, *15*, 130. [CrossRef] [PubMed]
126. Kempema, L.A.; Cui, X.; Holzer, F.M.; Walling, L.L. Arabidopsis Transcriptome Changes in Response to Phloem-Feeding Silverleaf Whitefly Nymphs. Similarities and Distinctions in Responses to Aphids. *Plant Physiol.* **2007**, *143*, 849–865. [CrossRef]
127. Agerbirk, N.; Olsen, C.E.; Nielsen, J.K. Seasonal Variation in Leaf Glucosinolates and Insect Resistance in Two Types of *Barbarea Vulgaris* ssp. *arcuata*. *Phytochemistry* **2001**, *58*, 91–100. [CrossRef]
128. Kroymann, J.; Donnerhacke, S.; Schnabelrauch, D.; Mitchell-Olds, T. Evolutionary Dynamics of an *Arabidopsis* Insect Resistance Quantitative Trait Locus. *Proc. Natl. Acad. Sci. USA* **2003**, *100*, 14587–14592. [CrossRef]
129. Ulmer, B.J.; Dosdall, L.M. Glucosinolate Profile and Oviposition Behavior in Relation to the Susceptibilities of Brassicaceae to the Cabbage Seedpod Weevil. *Entomol. Exp. Appl.* **2010**, *121*, 203–213. [CrossRef]
130. Bo, P.; Hopkins, R.; Rask, L.; Meijer, J. Infestation by Cabbage Aphid (*Brevicoryne brassicae*) on Oilseed Rape (*Brassica napus*) Causes a Long Lasting Induction of the Myrosinase System. *Entomol. Exp. Appl.* **2003**, *109*, 55–62.
131. Kusnierczyk, A.; Winge, P.; Jorstad, T.S.; Troczynska, J.; Rossiter, J.T.; Bones, A.M. Towards Global Understanding of Plant Defence against Aphids—Timing and Dynamics of Early Arabidopsis Defence Responses to Cabbage Aphid (*Brevicoryne brassicae*) Attack. *Plant Cell Environ.* **2008**, *31*, 1097–1115. [CrossRef]
132. Beukeboom, L.W. Editor's Choice: March 2020. *Entomol. Exp. Appl.* **2020**, *168*, 199. [CrossRef]
133. de Vos, M.; Jander, G. *Myzus Persicae* (green peach aphid) Salivary Components Induce Defence Responses in *Arabidopsis Thaliana*. *Plant Cell Environ.* **2009**, *32*, 1548–1560. [CrossRef] [PubMed]
134. Kuhlmann, F.; Muller, C. Independent Responses to Ultraviolet Radiation and Herbivore Attack in Broccoli. *J. Exp. Bot.* **2009**, *60*, 3467–3475. [CrossRef]
135. Pfalz, M.; Vogel, H.; Kroymann, J. The Gene Controlling the *Indole Glucosinolate Modifier1* Quantitative Trait Locus Alters Indole Glucosinolate Structures and Aphid Resistance in *Arabidopsis*. *Plant Cell* **2009**, *21*, 985–999. [CrossRef]
136. Hu, J.; Yang, J.J.; Liu, B.M.; Cui, H.Y.; Zhang, Y.J.; Jiao, X.G. Feeding Behavior Explains the Different Effects of Cabbage on MEAM1 and MED Cryptic Species of *Bemisia Tabaci*. *Insect Sci.* **2020**, *27*, 1276–1284. [CrossRef]
137. Anyanga, M.O.; Farman, D.I.; Ssemakula, G.N.; Mwanga, R.O.M.; Stevenson, P.C. Effects of Hydroxycinnamic Acid Esters on Sweetpotato Weevil Feeding and Oviposition and Interactions with *Bacillus Thuringiensis* Proteins. *J. Pest Sci.* **2020**. [CrossRef]
138. Yao, Q.; Peng, Z.; Tong, H.; Yang, F.; Xing, G.; Wang, L.; Zheng, J.; Zhang, Y.; Su, Q. Tomato Plant Flavonoids Increase Whitefly Resistance and Reduce Spread of Tomato yellow leaf curl virus. *J. Econ. Entomol.* **2019**, *112*, 2790–2796. [CrossRef] [PubMed]
139. Jost, R.; Altschmied, L.; Bloem, E.; Bogs, J.; Gershenzon, J.; Hahnel, U.; Hansch, R.; Hartmann, T.; Kopriva, S.; Kruse, C.; et al. Expression Profiling of Metabolic Genes in Response to Methyl Jasmonate Reveals Regulation of Genes of Primary and Secondary Sulfur-related Pathways in *Arabidopsis Thaliana*. *Photosynth. Res.* **2005**, *86*, 491–508. [CrossRef] [PubMed]
140. Pangesti, N.; Reichelt, M.; van de Mortel, J.E.; Kapsomenou, E.; Gershenzon, J.; van Loon, J.J.; Dicke, M.; Pineda, A. Jasmonic Acid and Ethylene Signaling Pathways Regulate Glucosinolate Levels in Plants During Rhizobacteria-Induced Systemic Resistance Against a Leaf-Chewing Herbivore. *J. Chem. Ecol.* **2016**, *42*, 1212–1225. [CrossRef] [PubMed]
141. Kong, W.; Li, J.; Yu, Q.; Cang, W.; Xu, R.; Wang, Y.; Ji, W. Two Novel Flavin-Containing Monooxygenases Involved in Biosynthesis of Aliphatic Glucosinolates. *Front. Plant Sci.* **2016**, *7*, 1292. [CrossRef]
142. Lu, C.; Han, M.H.; Guevara-Garcia, A.; Fedoroff, N.V. Mitogen-activated protein kinase signaling in postgermination arrest of development by abscisic acid. *Proc. Natl. Acad. Sci. USA* **2002**, *99*, 15812–15817. [CrossRef] [PubMed]

Article

Isolation and Characterization of Endophyte *Bacillus velezensis* KOF112 from Grapevine Shoot Xylem as Biological Control Agent for Fungal Diseases

Kazuhiro Hamaoka, Yoshinao Aoki and Shunji Suzuki *

Laboratory of Fruit Genetic Engineering, The Institute of Enology and Viticulture, University of Yamanashi, Kofu 400-0005, Japan; g19dia03@yamanashi.ac.jp (K.H.); yaoki@yamanashi.ac.jp (Y.A.)
* Correspondence: suzukis@yamanashi.ac.jp; Tel.: +81-55-220-8394

Citation: Hamaoka, K.; Aoki, Y.; Suzuki, S. Isolation and Characterization of Endophyte *Bacillus velezensis* KOF112 from Grapevine Shoot Xylem as Biological Control Agent for Fungal Diseases. *Plants* **2021**, *10*, 1815. https://doi.org/10.3390/plants10091815

Academic Editors: Carlos Agustí-Brisach and Eugenio Llorens

Received: 11 May 2021
Accepted: 27 August 2021
Published: 31 August 2021

Publisher's Note: MDPI stays neutral with regard to jurisdictional claims in published maps and institutional affiliations.

Abstract: As the use of chemical fungicides has raised environmental concerns, biological control agents have attracted interest as an alternative to chemical fungicides for plant-disease control. In this study, we attempted to explore biological control agents for three fungal phytopathogens causing downy mildew, gray mold, and ripe rot in grapevines, which are derived from shoot xylem of grapevines. KOF112, which was isolated from the Japanese indigenous wine grape *Vitis* sp. cv. Koshu, inhibited mycelial growth of *Botrytis cinerea*, *Colletotrichum gloeosporioides*, and *Phytophthora infestans*. The KOF112-inhibited mycelial tips were swollen or ruptured, suggesting that KOF112 produces antifungal substances. Analysis of the 16S rDNA sequence revealed that KOF112 is a strain of *Bacillus velezensis*. Comparative genome analysis indicated significant differences in the synthesis of non-ribosomal synthesized antimicrobial peptides and polyketides between KOF112 and the antagonistic *B. velezensis* FZB42. KOF112 showed biocontrol activities against gray mold caused by *B. cinerea*, anthracnose by *C. gloeosporioides*, and downy mildew by *Plasmopara viticola*. In the KOF112–*P. viticola* interaction, KOF112 inhibited zoospore release from *P. viticola* zoosporangia but not zoospore germination. In addition, KOF112 drastically upregulated the expression of genes encoding class IV chitinase and β-1,3-glucanase in grape leaves, suggesting that KOF112 also works as a biotic elicitor in grapevine. Because it is considered that endophytic KOF112 can colonize well in and/or on grapevine, KOF112 may contribute to pest-management strategies in viticulture and potentially reduce the frequency of chemical fungicide application.

Keywords: *Bacillus velezensis*; downy mildew; gray mold; ripe rot; PR protein; zoosporangia

1. Introduction

The production of Koshu wine in Japan started in 1874 in Yamanashi Prefecture. *Vitis* sp. cv. Koshu, a hybrid of *Vitis vinifera* L. and *V. davidii* Foex, is an indigenous wine grape in Japan [1]. Koshu was introduced from Europe to Japan through the Silk Road, and crossed with the Chinese wild species *V. davidii* en route to Japan. Koshu was recognized as a wine grape cultivar in 2010 by the International Organization for Vine and Wine, and was registered in *Vitis* International Variety Catalogue by Julius Kühn-Institut-Bundesforschungsinstitut für Kulturpflanzen (https://www.vivc.de/, accessed on 28 April 2021). At present, Koshu is one of the most widely cultivated wine grapes in Japan and one of the most important cultivars for white-wine making in Japan [2]. Whole genome analysis demonstrated that Koshu is susceptible to phytopathogenic attack as a result of deletions in genes associated with pathogen response, such as hypersensitive response [3]. For example, Koshu is more susceptible to downy mildew, which is caused by *Plasmopara viticola* (Berk and M.A. Curtis; Berl and De Toni), than other *V. vinifera* cultivars [4]. In addition, the humid climate of Japan, caused by the prolonged concentration of extremely moist airstreams over Japan as a result of global warming [5], has contributed to damaging Koshu grapevines by phytopathogenic fungal diseases. The high nighttime temperature resulting from the acceleration of global warming has promoted downy mildew infection in grapevines [6]. Thus, fungal disease control has become a subject of serious concern in viticulture for Koshu wine making.

A simple strategy against fungal disease is the use of chemical fungicides. However, there are two main problems with regard to the application of chemical fungicides. One is environmental pollution, and the other is the emergence of fungal phytopathogen populations resistant to the chemical fungicides. In particular, the latter has plagued vine growers in Japan, making it extremely difficult to control fungal diseases in grapevines. *P. viticola* is a high-risk phytopathogen that easily acquires chemical fungicide resistance [7]. Resistant genes, conferring resistance to quinone outside inhibitor and carboxylic acid amide, were detected in *P. viticola* populations in Japanese vineyards in 2010 [8] and in 2015 [9], respectively. Fungicide-resistant *Colletotrichum gloeosporioides* (Penzig) Penzig and Saccardo, which causes grape ripe rot, and *Botrytis cinerea* Pers. ex Fr., which causes grape gray mold, have already emerged in Japanese vineyards [10,11]. It is for these reasons that alternatives to chemical fungicides are attracting the interest of scientific communities.

One of the alternative disease control strategies to chemical fungicides is biological control using biofungicides. Biological control agents in biofungicides are isolated from nature, and are largely microorganisms [12]. A vast number of microorganisms isolated from nature have been identified as candidates for biological control agents in biofungicides [13,14]. Some microorganisms have been developed and launched as biofungicides in viticulture. For example, *Bacillus subtilis* QST-713 (product name, Serenade®) is available on the market, and is used to control gray mold in viticulture [15]. The introduction of biofungicide application in viticulture is expected to reduce the frequency of chemical fungicide application and to alleviate increasing environmental concerns on chemical fungicides and the emergence of resistance to the chemical fungicides in fungal phytopathogens.

Cyclic lipopeptides produced by biological control agents have received considerable attention as one of the tools for disease control in plants, because some cyclic lipopeptides also function as elicitors in plants as well as antimicrobial metabolites [16–18]. For example, fengycin and surfactin secreted by *B. subtilis* GLB191 contribute to protection against grape downy mildew by directly inhibiting *P. viticola* and inducing plant defense response [17]. Iturin A induces defense response in plants depending on its structure [19]. The cyclization of the seven amino acids and the β-hydroxy fatty acid chain of iturin A are required for the induction of plant defense response. Although evidence of how cyclic lipopeptides trigger plant defense response is lacking, microorganisms showing bifunctional activity against phytopathogens may be an innovative biological control agent in viticulture.

The objective of this study was to clarify the possibility of using endophytic bacteria as biological control agents in biofungicides used in viticulture. The colonization efficiency of biological control agents in biofungicides on/in plant tissues affects their antagonistic activities toward fungal phytopathogens [20]. In this study, we explored grapevine endophytic bacteria possessing in vitro antagonistic activities toward three fungal phytopathogens, *B. cinerea*, *C. gloeosporioides*, and *Phytophthora infestans* (a substitute of *P. viticola*), and isolated endophytic *Bacillus velezensis* KOF112 from the shoot xylem of Koshu grapevine. KOF112 showed in vivo biocontrol activities against gray mold caused by *B. cinerea*, anthracnose by *C. gloeosporioides*, and downy mildew by *P. viticola*. In addition, foliar application of KOF112 induced plant defense response in grapevine.

2. Results

2.1. Antagonistic Activity of KOF112 toward Phytopathogenic Fungi

Two hundred and forty-seven colonies were collected from Koshu shoot xylem and subjected to the in vitro bioassay using *B. cinerea*, *C. gloeosporioides*, and *P. infestans*. As *P. viticola* is an obligate biotrophic oomycete, *P. infestans* was used as the oomycete phytopathogen instead of *P. viticola*. As a result, one colony that exhibited strong antagonistic activity toward the three phytopathogenic fungi was successfully isolated and named KOF112. Large inhibition zones were formed between KOF112 and each phytopathogenic fungus (Figure 1A). The mycelial tips of growth-inhibited *B. cinerea*, *C. gloeosporioides*, and *P. infestans* by KOF112 were swollen or ruptured (Figure 1B). *Agrobacterium* sp. isolate CHB3, selected as the control isolate with no antifungal activity, exhibited no suppressive effect on the mycelial growth of all the fungi tested.

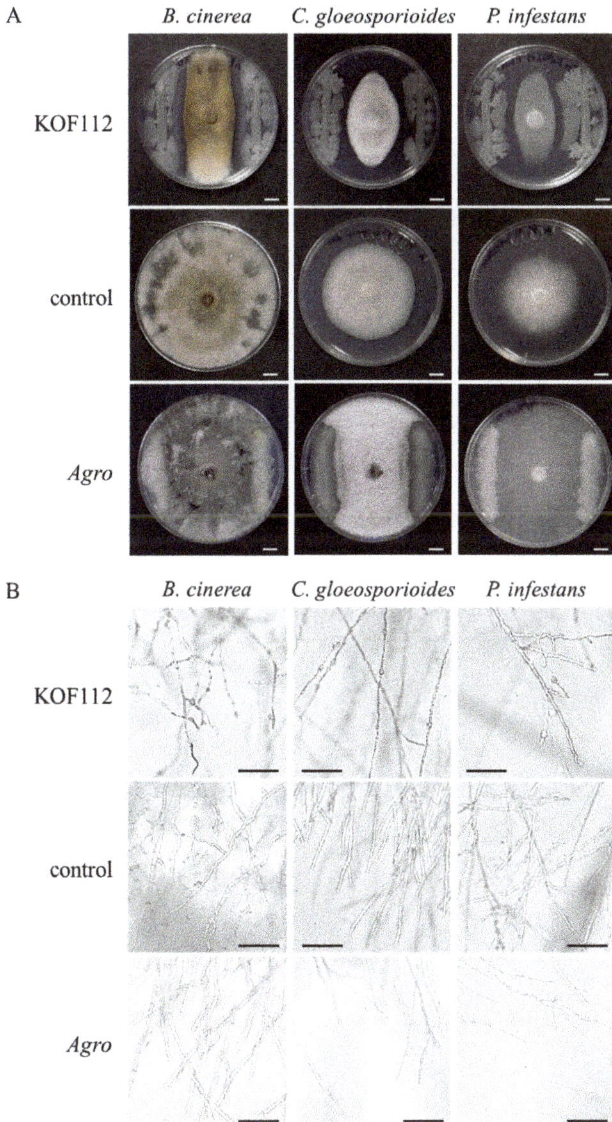

Figure 1. Antagonistic activity of KOF112 toward mycelial growth of phytopathogenic fungi. (**A**) KOF112 was streaked on the edge of SCD plates, and agar plugs cut from colonies of *B. cinerea*, *C. gloeosporioides* or *P. infestans* were placed at the center of the KOF112-growing SCD plates. After incubation for 5 d, a large inhibition zone was formed between KOF112 and fungal mycelia compared with control culture. Bar, 1 cm. (**B**) Microscopic observation. Mycelial tips of growth-inhibited fungi co-incubated with KOF112 were swollen or ruptured compared with control culture. Bar, 100 μm. *Agro, Agrobacterium* sp. isolate CHB3 used as control isolate with no antifungal activity.

These results suggest that KOF112 inhibits the mycelial growth of all the fungi tested, and that KOF112 produces antifungal substances.

2.2. KOF112 Is a Strain of Bacillus velezensis

The 16S rDNA nucleotide sequence of KOF112 had high homologies to the corresponding sequences of *B. velezensis* BIM B-1312D (100%), *B. velezensis* FJAT-52631 (100%), and *B. velezensis* CR-502 (99.9%). Phylogenetic analysis of the nucleotide sequences of KOF112 and *Bacillus* isolates showed that KOF112 formed a cluster with *B. velezensis* FZB42, which has a capacity to produce antimicrobial secondary metabolites (Figure 2) [21].

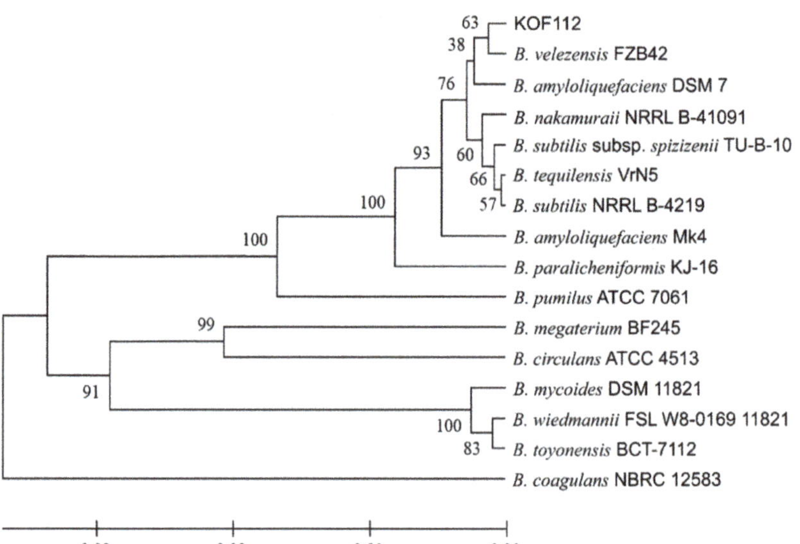

Figure 2. Phylogenetic analysis of 16S rDNA nucleotide sequence of KOF112 compared with those of *Bacillus* isolates. A phylogenetic tree was designed by means of the neighbor-joining (NJ) method using MEGA X program (bootstrap value with 1000 replicates). Bar indicates 1% band dissimilarity. Distance corresponds to the number of nucleotide substitutions per site. *B. velezensis* FZB42 (accession no. CP000560). *B. amyloliquefaciens* DSM7 (accession no. FN597644). *B. amyloliquefaciens* Mk4 (accession no. MT131178). *B. nakamuraii* NRRL B-41091 (accession no. KU836854). *B. subtilis* NRRL B-4219 (accession no. NR_116183). *B. tequilensis* VrN5 (accession no. LT986216). *B. subtilis* subsp. *spizenjii* TU-B-10 (accession no. CP602905). *B. paralicheniformis* KJ-16 (accession no. KY694465). *B. pumilus* ATCC 7061 (accession no. AY876289). *B. megaterium* BF245 (accession no. MK491077). *B. circulans* ATCC 4513 (accession no. AY724690). *B. coagulans* NBRC 12583 (accession no. KX261624). *B. mycoides* DSM 11821 (accession no. AB021199). *B. wiedmannii* FSL W8-0169 11821 (accession no. KU198626). *B. toyonensis* BCT-7112 (accession no. CP006863).

The draft genome of KOF112 includes a circular genome (3,916,789 bp) and a plasmid (13,003 bp). The annotation predicted 3746 coding sequences, 27 rRNA genes, and 86 tRNA genes. Comparison of KOF112 genome sequence with the genome sequences of antagonistic *Bacillus* isolates, *B. velezensis* FZB42 and *B. amyloliquefaciens* DSM7, which formed a cluster by 16S rDNA comparison (Figure 2), revealed significant differences among the genome sequences (Figure 3). For example, the gene clusters responsible for the non-ribosomal synthesized antimicrobial peptides and polyketides in KOF112 genome were different from those in FZB42 and DSM7 genomes. KOF112 genome has gene clusters for the biosynthesis of surfactin, bacillibactin, bacilysin, and bacillaene, but not gene clusters for the biosynthesis of iturin, bacillomycin, fengycin, difficidin, and macrolactin.

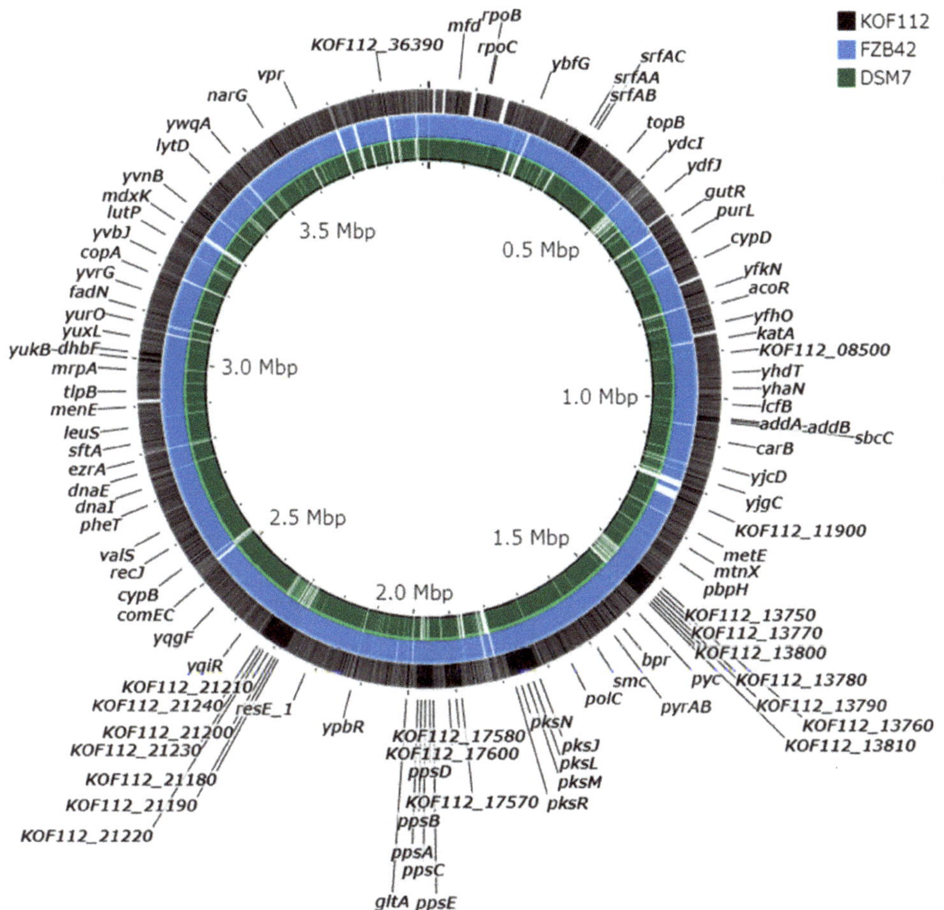

Figure 3. Genome comparison of KOF112, *B. velezensis* FZB42, and *B. amyloliquefaciens* DSM7. The whole genomes of KOF112 (outermost circle), FZB42 (middle circle), and DSM7 (innermost circle) were aligned using the CGView Server. The server used BLAST to compare the KOF112 genome sequence with FZB42 and DSM7 genome sequences. BLAST results and feature information for coding sequences were converted into a graphical map with black, blue, and green colors. Gene annotation of KOF112 coding sequences is partially shown in the figure.

Surfactin synthase subunit 1 (*srfAA*), surfactin synthase subunit 2 (*srfAB*), surfactin synthase subunit 3 (*srfAC*), and surfactin synthase thioesterase subunit (*srfAD*) were present in KOF112 genome (Figure 3). Surfactin biosynthetic gene cluster was selected as a non-ribosomal peptide synthase cluster and compared among KOF112, *B. velezensis* FZB42, and *B. amyloliquefaciens* DSM7 (Figure 4). The organization of *srfAA*, *srfAB*, *srfAC*, and *srfAD* showed a perfect match among the isolates (Figure 4A). Phylogenetic analysis of each gene indicated that KOF112 surfactin biosynthetic genes formed a cluster with *B. velezensis* FZB42 surfactin biosynthetic genes (Figure 4B).

Taken together, the results suggest that KOF112 is a strain of *B. velezensis*.

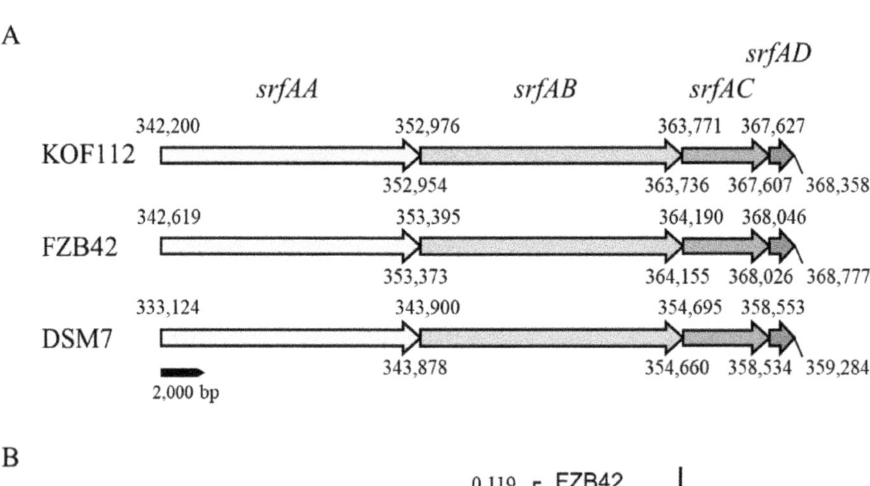

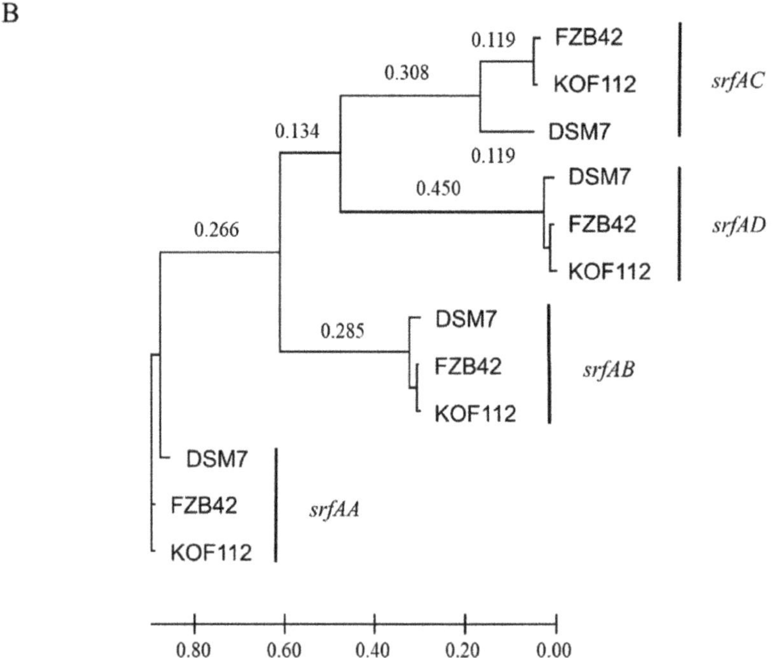

Figure 4. Comparison of surfactin biosynthetic gene clusters among KOF112, *B. velezensis* FZB42, and *B. amyloliquefaciens* DSM7. (**A**) Organization of surfactin biosynthesis genes. (**B**) Phylogenetic analysis of *srfAA*, *srfAB*, *srfAC*, and *srfAD* of KOF112 compared with those of FZB42 and DSM7. A phylogenetic tree was designed by means of the NJ method using MEGA X program (bootstrap value with 1000 replicates). Bar indicates 1% band dissimilarity. Distance corresponds to the number of nucleotide substitutions per site.

2.3. Biocontrol Activity of KOF112 against Downy Mildew in Grapevine

Because KOF112 inoculum contained 10% soybean casein digest (SCD) medium, 10% SCD medium was used as a control in the assay for biocontrol activity of KOF112. A large number of *P. viticola* white symptoms were observed on the untreated disks and the disks treated with SCD, whereas growth of symptoms on the disks treated with KOF112 was apparently inhibited (Figure 5A). In particular, 1×10^8 cfu/mL KOF112 completely controlled *P. viticola* symptoms. Disease severity was significantly reduced by KOF112 compared with control or SCD treatment (Figure 5B). The inhibitory effect of KOF112 on *P. viticola* infection was dose-dependent. No symptoms were observed on the disks treated with 1×10^8 cfu/mL KOF112, whereas a large number of white symptoms were visible on the disks treated with 1×10^5 cfu/mL KOF112. *Agrobacterium* sp. isolate CHB3 exhibited no inhibitory effect on downy mildew.

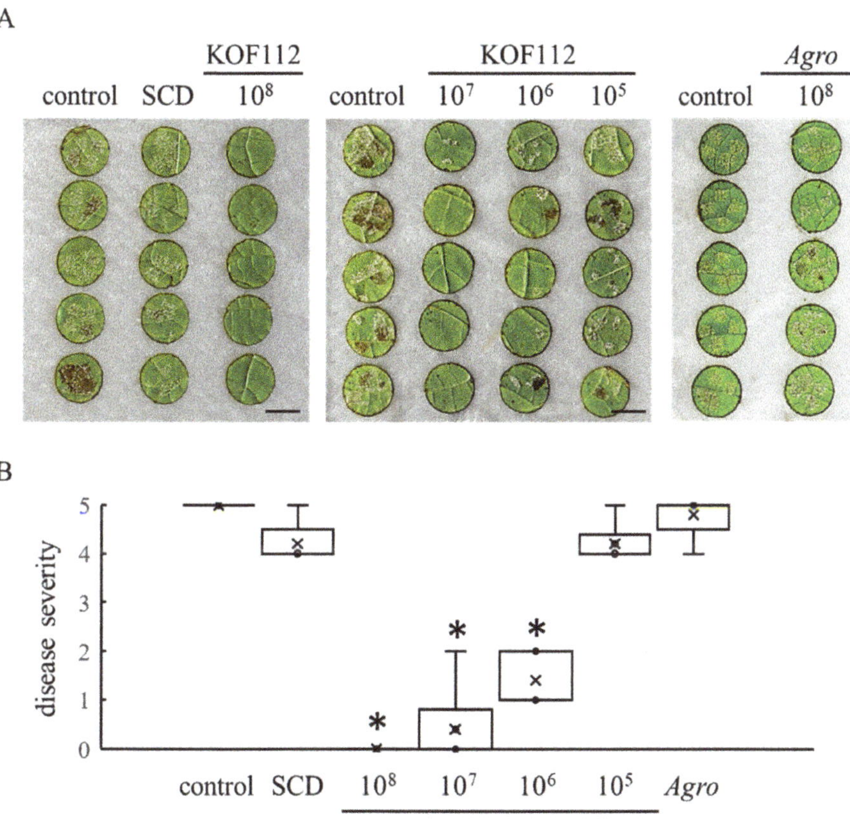

Figure 5. Suppression of grape downy mildew by KOF112. (**A**) Representative symptoms of grape downy mildew on leaf disks treated with KOF112 (1×10^8, 1×10^7, 1×10^6 or 1×10^5 cfu/mL) or *Agrobacterium* sp. isolate CHB3 (1×10^8 cfu/mL). Bar, 1 cm. (**B**) Disease severity was evaluated as described in Materials and Methods. Crosses ($\times$) indicate means of two independent experiments with five leaf disks. * $p < 0.05$ compared with control and SCD. Control, untreated. SCD, treated with 10% SCD. *Agro*, treated with *Agrobacterium* sp. isolate CHB3 used as control isolate with no antifungal activity.

2.4. Biocontrol Activity of KOF112 against Gray Mold in Cucumber

Cucumber leaves were used as the alternative host of *B. cinerea* because it was difficult to evaluate *B. cinerea* symptoms on grapevine leaves. The cucumber-*B. cinerea* pathosystem was used. Large and severe symptoms caused by *B. cinerea* were clearly observed on control or SCD-treated leaves (Figure 6). KOF112 (1×10^8 cfu/mL) significantly reduced the disease severity caused by *B. cinerea* compared with control and SCD treatment (Figure 6). *Agrobacterium* sp. isolate CHB3 exhibited no inhibitory effect on gray mold.

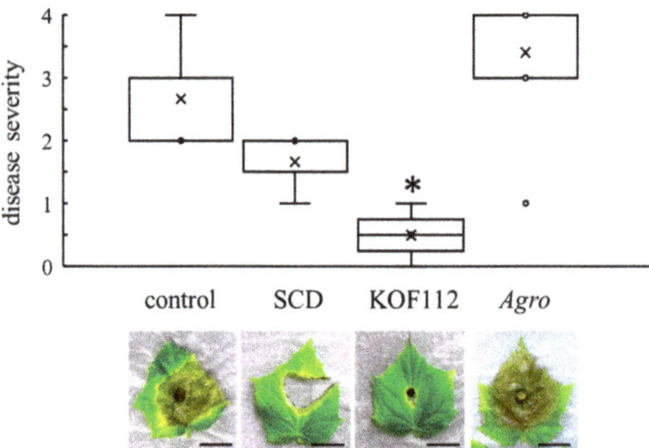

Figure 6. Suppression of cucumber gray mold by KOF112. Disease severity was evaluated as described in Materials and Methods. Crosses (×) indicate means of two independent experiments with three true leaves. * $p < 0.05$ compared with control and SCD. The representative symptom of cucumber gray mold on true leaves is shown below the graph. Control, untreated. SCD, treated with 10% SCD. KOF112, treated with 1×10^8 cfu/mL KOF112. *Agro*, treated with 1×10^8 cfu/mL *Agrobacterium* sp. isolate CHB3 used as control isolate with no antifungal activity. Bar, 2 cm.

2.5. Biocontrol Activity of KOF112 against Anthracnose in Strawberry

Strawberry leaves were used as the alternative host of *C. gloeosporioides* because it was difficult to evaluate *C. gloeosporioides* symptoms on grapevine leaves. The strawberry-*C. gloeosporioides* pathosystem was used. Large symptoms caused by *C. gloeosporioides* were clearly observed on control or SCD-treated leaves (Figure 7). KOF112 (1×10^8 cfu/mL) significantly reduced the disease severity caused by *C. gloeosporioides* compared with control and SCD treatment (Figure 7). *Agrobacterium* sp. isolate CHB3 exhibited no inhibitory effect on anthracnose.

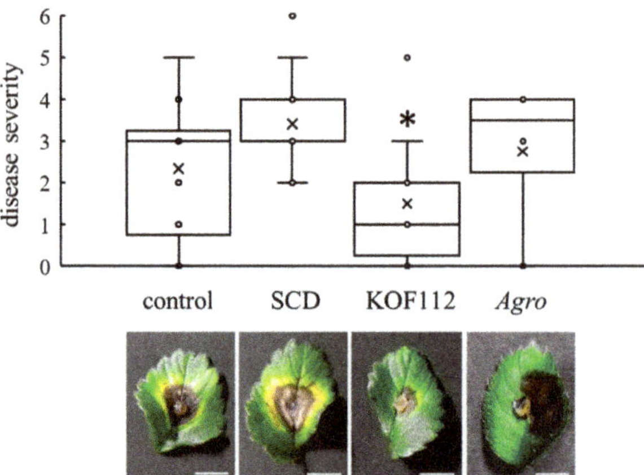

Figure 7. Suppression of strawberry anthracnose by KOF112. Disease severity was evaluated as described in Materials and Methods. Crosses (×) indicate means of four independent experiments with three leaves. * $p < 0.05$ compared with control and SCD. Each representative symptom of strawberry anthracnose on detached leaves is shown below the graph. Control, untreated. SCD, treated with 10% SCD. KOF112, treated with 1×10^8 cfu/mL KOF112. *Agro*, treated with 1×10^8 cfu/mL *Agrobacterium* sp. isolate CHB3, used as a control isolate with no antifungal activity. Bar, 1 cm.

2.6. Inhibition of Zoospore Release from P. viticola Zoosporangia by KOF112

As KOF112 inhibited the mycelial growth of *B. cinerea*, *C. gloeosporioides*, and *P. infestans* (Figure 1), mycelial disks of *B. cinerea* and *C. gloeosporioides* were used as inocula in the bioassays for biocontrol activity of KOF112 against gray mold and anthracnose, respectively (Figures 6 and 7). On the other hand, we used *P. viticola* zoosporangia as inoculum in the bioassay for biocontrol activity of KOF112 against downy mildew (Figure 5). Zoosporangia release many zoospores after inoculation, and the zoospores penetrate leaves through stomata [22]. Microscopic observation was performed to evaluate the effect of KOF112 on the early infection behaviors of *P. viticola* zoosporangia and zoospores. Empty zoosporangia were observed in the case of zoosporangia not treated or treated with SCD within 24 h after the treatment, whereas a large number of zoosporangia treated with KOF112 still had zoospores inside them after 24 h (Figure 8A). Interestingly, KOF112 seemed to surround zoosporangia that did not release zoospores (Figure 8A). Approximately 50% of zoosporangia not treated or treated with SCD released zoospores 24 h after the treatment, whereas approximately 24% of zoosporangia treated with KOF112 released zoospores after the same treatment period (Figure 8B). On the other hand, approximately 45% of zoospores germinated in spore suspension not treated or treated with 10% SCD after incubation for 20 h, whereas approximately 50% of zoospores germinated in the presence of KOF112 (Figure 8C). *Agrobacterium* sp. isolate CHB3 exhibited no effect on both zoospore release from zoosporangia and zoospore germination.

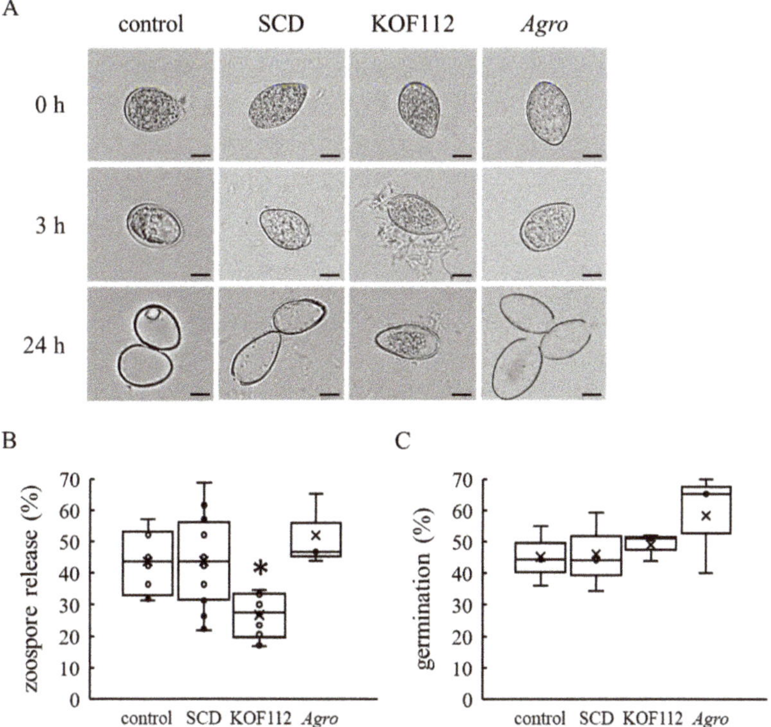

Figure 8. KOF112 inhibits zoospore release from zoosporangia but not zoospore germination. (**A**) Zoosporangia at 0, 3, and 24 h after preparation. Bar, 20 μm. (**B**) Zoospore release. The rates of

zoospore release from zoosporangia were calculated as described in Materials and Methods. Crosses (×) indicate means of eight independent preparations. * $p < 0.05$ compared with control and SCD. (C) Zoospore gemination. The rates of zoospore germination were calculated as described in Materials and Methods. Crosses (×) indicate means of three independent preparations. Control, untreated. SCD, treated with 10% SCD. KOF112, treated with 1×10^8 cfu/mL KOF112. Agro, treated with 1×10^8 cfu/mL Agrobacterium sp. isolate CHB3, used as a control isolate with no antifungal activity.

These results suggest that KOF112 inhibits zoospore release from *P. viticola* zoosporangia but not zoospore germination.

2.7. KOF112 Induces Plant Defense Response in Grapevine

The possibility that KOF112 induced plant defense response in grapevine was evaluated. We selected genes encoding class IV chitinase and β-1,3-glucanase as indicators of plant defense response, because chitinase and β-1,3-glucanase gene expression is induced through jasmonic acid (JA) and salicylic acid (SA)-dependent defense pathways, respectively. KOF112 drastically upregulated the expression of a gene-encoding class IV chitinase in grape leaf disks 48 h after KOF112 treatment, compared with untreated control and SCD treatment (Figure 9). The transcripts of β-1,3-glucanase gene were also increased 24 and 48 h after KOF112 treatment compared with untreated control and SCD treatment (Figure 9). *Agrobacterium* sp. isolate CHB3 did not upregulate the transcription of both genes.

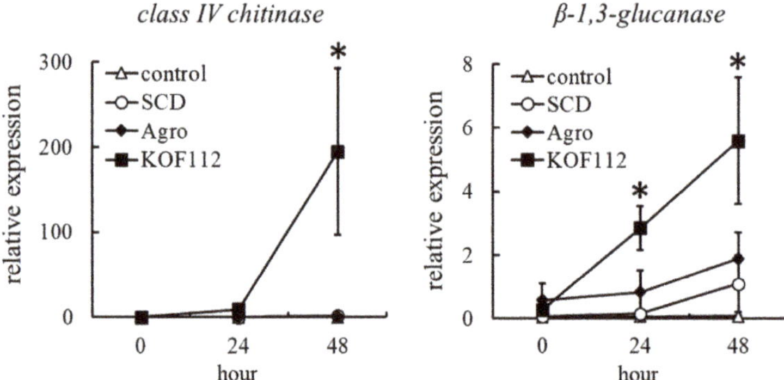

Figure 9. Transcription profiles of genes encoding class IV chitinase and β-1,3-glucanase in grape leaves treated with KOF112. Transcription levels of the genes in grape leaf disks 0, 24, and 48 h after KOF112 treatment were estimated by real-time RT-PCR. Data were calculated as gene expression relative to ubiquitin gene expression. Bars indicate means ± standard deviations of three independent experiments with three leaf disks. * $p < 0.05$ compared with control and SCD. Control, untreated. SCD, treated with 10% SCD. KOF112, treated with 1×10^8 cfu/mL KOF112. *Agro*, treated with 1×10^8 cfu/mL *Agrobacterium* sp. isolate CHB3, used as a control isolate with no antifungal activity.

These results suggest that KOF112 also works as a biotic elicitor and induces plant defense response in grapevine through both JA- and SA-dependent defense pathways.

3. Discussion

Grapes are one of the most important fruits cultivated worldwide. However, their high susceptibility to pre- and post-harvest pathogens has resulted in significant economic losses. The main diseases in viticulture are gray mold [23], ripe rot [24], and downy mildew [25]. As a biological control agent, KOF112 is able to suppress those three diseases. Although the disease-suppressing effect of biological control agents is weaker than that of chemical fungicides, the application of KOF112 having antagonistic activities toward a wide range of phytopathogenic fungi, including *Ascomycetes* and *Oomycetes*, may contribute to reducing the frequency of chemical fungicide application in viticulture, as well as to inhibiting the development of chemical fungicide resistance. The supernatants of biological control agents could be used as a new biostimulant in sustainable agriculture [26]. The supernatant of *B. subtilis* GLB191, which contains surfactin and fengycin, exerted a direct antifungal effect and induced plant defense response, thereby contributing to protection against grape downy mildew [17]. Future laboratory and field trials are necessary to evaluate whether the supernatant of KOF112 can be used as a biostimulant against gray mold, ripe rot, and downy mildew, and to identify the optimum conditions, including the dose of KOF112 and the timing of KOF112 application, to ensure that the combination treatment of KOF112 and chemical fungicides works effectively in viticulture.

Most commercially available biological control agents are *Bacillus*, which produces a large number of antibiotics [27,28]. *B. velezensis* FZB42 (former name *B. amyloliquefaciens* subsp. *plantarum*) [29] is used commercially as a biological control agent in agriculture [30] and produces such metabolites as peptides and polyketides, having antifungal, antibacterial, and nematocidal activities [31]. However, doubtful evidence related to the direct antifungal activity of FZB42 metabolites against competing plant pathogens has been presented. The concentrations of the metabolites produced by FZB42 in plants [32] were relatively low and/or undetectable, with the exception of surfactin. On the other hand, FZB42 compensated changes in microbial community structure caused by pathogens and helped plant-associated *Bacilli* contribute to plant protection [33]. Consequently, the sublethal concentrations of cyclic lipopeptides produced by *Bacilli* triggered plant defense response systemically.

Cyclic lipopeptides were found to enhance plant defense response to phytopathogenic fungi [34,35]. Surfactin and fengycin protected grape against downy mildew by exerting a direct antifungal effect and inducing plant defense response [17]. Cyclic lipopeptides activated plant defense response through distinct defense pathways [36]. Mycosubtilin (iturin family) activated both JA- and SA-dependent defense pathways, whereas surfactin mainly induced an SA-dependent plant defense response [36]. In the present study, gene expression analysis suggested that KOF112 induced plant defense response in grapevine through both JA- and SA-dependent defense pathways. Thus, enhancing plant defense response by antifungal metabolites produced by biological control agents is one of the mechanisms for protecting plants. Future studies employing liquid chromatography-tandem mass spectrometry analysis would identify the cyclic lipopeptides produced by KOF112 and reveal which cyclic lipopeptides contribute to the protective effect of KOF112 against fungal diseases.

In this study, we found that KOF112 also induced plant defense response in grapevine. Chitinase inhibited infection by *B. cinerea* [37] and *C. gloeosporioides* [38], whereas β-1,3-glucanase exerted direct antimicrobial activity against *B. cinerea* [39], *C. gloeosporioides* [40], and *P. viticola* [41]. Genome sequencing was attempted to clarify the biocontrol activity of the biological control agents [42]. Comparative genome analysis of KOF112 and FZB42 demonstrated that KOF112 might produce fewer antimicrobial peptides and polyketides than FZB42. The direct inhibition of phytopathogenic fungal mycelial growth in vitro is an indirect indication that KOF112 produces some antibiotics. Although the inhibition of zoospore release from *P. viticola* zoosporangia by KOF112 may be an interesting mechanism underlying the biocontrol activity against grape downy mildew, the probable antibiotic production by KOF112 may confer enhanced plant defense response in the same way as FZB42. Further studies employing the genomic analysis of KOF112 for exploring genes encoding bioactive substances, as well as the qualitative and quantitative analyses of antibiotics produced by KOF112, would reveal the main mechanism underlying plant disease control by KOF112. Because we were unable to analyze the biocontrol activity of KOF112 by field trials in vineyards, we could not verify whether our laboratory experiments using leaf disks support our hypothesis of increased plant resistance. We need to conduct field trials in vineyards to verify our hypothesis that KOF112 also works as a biotic elicitor in pest management strategies.

Grapevines are a rich source of potential biological control agents for fungal and oomycete pathogens. Bacterial isolates collected from endophytic and epiphytic communities living in grapevine leaves inhibited *B. cinerea* and *P. infestans* mycelial growth [43]. The colonization of biological control agents *in planta* is one of the driving forces for protecting plants against diseases. The suppressive activities of biological control agents against phytopathogenic fungi are influenced by the capacity of those agents to colonize in planta [20,44]. Endophytes are considered potential biological control agents because of their colonization ability. In plant root endospheres, high motility as well as enhanced plant cell-wall degradation and reactive oxygen-species-scavenging abilities seem to be important traits for successful endophytic colonization [45]. Because endophytic bacteria play a role in the resistance to biotic and abiotic stresses as well as plant growth and development, manipulating endophytic bacteria would help us develop novel and innovative techniques for improving agricultural production. In fact, plants have unique endophytes, some of which confer resistance to the plants themselves [46]. In this study, we focused on endophytic bacteria and isolated antifungal endophyte KOF112 from grapevine shoot xylem. It is reasonable to assume that endophytes show greater affinity for plants than soil microorganisms. Although KOF112 can induce plant defense response in grapevine, we cannot present concrete results of KOF112 colonization in grapevine at the moment. Further investigation by scanning electron microscopy would reveal whether KOF112 colonizes foliar-sprayed leaves. If KOF112 can colonize well in and/or on grapevine, KOF112 would have an edge over commercial biological fungicides that are generally considered to have a short life span on grapevine. Future studies involving field trials of KOF112 application to bunches and leaves and/or KOF112 injection into grapevine shoot xylems would reveal whether KOF112 colonization in grapevine results in the optimal biocontrol activity of endophytic KOF112 against fungal diseases.

4. Materials and Methods

4.1. Plant Materials

Grapevines (*Vitis* sp. cv. Koshu, *V. vinifera* cvs. Pinot Noir, Chardonnay, and Cabernet Sauvignon) were cultivated in the experimental vineyard of The Institute of Enology and Viticulture, University of Yamanashi, Yamanashi, Japan (latitude 35.680528, longitude 138.569268, elevation 250 m). The grapevines were trained in double cordon style and were approximately 30 years old.

4.2. Isolation of Endophytic Bacteria from Grapevine Shoot Xylem

Grapevine shoots were collected on 9 December 2019. The surface of the shoots was sterilized with 0.1% (*v/v*) sodium hypochlorite solution for 3 min at room temperature. Bark and epidermal tissue were peeled off from the shoots using a knife sterilized with 70% ethanol. Xylem was shaved using a grater sterilized with 70% ethanol. One gram of shaved xylem was placed in a sterilized 100 mL flask containing 40 mL of phosphate buffer solution (pH 7.4). After shaking at 130 rpm for 3 h at 25 °C, the phosphate buffer solution with shaved xylem was filtered through sterilized cotton gauze. One hundred microliters of filtrate was plated on SCD, Luria–Bertani broth, and potato dextrose broth agar (PDA) plates, and the plates were incubated at 25 °C for 3 d. Bacterial isolates that grew on the plates were used in the in vitro bioassay.

4.3. In Vitro Bioassay

To screen bacterial isolates for their antagonistic activity toward three fungal phytopathogens, *B. cinerea*, *C. gloeosporioides*, and *P. infestans* (an oomycete pathogen of late blight), the dual culture technique was performed as described previously (Figure 1) [47]. *B. cinerea* and *C. gloeosporioides* laboratory strains isolated from the experimental vineyard of The Institute of Enology and Viticulture were used [11,48]. *P. infestans* isolate (NBRC 9173) was obtained from the NITE Biological Resource Center. Briefly, bacterial isolates were streaked on the edge of SCD plates. An agar plug (6.5 mm diameter) cut from colonies of the fungal phytopathogens was placed at the center of the bacteria-growing SCD plates, and the plates were incubated at 25 °C for 5 d. Bacterial isolates that suppressed the mycelial growth of all the fungal phytopathogens by forming a growth inhibition zone were selected as potent antagonists. From the assay, we obtained KOF112 as the best candidate strain. *Agrobacterium* sp. isolate CHB3 was selected as the control isolate, with no antifungal activity to evaluate the biocontrol activity of KOF112 against the three fungal phytopathogens.

4.4. Identification of KOF112 by 16S rDNA Sequence Analysis

Genomic DNA was extracted from the one-day culture of KOF112 in SCD medium using a DNeasy PowerSoil Kit (Qiagen, Hilden, Germany) in accordance with the manufacturer's instructions. PCR conditions for amplifying partial 16S rDNA were as follows: after incubation at 94 °C for 5 min, PCR amplification was performed for 30 cycles at 94 °C for 30 s, 55 °C for 30 s, and 72 °C for 1 min, with a final extension step at 72 °C for 7 min. The nucleotide sequences of the primer set for amplifying partial 16S rDNA from bacterial genome were as follows: 795F (5′-GGATTAGATACCCTGGTA-3′) and 1492R (5′-GGYTACCTTGTTACGACTT-3′). The nucleotide sequences of the amplicons were analyzed using the dye terminator method and subjected to the Basic Local Alignment Search Tool (BLAST, NCBI). Phylogenetic analysis was performed using the partial 16S rDNA of KOF112 and *Bacillus* isolates deposited in the NCBI database. The nucleotide sequences were subjected to the NJ method using Molecular Evolutionary Genetics Analysis software, MEGA10 (www.megasoftware.net, accessed on 7 April 2021).

4.5. Genome Sequencing, Assembly, and Annotation

DNA extraction from KOF112 was performed using a DNeasy PowerSoil Kit (Qiagen). Genome sequencing, assembly, and annotation were performed as described previously [49]. Briefly, short-read sequencing using a DNBSEQ-G400 sequencer (MGI Tech, Shenzhen, China) and long-read sequencing using the GridION nanopore sequencing platform (Oxford Nanopore Technologies, Oxford, UK) were performed in accordance with the manufacturers' instructions. The high-quality short-read and long-read sequences were assembled using Unicycler (ver. 0.4.7) under default conditions. The annotation was performed by Prokka (ver. 1.13). The draft genome sequence of KOF112 was deposited in DDBJ/ENA/GenBank under accession no. AP024603 (genome) and AP024604 (plasmid).

4.6. Comparative Genome Analysis

Genome sequences of *B. velezensis* FZB42 (former name *B. amyloliquefaciens* subsp. plantarum, accession no. CP000560) and *B. amyloliquefaciens* DSM7 (accession no. FN597644) were used as representative antagonistic *Bacillus* species for comparative genome analysis. Comparative genome analysis was performed using the CGView Server [50].

4.7. Biocontrol Activity of KOF112 against Downy Mildew

An in vivo bioassay using grape leaf disks was performed as described previously, with minor modifications [51]. Briefly, Koshu leaves were collected from potted seedlings. Five leaf disks each having a diameter of 13 mm were cut out from the leaves using a cork polisher and placed upside down on moistened filter paper in square Petri dishes (140 mm × 100 mm). KOF112 was incubated in SCD medium overnight and adjusted to 1×10^8, 1×10^7, 1×10^6, and 1×10^5 cfu/mL with SCD medium and sterile water (all KOF112 solutions contained 10% SCD medium). Ten microliters of KOF112 solution was dropped onto four locations on the abaxial surface of the leaf disk. Ten percent SCD medium and *Agrobacterium* sp. isolate CHB3 (1×10^8 cfu/mL, containing 10% SCD) were used as control. The leaf disks were dried at room temperature in a flow cabinet for 3 h.

Inoculation of *P. viticola* zoosporangia was performed as described previously with minor modifications [6]. Briefly, field-isolated *P. viticola* was maintained on leaves of potted Koshu seedlings in a growth chamber under light irradiation ($11.8\ \mathrm{Wm}^{-2}/16$

h/d) at 21 °C. Fresh zoosporangia of *P. viticola* were washed off with sterile water from the symptoms on the Koshu leaves. Ten microliters of the zoosporangium suspension (1×10^4 zoosporangia/mL) was dropped onto the same locations as the pretreated KOF112 solutions on the leaf disks. The Petri dishes containing the leaf disks were placed in a plastic box containing moistened paper towel. The box was incubated in the dark for 24 h and then in an incubator (21 °C, 11.8 Wm^{-2}/16 h/d). Downy mildew symptoms on each disk were assessed 7 days post *P. viticola* inoculation. Disease severity was scored by evaluating the symptom on each disk as described previously [51]: 0, no symptoms; 1, white symptom occupies up to 1/6 of the disk; 2, white symptom occupies up to 1/3 of the disk; 3, white symptom occupies up to half of the disk; 4, white symptom occupies up to two-thirds of the disk; 5, white symptom occupies more than two-thirds of the disk. Two independent experiments were performed with five leaf disks.

4.8. Biocontrol Activity of KOF112 against Gray Mold

Three true leaves (approximately 4 cm in size) were detached from cucumber (*Cucumis sativus* cv. Sharp-1) seedlings and sprayed with 500 µL of KOF112 solution (1×10^8 cfu/mL, containing 10% SCD) with a hand sprayer. As the control experiment, sterile water, 10% SCD medium, or *Agrobacterium* sp. isolate CHB3 (1×10^8 cfu/mL, containing 10% SCD) was sprayed. After air drying the cotyledons, the center of each cotyledon was punctured with a sterile needle. Mycelial disks (6.5 mm diameter) of *B. cinerea* laboratory strain were excised from the leading edge of the colonies on PDA plates, and a disk was placed on the wound of each cotyledon. After incubation in a moisture chamber at 25 °C for 6 d in an incubator (11.8 Wm^{-2}/16 h/d), disease severity was scored by evaluating the symptom of each cotyledon as described previously [11]: 0, no symptoms; 0.5, rot only under inoculum; 1, rot two times larger than the disk; 2, rot three times larger than the disk; 3, rot four times larger than the disk; 4, rot more than five times larger than the disk. Two independent experiments were performed with three leaves.

4.9. Biocontrol Activity of KOF112 against Anthracnose

Strawberry leaves (*Fragaria ananassa* cv. Hokowase) were detached from seedlings cultivated in a growth chamber. Three detached leaves were sprayed with 500 µL of KOF112 solution (1×10^8 cfu/mL) with a hand sprayer. As the control experiment, sterile water, 10% SCD medium, or *Agrobacterium* sp. isolate CHB3 (1×10^8 cfu/mL, containing 10% SCD) was sprayed. After air drying the leaves, the center of each leaf was punctured with a sterile needle. Mycelial disks (6.5 mm diameter) of *C. gloeosporioides* laboratory strain were excised from the leading edge of the colonies on PDA plates, and a disk was placed on the wound of each leaf. After incubation in a moisture chamber at 25 °C for 8 d in the dark, disease severity was scored by evaluating the symptom of each leaf as described previously [48]: 0, no symptoms; 1, rot only under the disk; 2, rot two times larger than the disk; 3, rot three times larger than the disk; 4, rot four times larger than the disk; 5, rot more than five times larger than the disk; 6, rot more than six times larger than the disk; 7, rot more than seven times larger than the disk. Four independent experiments were performed with three leaves.

4.10. Light Microscope Observation of P. viticola Zoospore Release from Zoosporangia and Zoospore Germination

A zoosporangium suspension of *P. viticola* (1×10^4 zoosporangia/mL) was prepared as mentioned above. Nine hundred microliters of the zoosporangium suspension was incubated with 100 µL of KOF112 solution (1×10^8 cfu/mL, containing 10% SCD), 10% SCD medium, or *Agrobacterium* sp. isolate CHB3 (1×10^8 cfu/mL, containing 10% SCD) in a microtube at 22 °C for 1 h under light irradiation (11.8 Wm^{-2}), and then for 23 h in the dark. Zoospore release from zoosporangia was observed under a light microscope (Olympus BX51, Tokyo, Japan). Zoospore release rate was calculated using the following formula:

$$\text{zoospore release (\%)} = \text{number of empty zoosporangia/number of total zoosporangia} \times 100$$

Zoospore germination was counted as described previously, with minor modifications [6]. Briefly, a zoosporangium suspension of *P. viticola* (1×10^4 zoosporangia/mL) was prepared as mentioned above. Five hundred microliters of the zoosporangium suspension was incubated in a microtube at 30 °C for 4 h under light irradiation ($11.8\ \mathrm{Wm}^{-2}$). One hundred microliters of KOF112 solution (1×10^8 cfu/mL, containing 10% SCD), 10% SCD, or *Agrobacterium* sp. isolate CHB3 (1×10^8 cfu/mL, containing 10% SCD) was added into the microtube, and then incubation was carried out at 30 °C for 12 h under light irradiation ($11.8\ \mathrm{Wm}^{-2}$), followed by incubation at 30 °C for 8 h in the dark. The suspension was stained with 0.05% aniline blue in 0.0067M K_2HPO_4 (pH 9–9.5) at room temperature for 20 min. Zoospore germination was observed using a fluorescence microscope (Olympus BX51). Zoospore germination rate was calculated using the following formula:

$$\text{germination (\%)} = \text{number of germinated zoospores/number of total zoospores} \times 100$$

4.11. Real-Time RT-PCR

To determine whether KOF112 induces plant defense response in grapevines, transcriptional alteration of genes encoding two PR proteins, class IV chitinase and β-1,3-glucanase, was evaluated in KOF112-treated grapevine leaves. Three leaf disks each having a diameter of 13 mm were cut out from Koshu leaves using a cork polisher and placed upside down on moistened filter paper in square Petri dishes (140 mm × 100 mm). Ten microliters of KOF112 solution (1×10^8 cfu/mL, containing 10% SCD) was dropped onto four locations on the abaxial surface of the leaf disk. Ten percent SCD medium or *Agrobacterium* sp. isolate CHB3 (1×10^8 cfu/mL, containing 10% SCD) was used as control. After incubation at 22 °C for 24 h and 48 h in an incubator ($11.8\ \mathrm{Wm}^{-2}$/16 h/d), the disks were homogenized in a mortar containing liquid nitrogen using a pestle. Total RNA isolation from the pulverized samples was performed using NucleoSpin RNA Plant (Takara, Shiga, Japan), and purification was carried out using Fruit-mate for RNA Purification (Takara). First-strand cDNA was synthesized from the total RNA using PrimeScript RT Master Mix (Perfect Real Time) (Takara). Real-time RT-PCR was performed using an SYBR Premix Ex Taq II (Perfect Real Time) (Takara) with a Thermal Cycler Dice Real Time System (Takara). PCR conditions were as follows: incubation at 95 °C for 30 s, followed by 40 cycles at 95 °C for 5 s and at 60 °C for 45 s. The primers used for amplification were as follows: ubiquitin primers (5′-GTGGTATTATTGAGCCATCCTT-3′ and 5′-AACCTCCAATCCAGTCATCTAC-3′, GenBank accession no. BN000705); class IV chitinase primers (5′-CAATCGGGTCCTTGTGATTC-3′ and 5′-CAAGGCACTGAGAAACGCT-3′, GenBank accession no. U97522), and β-1,3-glucanase primers (5′-GAATCTGTTCGATGCCATGC-3′ and 5′-GCATTATCAACCGTAGTCCC-3′, GenBank accession no. DQ267748). Ubiquitin primers were used as the reference gene to normalize each gene expression because ubiquitin gene expression was stable in the grapevine leaves [52]. Using the standard curve method of Thermal Cycler Dice Real Time System Single Software ver. 3.00 (Takara), gene expression levels were determined as the number of amplification cycles needed to reach a fixed threshold and are expressed as relative values to ubiquitin.

4.12. Statistical Analysis

Data are presented as means ± standard deviations of biological replicates. Statistical analysis was performed by the Student's *t*-test or Dunnett's multiple comparison test using Excel Statistics software 2012 (Social Survey Research Information, Tokyo, Japan).

Author Contributions: Conceptualization, K.H. and S.S.; methodology, K.H. and Y.A.; investigation, K.H.; data curation, K.H. and Y.A.; writing—original draft preparation, K.H.; writing—review and editing, S.S.; project administration, S.S. All authors have read and agreed to the published version of the manuscript.

Funding: This research received no external funding.

Data Availability Statement: KOF112 genome was deposited in DDBJ/ENA/GenBank under accession numbers PRJDB11468 (BioProject), SAMD00293740 (BioSample), AP024603 (genome), AP024604 (plasmid), and DRA011804 (raw sequencing reads).

Acknowledgments: We are grateful to Keisuke Sugiyama of the University of Yamanashi for helpful discussion.

References

1. Goto-Yamamoto, N.; Sawler, J.; Myles, S. Genetic analysis of East Asian grape cultivars suggests hybridization with wild *Vitis*. *PLoS ONE* **2015**, *10*, e0140841. [CrossRef]
2. Koshu of Japan. Available online: www.koshuofjapan.com/index.html (accessed on 4 April 2021).
3. Tanaka, K.; Hamaguchi, Y.; Suzuki, S.; Enoki, S. Genomic characterization of Japanese indigenous wine grape *Vitis* sp. cv. Koshu. *Front. Plant Sci.* **2020**, *11*, 532211. [CrossRef]
4. Kono, A.; Ban, Y.; Sato, A.; Mitani, N. Evaluation of 17 table grape accessions for foliar resistance to downy mildew. *Acta Hortic.* **2015**, *1082*, 207–211. [CrossRef]
5. Tsuguti, H.; Seino, N.; Kawase, H.; Imada, Y.; Nakaegawa, T.; Takayabu, I. Meteorological overview and mesoscale characteristics of the heavy rain event of July 2018 in Japan. *Landslides* **2019**, *16*, 363–371. [CrossRef]
6. Aoki, Y.; Usujima, A.; Suzuki, S. High night temperature promotes downy mildew in grapevine via attenuating plant defense response and enhancing early *Plasmopara viticola* infection. *Plant Prot. Sci.* **2021**, *57*, 21–30. [CrossRef]
7. Fungicide Resistance Action Committee. *CAA Working Group Report*. Available online: http://www.frac.info/frac/index.html (accessed on 11 April 2021).
8. Furuya, S.; Mochizuki, M.; Saito, S.; Kobayashi, H.; Takayanagi, T.; Suzuki, S. Monitoring of QoI fungicide resistance in *Plasmopara viticola* populations in Japan. *Pest Manag. Sci.* **2010**, *66*, 1268–1272. [CrossRef] [PubMed]
9. Aoki, Y.; Kawagoe, Y.; Fujimori, N.; Tanaka, S.; Suzuki, S. Monitoring of a single point mutation in the *PvCesA3* allele conferring resistance to carboxylic acid amide fungicides in *Plasmopara viticola* populations in Yamanashi prefecture, Japan. *Plant Health Prog.* **2015**, *16*, 84–87. [CrossRef]
10. Chung, W.H.; Ishii, H.; Nishimura, K.; Fukaya, M.; Yano, K.; Kajitani, Y. Fungicide sensitivity and phylogenetic relationship of anthracnose fungi isolated from various fruit crops in Japan. *Plant Dis.* **2006**, *90*, 506–512. [CrossRef]
11. Saito, S.; Suzuki, S.; Takayanagi, T. Nested PCR-RFLP is a high-speed method to detect fungicide-resistant *Botrytis cinerea* at an early growth stage of grapes. *Pest Manag. Sci.* **2009**, *65*, 197–204. [CrossRef] [PubMed]
12. Otoguro, M.; Suzuki, S. Status and future of disease protection and grape berry quality alteration by microorganisms in viticulture. *Lett. Appl. Microbiol.* **2018**, *67*, 106–112. [CrossRef] [PubMed]
13. Aoki, T.; Aoki, Y.; Ishiai, S.; Otoguro, M.; Suzuki, S. Impact of *Bacillus cereus* NRKT on grape ripe rot disease through resveratrol synthesis in berry skin. *Pest Manag. Sci.* **2017**, *73*, 174–180. [CrossRef] [PubMed]
14. Aoki, Y.; Haga, S.; Suzuki, S. Direct antagonistic activity of chitinase produced by *Trichoderma* sp. SANA20 as biological control agent for gray mould caused by *Botrytis cinerea*. *Cogent Biol.* **2020**, *6*, 1747903. [CrossRef]
15. Rotolo, C.; De Miccolis Angelini, R.M.; Dongiovanni, C.; Pollastro, S.; Fumarola, G.; Di Carolo, M.; Perrelli, D.; Natale, P.; Faretra, F. Use of biocontrol agents and botanicals in integrated management of *Botrytis cinerea* in table grape vineyards. *Pest Manag. Sci.* **2018**, *74*, 715–725. [CrossRef]
16. Heyman, L.; Ferrarini, E.; Omoboye, O.O.; Sanchez, L.; Barka, E.A.; Höfte, M. Potential of *Pseudomonas* cyclic lipopeptides to control downy mildew in grapevine by induced resistance and direct antagonism. In *Molecular Plant-Microbe Interactions*; American Phytopathological Society: St Paul, MN, USA, 2019; p. 95.
17. Li, Y.; Héloir, M.C.; Zhang, X.; Geissler, M.; Trouvelot, S.; Jacquens, L.; Henkel, M.; Su, X.; Fang, X.; Wang, Q.; et al. Surfactin and fengycin contribute to the protection of a *Bacillus subtilis* strain against grape downy mildew by both direct effect and defence stimulation. *Mol. Plant Pathol.* **2019**, *20*, 1037–1050. [CrossRef]
18. Omoboye, O.O.; Oni, F.E.; Batool, H.; Yimer, H.Z.; De Mot, R.; Höfte, M. *Pseudomonas* cyclic lipopeptides suppress the rice blast fungus *Magnaporthe oryzae* by induced resistance and direct antagonism. *Front. Plant Sci.* **2019**, *10*, 901. [CrossRef]
19. Kawagoe, Y.; Shiraishi, S.; Kondo, H.; Yamamoto, S.; Aoki, Y.; Suzuki, S. Cyclic peptide iturin A structure-dependently induces defense response in *Arabidopsis* plants by activating SA and JA signaling pathways. *Biochem. Biophys. Res. Commun.* **2015**, *460*, 1015–1020. [CrossRef]
20. Romero, D.; Pérez-García, A.; Rivera, M.E.; Cazorla, F.M.; de Vicente, A. Isolation and evaluation of antagonistic bacteria towards the cucurbit powdery mildew fungus *Podosphaera fusca*. *Appl. Microbiol. Biotechnol.* **2004**, *64*, 263–269. [CrossRef] [PubMed]
21. Scholz, R.; Vater, J.; Budiharjo, A.; Wang, Z.; He, Y.; Dietel, K.; Borriss, R. Amylocyclicin, a novel circular bacteriocin produced by *Bacillus amyloliquefaciens* FZB42. *J. Bacteriol.* **2014**, *196*, 1842–1852. [CrossRef] [PubMed]
22. Kiefer, B.; Riemann, M.; Büche, C.; Kassemeyer, H.-H.; Nick, P. The host guides morphogenesis and stomatal targeting in the grapevine pathogen *Plasmopara viticola*. *Planta* **2002**, *215*, 387–393. [CrossRef]
23. McClellan, W.D.; Hewitt, W.B. Early botrytis rot of grapes: Time of infection and latency of *Botrytis cinerea* Pers. in *Vitis vinifera* L. *Phytopathology* **1973**, *63*, 1151–1156. [CrossRef]

24. Whitelaw-Weckert, M.A.; Curtin, S.J.; Huang, R.; Steel, C.C.; Blanchard, C.L.; Roffey, P.E. Phylogenetic relationships and pathogenicity of *Colletotrichum acutatum* isolates from grape in subtropical Australia. *Plant Pathol.* **2007**, *56*, 448–463. [CrossRef]
25. Emmett, R.W.; Wicks, T.J.; Magarey, P.A. Downy mildew of grapes. In *Diseases of Fruit Crops. Plant Diseases of International Importance*; Kumar, J., Chaube, H.S., Singh, U.S., Mukhopadhyay, A.N., Eds.; Prentice Hall: Englewood Cliffs, NJ, USA, 1992; pp. 90–128.
26. Pellegrini, M.; Pagnani, G.; Bernardi, M.; Mattedi, A.; Spera, D.M.; Gallo, M.D. Cell-free supernatants of plant growth-promoting bacteria: A review of their use as biostimulant and microbial biocontrol agents in sustainable agriculture. *Sustainability* **2020**, *12*, 9917. [CrossRef]
27. Ongena, M.; Jacques, P. *Bacillus* lipopeptides: Versatile weapons for plant disease biocontrol. *Trends Microbiol.* **2008**, *16*, 115–125. [CrossRef] [PubMed]
28. Stein, T. *Bacillus subtilis* antibiotics: Structures, syntheses and specific functions. *Mol. Microbiol.* **2005**, *56*, 845–857. [CrossRef] [PubMed]
29. Dunlap, C.; Kim, S.J.; Kwon, S.W.; Rooney, A. *Bacillus velezensis* is not a later heterotypic synonym of *Bacillus amyloliquefaciens*, *Bacillus methylotrophicus*, *Bacillus amyloliquefaciens* subsp. *plantarum* and '*Bacillus oryzicola*' are later heterotypic synonyms of *Bacillus velezensis* based on phylogenomics. *Int. J. Syst. Evol. Microbiol.* **2016**, *66*, 1212–1217.
30. Borriss, R. Use of plant-associated *Bacillus* strains as biofertilizers and biocontrol agents. In *Bacteria in Agrobiology: Plant Growth Responses*; Maheshwari, D.K., Ed.; Springer: Berlin/Heidelberg, Germany, 2011; pp. 41–76.
31. Chen, X.H.; Koumoutsi, A.; Scholz, R.; Schneider, K.; Vater, J.; Süssmuth, R.; Piel, J.; Borriss, R. Genome analysis of *Bacillus amyloliquefaciens* FZB42 reveals its potential for biocontrol of plant pathogens. *J. Biotechnol.* **2009**, *140*, 27–37. [CrossRef] [PubMed]
32. Chowdhury, S.P.; Uhl, J.; Grosch, R.; Alquéres, S.; Pittroff, S.; Dietel, K.; Schmitt-Kopplin, P.; Borriss, R.; Hartmann, A. Cyclic lipopeptides of *Bacillus amyloliquefaciens* subsp. plantarum colonizing the lettuce rhizosphere enhance plant defense responses toward the bottom rot pathogen *Rhizoctonia solani*. *Mol. Plant Microbe Interact.* **2015**, *28*, 984–995. [CrossRef]
33. Chowdhury, S.P.; Hartmann, A.; Gao, X.; Borriss, R. Biocontrol mechanism by root-associated *Bacillus amyloliquefaciens* FZB42—a review. *Front. Microbiol.* **2015**, *6*, 780. [CrossRef]
34. Yamamoto, S.; Shiraishi, S.; Suzuki, S. Are cyclic lipopeptides produced by *Bacillus amyloliquefaciens* S13-3 responsible for the plant defense response in strawberry against *Colletotrichum gloeosporioides*? *Lett. Appl. Microbiol.* **2015**, *60*, 379–386. [CrossRef]
35. Pršić, J.; Ongena, M. Elicitors of plant immunity triggered by beneficial bacteria. *Front. Plant Sci.* **2020**, *11*, 594530. [CrossRef]
36. Farace, G.; Fernandez, O.; Jacquens, L.; Coutte, F.; Krier, F.; Jacques, P.; Clément, C.; Barka, E.A.; Jacquard, C.; Dorey, S. Cyclic lipopeptides from *Bacillus subtilis* activate distinct patterns of defence responses in grapevine. *Mol. Plant Pathol.* **2015**, *16*, 177–187. [CrossRef]
37. Aziz, A.; Heyraud, A.; Lambert, B. Oligogalacturonide signal transduction, induction of defense-related responses and protection of grapevine against *Botrytis cinerea*. *Planta* **2004**, *218*, 767–774. [CrossRef]
38. Ali, M.; Tumbeh Lamin-Samu, A.; Muhammad, I.; Farghal, M.; Khattak, A.M.; Jan, I.; ul Haq, S.; Khan, A.; Gong, Z.-H.; Lu, G. Melatonin mitigates the infection of *Colletotrichum gloeosporioides* via modulation of the chitinase gene and antioxidant activity in *Capsicum annuum* L. *Antioxidants* **2001**, *10*, 7. [CrossRef] [PubMed]
39. Fujimori, N.; Enoki, S.; Suzuki, A.; Naznin, H.A.; Shimizu, M.; Suzuki, S. Grape apoplasmic β-1,3-glucanase confers fungal disease resistance in *Arabidopsis*. *Sci. Hortic.* **2016**, *200*, 105–110. [CrossRef]
40. Bautista-Rosales, P.U.; Calderon-Santoyo, M.; Servín-Villegas, R.; Ochoa-Álvarez, N.A. Action mechanisms of the yeast *Meyerozyma caribbica* for the control of the phytopathogen *Colletotrichum gloeosporioides* in mangoes. *Biol. Control* **2013**, *65*, 293–301. [CrossRef]
41. Mestre, P.; Arista, G.; Piron, M.C.; Rustenholz, C.; Ritzenthaler, C.; Merdinoglu, D.; Chich, J.F. Identification of a *Vitis vinifera* endo-β-1,3-glucanase with antimicrobial activity against *Plasmopara viticola*. *Mol. Plant Pathol.* **2017**, *18*, 708–719. [CrossRef] [PubMed]
42. Leal, C.; Fontaine, F.; Aziz, A.; Egas, C.; Clément, C.; Trotel-Aziz, P. Genome sequence analysis of the beneficial *Bacillus subtilis* PTA-271 isolated from a *Vitis vinifera* (cv. Chardonnay) rhizospheric soil: Assets for sustainable biocontrol. *Environ. Microbiome* **2021**, *16*, 3. [CrossRef]
43. Bruisson, S.; Zufferey, M.; L'Haridon, F.; Trutmann, E.; Anand, A.; Dutartre, A.; De Vrieze, M.; Weisskopf, L. Endophytes and epiphytes from the grapevine leaf microbiome as potential biocontrol agents against phytopathogens. *Front Microbiol.* **2019**, *10*, 2726. [CrossRef]
44. Shimizu, M.; Yazawa, S.; Ushijima, Y. A promising strain of endophytic *Streptomyces* sp. for biological control of cucumber anthracnose. *J. Gen. Plant Pathol.* **2009**, *75*, 27–36. [CrossRef]
45. Liu, H.; Carvalhais, L.C.; Crawford, M.; Singh, E.; Dennis, P.G.; Pieterse, C.M.J.; Schenk, P.M. Inner plant values: Diversity, colonization and benefits from endophytic bacteria. *Front. Microbiol.* **2017**, *8*, 2552. [CrossRef]
46. Rajendran, L.; Samiyappan, R.; Raguchander, T.; Saravanakumar, D. Endophytic bacteria mediate plant resistance against cotton bollworm. *J. Plant Interact.* **2007**, *2*, 1–10. [CrossRef]
47. Furuya, S.; Mochizuki, M.; Aoki, Y.; Kobayashi, H.; Takayanagi, T.; Shimizu, M.; Suzuki, S. Isolation and characterization of *Bacillus subtilis* KS1 for the biocontrol of grapevine fungal diseases. *Biocontrol Sci. Technol.* **2011**, *21*, 705–720. [CrossRef]
48. Mochizuki, M.; Yamamoto, S.; Aoki, Y.; Suzuki, S. Isolation and characterization of *Bacillus amyloliquefaciens* S13-3 as a biological control agent for anthracnose caused by *Colletotrichum gloeosporioides*. *Biocontrol Sci. Technol.* **2012**, *22*, 697–709. [CrossRef]

49. Hamaoka, K.; Suzuki, S. Draft genome sequence of *Bacillus velezensis* KOF112, an antifungal endophytic isolate from a shoot xylem of Japanese indigenous wine grape *Vitis* sp. cv. Koshu. *Microbiol. Resour. Announc.* **2021**, *10*, e00422-21. [CrossRef] [PubMed]

50. Grant, J.R.; Stothard, P. The CGView Server: A comparative genomics tool for circular genomes. *Nucleic Acids Res.* **2008**, *36*, W181–W184. [CrossRef] [PubMed]

51. Aoki, Y.; Trung, N.V.; Suzuki, S. Impact of *Piper betle* leaf extract on grape downy mildew: Effects of combining 4-allylpyrocatechol with eugenol, α-pinene or β-pinene. *Plant Prot. Sci.* **2019**, *55*, 23–30.

52. Bogs, J.; Downey, M.O.; Harvey, J.S.; Ashton, A.R.; Tanner, G.J.; Robinson, S.P. Proanthocyanidin synthesis and expression of genes encoding leucoanthocyanidin reductase and anthocyanidin reductase in developing grape berries and grapevine leaves. *Plant Physiol.* **2005**, *139*, 652–663. [CrossRef]

Article

Combining an HA + Cu (II) Site-Targeted Copper-Based Product with a Pruning Wound Protection Program to Prevent Infection with *Lasiodiplodia* spp. in Grapevine

Pedro Reis [1,*], Ana Gaspar [1], Artur Alves [2], Florence Fontaine [3] and Cecília Rego [1]

[1] LEAF—Linking Landscape, Environment, Agriculture and Food-Research Center, Associated Laboratory TERRA, Instituto Superior de Agronomia, Universidade de Lisboa, Tapada da Ajuda, 1349-017 Lisboa, Portugal; anapatriciagaspar@hotmail.com (A.G.); crego@isa.ulisboa.pt (C.R.)

[2] CESAM—Centre for Environmental and Marine Studies, Department of Biology, University of Aveiro, 3810-193 Aveiro, Portugal; artur.alves@ua.pt

[3] SFR Condorcet FR CNRS 3417, Université de Reims Champagne-Ardenne, Résistance Induite et Bioprotection des Plantes EA 4707, BP 1039, CEDEX 2, 51687 Reims, France; florence.fontaine@univ-reims.fr

* Correspondence: pedroreis@isa.ulisboa.pt

Abstract: The genus *Lasiodiplodia* has been reported from several grape growing regions and is considered as one of the fastest wood colonizers, causing Botryosphaeria dieback. The aim of this study was to (i) evaluate the efficacy of Esquive®, a biocontrol agent, on vineyard pruning wound protection, applied single or, in a combined protection strategy with a new site-targeted copper-based treatment (LC2017), and (ii) compare their efficacy with chemical protection provided by the commercially available product, Tessior®. For two seasons, protectants were applied onto pruning wounds, while LC2017 was applied throughout the season according to the manufacturer's instructions. Pruning wounds of two different cultivars were inoculated with three isolates of *Lasiodiplodia* spp. Efficacy of the wound protectants, varied between both years of the assay and according to the cultivar studied but were able to control the pathogen to some extent. The application of LC2017 did not show clear evidence of improving the control obtained by the sole application of the other products tested. Nevertheless, LC2017 showed a fungistatic effect against *Lasiodiplodia* spp., in vitro, and has previously shown an elicitor effect against grapevine trunk diseases. Therefore, this combination of two protection strategies may constitute a promising long-term approach to mitigate the impact of Botryosphaeria dieback.

Keywords: Botryosphaeriaceae; *Trichoderma atroviride* strain I-1237; Hydroxyapatite-Copper; *Vitis vinifera*; combined strategy

Citation: Reis, P.; Gaspar, A.; Alves, A.; Fontaine, F.; Rego, C. Combining an HA + Cu (II) Site-Targeted Copper-Based Product with a Pruning Wound Protection Program to Prevent Infection with *Lasiodiplodia* spp. in Grapevine. *Plants* **2021**, *10*, 2376. https://doi.org/10.3390/plants10112376

Academic Editors: Carlos Agustí-Brisach and Eugenio Llorens

Received: 30 September 2021
Accepted: 1 November 2021
Published: 4 November 2021

Publisher's Note: MDPI stays neutral with regard to jurisdictional claims in published maps and institutional affiliations.

1. Introduction

Botryosphaeria dieback is currently among the most significant grapevine trunk diseases (GTDs) in all the grape-growing regions of the world [1]. It represents one of the major threats to sustainable and economically viable viticulture due to the reduction in yield, increased crop management costs, and shortened life span of vines and vineyards [1–3]. This disease is caused by fungi of the family Botryosphaeriaceae and more than 26 taxa of this family have been associated with Botryosphaeria dieback in grapevine [4–8]. *Lasiodiplodia* spp. and *Neofusicoccum* spp. were previously proven as being amongst the fastest wood colonizing genera, and therefore considered as some of the most virulent GTD fungi [9–12]. The genus *Lasiodiplodia* comprises 34 species [13,14], from which ten have been reported from the grapevine [13–19]. *Lasiodiplodia theobromae* is the most commonly isolated species in grapevine and, although it is most common in tropical and sub-tropical regions, it can be found in vineyards around the world, such as Australia [20], Algeria [21], Brazil [17], Bolivia [22], China [19], Italy [23], Mexico [9], Peru [24], Portugal [25], Spain [26], Turkey [27], and the USA [9]. Common external symptoms caused by infection with botryosphaeria-ceous fungi on grapevine, include leaf spots and wilting, dry fruit rots, bud necrosis and perennial cankers, cordon dieback, and eventually the sudden death of the plant [7,28,29]. The internal wood symptoms usually consist of wedge-shaped necrotic sectors and brown streaking below the bark, sometimes beginning in the pruning wounds [30]. These fungi are air-borne and infect grapevine through any type of wound, but primarily the infection occurs through the pruning ones [1]. Pycnidia of different Botryosphaeriaceae species associated with Botryosphaeria dieback can be found within old pruning wounds, in dead or cankered wood, embedded on the bark of cordons and trunks of infected grapevines. They can also be found on pruning debris left in the vineyard, constituting a potential source of inoculum for new infections [7,31,32]. Up to now, there are no curative methods to mitigate infection by Botryosphaeriaceae species. Preventive control methods such as pruning wound protection are currently the practice proving to be more efficient [1,28], especially if carried out from the early stages of the vineyard lifespan [3,33]. The field efficacy of chemical wound protectants against botryosphaeriaceous fungi has been demon-strated in several grape-growing regions of the world, namely Australia [34], Chile [35], New Zealand [36,37], Portugal [38], South Africa [39,40], Spain [41], and the USA [42]. An-other method considered to be the most effective strategy for controlling infection by GTD pathogens is the application of pastes and paints amended, or not, with fungicides. This latter can provide a physical barrier, preventing spore germination to occur in the wound, but if this barrier is altered by any external factor, the supplementation with a fungicide will act on the pathogen, inhibiting its growth [1]. Their efficacy has also been shown, specifi-cally against Botryosphaeriaceae [34,35,41–43]. The currently existing active ingredients (AIs) are effective in protecting pruning wounds but have limited systemic activity. These AIs usually do not penetrate well enough into the grapevine tissues to effectively control pathogens inside the vascular system [28,44] and limit the colonization by pathogens. Therefore, new methods to efficiently deliver fungicides to specific targeted areas of the plant are considered to have great potential to improve GTD control. The application of site-targeted fungicides to protect vascular tissues against GTD pathogen colonization has been recently investigated by several authors [44–46]. Lignin nanocarriers loaded with pyr-aclostrobin, as a targeted drug delivery system [45], phloem mobile derivative of fenpiclonil in combination with beneficial endophyte [44], and copper-based treatments, formulated with hydroxyapatite (HA) as co-adjuvant with innovative delivery properties [47,48] have shown promising results in controlling both esca and Botryosphaeria dieback pathogens. This last formulation has also shown an elicitation ability towards several genes related to plant defense [48]. However, increasing consumer demands for reduced chemical use and growing restrictions on the use of synthetic pesticides have increased interest in the use of natural active ingredients such as biocontrol agents (BCAs). Research on BCAs has greatly increased over the last years, being *Trichoderma* species the most studied as bio-pesticides [49]. Several *Trichoderma*-based products are currently used as pruning wound protectants against GTDs (for review see [1] and [28]) and several studies have been conducted using these products, showing encouraging results, against Botryosphaeriaceae species [28,40,50–56]. However, although several promising solutions currently exist, it seems impossible to manage Botryosphaeria dieback using only a single approach. An integrated pest management (IPM) strategy has been recommended including cultural practices, organic products, BCAs, responsible use of chemical fungicides, and control management that may combine both chemical and biological products [28,33].

To the best of our knowledge, no studies on strategies integrating pruning wound pro-tection with chemical or biological products, and the application of site-targeted fungicides, with elicitor properties, have been conducted against Botryosphaeria dieback pathogens, on established vineyards. The main objectives of this work were thus to (i) test the efficacy of one *Trichoderma* based formulation, namely Esquive, single or in combination with LC2017, a new site-targeted copper-based treatment, formulated with hydroxyapatite loaded with copper (II) sulphate pentahydrate (CuSPHy + HA) which has a fungistatic and elicitor effect and, (ii) to compare their efficacy with a strategy of combining the application of commercially available chemical pruning wound protectant, Tessior®, a liquid polymer containing boscalid and pyraclostrobin alone and with LC2017, to prevent both infection and colonization by *Lasiodiplodia* spp. in field trials.

2. Results

2.1. In Vitro Assays

2.1.1. Mycelial Growth Inhibition Assays

The inhibition of mycelial growth for all fungi under study ranged from 3.7–100% (Table 1). Significant differences were observed amongst several concentrations of fungicide used. Independently of the fungal species, the 12.5 mL/L concentration was able to completely inhibit mycelial growth. For all *Lasiodiplodia* spp. Isolates, the remaining concentrations were able to cause a similar growth inhibition rate. It is noteworthy that a significant decrease in mycelial inhibition percentage was verified from the highest concentration of fungicide (12.5 mL/L) to the second-highest concentration (2.5 mL/L) with inhibition percentage values decreasing from 100% to 14–15%. Higher inhibition percentages could be observed for *Trichoderma atroviride* strain I-1237, with a lower decrease between the two referred concentrations, from 100% to 44.6%. For the *Lasiodiplodia* spp. isolates, similar values were observed for the remaining concentrations. Moreover, for the three isolates, no significant differences were reported for the two lowest concentrations tested of 0.125 mL/L and 0.025 mL/L. *T. atroviride* strain I-1237 showed a slightly different behavior for the 1.25 mL/L concentration compared to the *Lasiodiplodia* spp. isolates, with a higher inhibition percentage, but showed similar behavior for the lowest concentration tested. In this case, *T. atroviride* strain I-1237 was the only fungi under study that showed significant differences for the inhibition percentage at the two lowest fungicide concentrations recording the second-lowest inhibition percentage for the 0.025 mL/L concentration. To verify if the effect observed on plates with 100% mycelial growth inhibition was fungistatic and not fungicidal, mycelium disks were transferred to fresh PDA and allowed to grow for 48 h. For all fungi, mycelial growth was observed with average values of 7.0 cm for isolate Bt105, 5.36 cm for isolate LA-SOL3, 4.54 cm for CBS124060, and finally 5.08 cm for *T. atroviride* strain I-1237. The renewed growth for all fungi is an apparent indicator that, during these in vitro assays, product LC2017 merely showed a fungistatic instead of a fungicidal effect.

Table 1. Mycelial growth inhibition rate (%) of product LC2017 on all the *Botryosphaeriaceae* isolates under study (Bt105, LA-SOL3, CBS124060) and the *Trichoderma atroviride* strain I-1237 (Esquive).

LC 2017 Concentration	Mycelial Growth Inhibition (%) [z]			
	Bt105	**LA-SOL3**	**CBS124060**	***T. atroviride* (I-1237)**
A (12.5 mL/L)	100.0 [a]	100.0 [a]	100.0 [a]	100.0 [a]
B (2.5 mL/L)	14.5 [b]	13.9 [b]	15.0 [b]	44.6 [b]
C (1.25 mL/L)	10.4 [c]	9.4 [c]	10.6 [c]	26.0 [c]
D (0.25 mL/L)	8.0 [cd]	6.2 [d]	9.3 [c]	9.9 [d]
E (0.125 mL/L)	6.1 [d]	5.1 [d]	8.1 [c]	9.2 [d]
F (0.025 mL/L)	5.4 [d]	3.7 [d]	4.5 [c]	3.9 [e]

[z] Values in the same column followed by the same letter (a, b, c, d, e) do not significantly differ according to Tukey's test ($p = 0.05$).

2.1.2. Dual Culture Antagonism Assays

The levels of antagonism of *T. atroviride* strain I-1237 against the *Lasiodiplodia* spp. isolates evaluated, ranged from 14.0% to 51.2% (Figure 1). The lowest mean percentage of inhibition of radial mycelial growth was observed for isolate LA-SOL3 (19.8%) (Figure 2B) while the highest mean value of inhibition was recorded for isolate CBS124060 (30.2%) (Figure 2C). Significant statistical differences were observed between the mean inhibition percentage of radial growth of these two isolates. Isolate Bt105 showed values between the remaining two isolates with a mean inhibition percentage of radial growth of 27.3%, not significantly different from any of the referred isolates. Overall, the results show an apparent higher antagonism by *T. atroviride* strain I-1237 against the mycelial growth of the *L. mediterranea* isolate than for *L. theobromae* isolates (Figure 2).

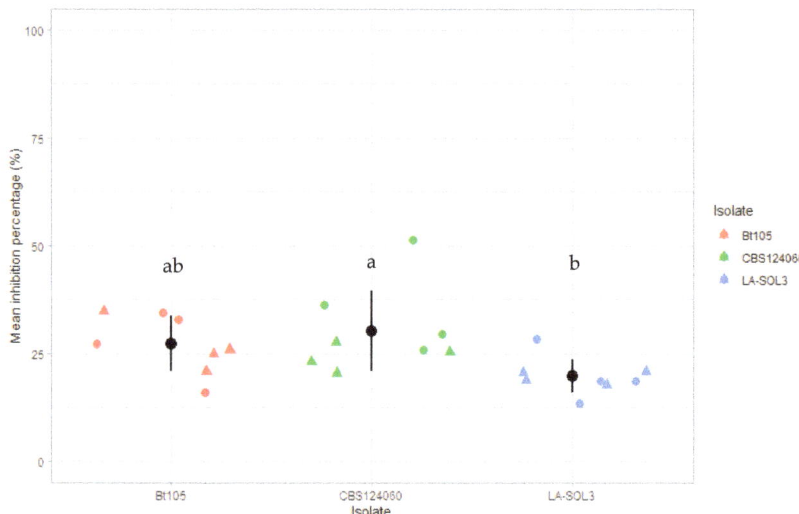

Figure 1. Mean inhibition percentage of growth radius for *Trichoderma atroviride* strain I-1237 against *Lasiodiplodia theobromae* isolates Bt105 and LA-SOL3 and *Lasiodiplodia mediterranea* isolate CBS124060. Dots and triangles represent data from both experiments, black dots represent the mean percentage of growth inhibition for the total replicates of both experiments and black bars represent the standard error of means. Columns with the same letter (a,b) are not significantly different according to Tukey's test ($p = 0.05$).

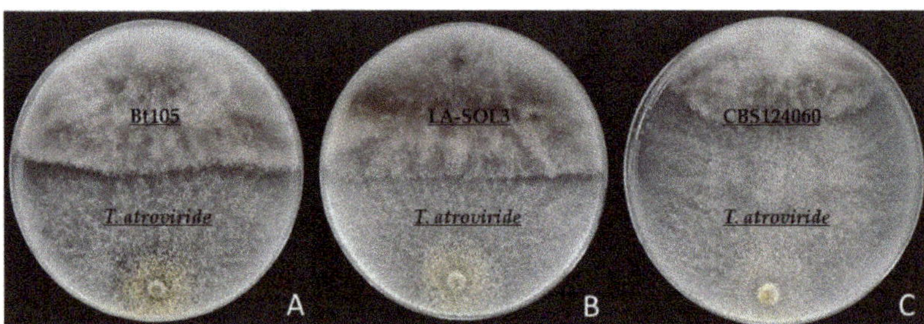

Figure 2. Dual culture antagonism assay of *T. atroviride* strain I-1237 and the three *Lasiodiplodia* spp. isolates under study. (**A**) isolate Bt105; (**B**) isolate LA-SOL3; (**C**) isolate CBS124060.

2.2. Field Assays

The efficacy of the treatments was assessed based on the mean percentage of pathogen recovery (MPR) and, the mean percentage of disease control (MPDC) was calculated according to Sosnowski et al. [57,58] and Martínez-Diz, et al. [41]. Statistically significant differences were found amongst the treatments for both cultivars and years of the study (Table 2) when compared to the inoculated controls. For cultivar Cabernet Sauvignon, during the 2019 assay, treatments with Esquive + LC2017 reduced the MPR of isolate CBS124060 showing an MPDC of 70.8%, while Tessior + LC2017 was able to significantly reduce the MPR for both isolates Bt105 and CBS124060, with MPDCs of 81.5% and 75.0%, respectively. Tessior alone significantly reduced the MPR of isolate LA-SOL3 with an MPDC of 78.3%, and although an MPDC of 34.8% for isolate LA-SOL3 was recorded for the treatment with Esquive alone, no significant reduction was observed for any isolate. During 2020, a significant reduction in MPR of isolate Bt105 was verified for all the treatments, being the highest MPDC of 67.7% for the treatment with Tessior + LC2017. The only remaining significant reduction in MPR was on the pruning wound treatment with Tessior challenged with the isolate CBS124060 (MPDC 55%). Regarding cultivar Touriga Nacional, in 2019, no significant reduction in MPR was found for any treatment, when compared with the inoculated control, with the sole exception of Esquive inoculated with isolate CBS124060 (MPDC 53.3%). The 2020 assay showed a significant reduction in the MPR on treatment Esquive + LC2017 for isolates LA-SOL3 and CBS124060, with MPDCs of 48.4% and 58.6%, respectively. Isolate Bt105 provided a significant reduction in MPR with an MPDC of 52.5% on treatment with Tessior + LC2017, while treatment with Tessior was able to provide an MPDC of 45.2%. No significant reduction in MPR was verified for none of the remaining treatments when compared to the inoculated controls. *Trichoderma* spp. was only isolated from samples treated with Esquive. In 2019, recovery rates on Cabernet Sauvignon ranged from 24% to 16% for Esquive + LC2017 and Esquive treatments, respectively. For Touriga Nacional, the recovery rate for Esquive + LC2017 treatment was 20%, while a 30% recovery rate was recorded for Esquive applied alone. During 2020, *Trichoderma* spp. recovery rates for treatments with Esquive + LC2017 and Esquive alone were 38% and 28% for Cabernet Sauvignon and 25% and 22% for Touriga Nacional.

Table 2. Efficacy of treatments used solely as pruning wound protectants and in a combination with the copper-based product, LC2017, to reduce *Lasiodiplodia* sp. as part of an integrated disease management.

Treatment	Product	Isolate	Cabernet Sauvignon				Touriga Nacional			
			2019		2020		2019		2020	
			MPR [y]	MPDC [z]	MPR	MPDC	MPR	MPDC	MPR	MPDC
1	Esquive + LC2017	Bt105	55.0 abc	18.5	40.0 bc	48.4	52.5 abc	25.0	57.5 abc	9.0
2	Esquive + LC2017	LA-SOL3	42.5 abcde	26.1	30.0 bc	53.8	57.5 abc	8.0	30.0 c	58.6
3	Esquive + LC2017	CBS124060	17.5 de	70.8	20.0 c	60.0	60.0 abc	20.0	40.0 bc	48.4
4	Tessior + LC2017	Bt105	12.5 e	81.5	25.0 c	67.7	42.5 bc	39.3	30.0 c	52.5
5	Tessior + LC2017	LA-SOL3	22.5 bcde	60.9	45.0 bc	30.8	60.0 abc	4.0	60.0 abc	17.2
6	Tessior + LC2017	CBS124060	15.0 de	75.0	35.0 bc	30.0	57.5 abc	23.3	62.5 ab	19.4
7	Tessior	Bt105	42.5 abcde	37.0	37.5 bc	51.6	52.5 abc	25.0	42.5 bc	32.8
8	Tessior	LA-SOL3	12.5 e	78.3	40.0 bc	38.5	50.0 abc	20.0	55.0 abc	24.1
9	Tessior	CBS124060	20.0 cde	66.7	22.5 c	55.0	65.0 abc	13.3	42.5 bc	45.2
10	Esquive	Bt105	62.5 ab	7.4	35.0 bc	54.8	50.0 abc	28.6	40.0 bc	36.7
11	Esquive	LA-SOL3	37.5 abcde	34.8	40.0 bc	38.5	32.5 c	48.0	52.5 abc	27.6
12	Esquive	CBS124060	50.0 abcd	16.7	20.0 c	60.0	35.0 c	53.3	65.0 ab	15.0
13	Inoculated Control	Bt105	67.5 a	-	75.0 a	-	70.0 ab	-	65.0 ab	-

Table 2. *Cont.*

Treatment	Product	Isolate	Cabernet Sauvignon				Touriga Nacional			
			2019		2020		2019		2020	
			MPR [y]	MPDC [z]	MPR	MPDC	MPR	MPDC	MPR	MPDC
14	Inoculated Control	LA-SOL3	57.5 [abc]	-	65.0 [ab]	-	62.5 [abc]	-	72.5 [a]	-
15	Inoculated Control	CBS124060	60.0 [abc]	-	50.0 [abc]	-	75.0 [a]	-	77.5 [a]	-

[y] Efficacy of all the treatments based on the mean percentage of recovery (MPR) of all the isolates used in this study, from the inoculated pruning wounds. Values in the same column with the same letter (a, b, c, d, e) do not significantly differ according to Tukey's test ($p = 0.05$). [z] Mean percentage of disease control (MPDC) of all the treatments calculated according to the formula MPDC = 100 × [1 − (MPR treatment/MPR inoculated control)].

Meteorological data recorded during the time of pruning treatments and pathogen inoculation for both years are presented in Figure 3. In 2019, the average daily temperature in the week of treatment and inoculation (11–15 February) was 10.05 °C with no rain events recorded. In 2020, treatments and pathogens inoculations were performed during the week from the 16–20 February, with an average temperature of 13.3 °C and only one rain event recorded on the 17th with 1.1 mm of precipitation. Regarding the total rainfall and average temperature for the month of February of both years, 2019 recorded an average temperature of 12.2 °C and a total precipitation of 12.7 mm spread throughout five rain events. In 2020, the average temperature was 14.0 °C, and a total precipitation of 8.4 mm with eight rain events.

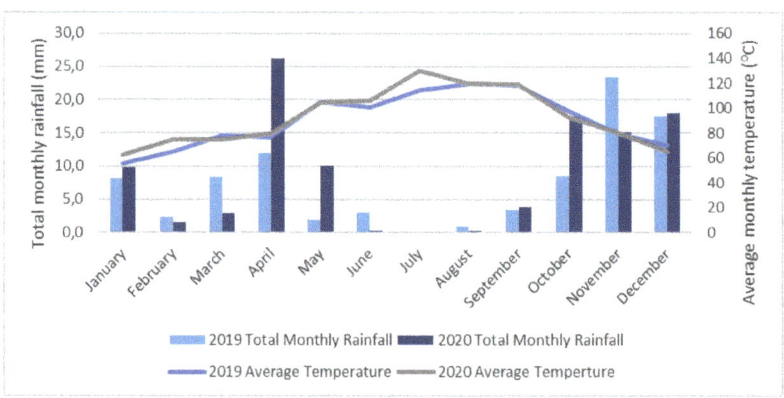

Figure 3. Monthly rainfall and average temperature for the two years of the trial.

The principal component analysis (PCA) was performed to better visualize the relationship between all the variables (Figure 4A,B). For 2019 (Figure 4A), the two first principal components, accounted for 59.9% of the total variability of the data (PC1—41.78%; PC2—18.12%), while for 2020 (Figure 4B) the two first principal components represented 53.98% of the total variability of the data (PC1—35.34%; PC2—18.64%). For both years, we could also observe a clear distribution of the RP of the treatments according to the cultivar, which probably results from the influence of the different susceptibility that both cultivars may have towards the different pathogens studied. The relationship between the treatments and the meteorological variables could be observed in the vector plots of Figure 4A,B. For both years, meteorological variables showed similar behavior with precipitation being more related to component 1, while temperature showed a higher relation with component 2. In the vector plots, it was observed that the two meteorological variables were very weakly correlated, especially in 2020, where almost no correlation occurred between the variables.

Since the vineyard was in an area with a Köppen climatic classification of Csa (hot-summer Mediterranean climate), it makes sense that these variables are negatively correlated since higher temperatures are equivalent to less precipitation. This is again evident for the lower correlation verified in 2020, as less rainfall was recorded during the summer months compared to 2019 (Figure 3). For 2019 (Figure 4A), most of the treatments showed a much higher correlation with precipitation than with the average temperature recorded throughout the duration of the assay. The only exception was for treatments 2, 12, and 13 which showed a higher correlation to the average temperature recorded. A negative correlation was verified between some treatments, such as T4 and T12 and, T2 and T10. T4 was a treatment with Tessior + LC2017 inoculated with isolate Bt105 while T12 was a treatment with Esquive inoculated with isolate CBS124060. This negative correlation may be due to not only these treatments having very different modes of action (chemical combination vs BCA), but also because the behavior of the two cultivars was completely opposite. T4 appeared to have higher efficiency than T12 on Cabernet Sauvignon, with MDPC values of 81.5% and 20% respectively, while the exact opposite was verified for Touriga Nacional, with 39.3% and 53.3% (Table 2). Treatments T2 and T10 also showed a negative correlation with each other, and although they are both treatments with Esquive (T2 is Esquive + LC2017 inoculated with isolate LA-SOL3 and T10 is Esquive applied alone inoculated with isolate Bt105) again the different efficacy of the treatments in both cultivars seems to have significance, since T2 seems to be more effective in Cabernet Sauvignon with higher MPDC values than T10, but the opposite was recorded for Touriga Nacional (Table 2). This trend is also verified for treatments T2 and T1, which also show a negative correlation. Both treatments are Esquive + LC2017 but inoculated with different isolates (T1-Bt105 and T2-LA-SOL3). MPDC values of T2 (26.1%) appear to show once again a higher efficacy in Cabernet Sauvignon than in Touriga Nacional (18.5%), while T1 appears to have a better impact in protecting Touriga Nacional (25%) than Cabernet Sauvignon (18%). This behavior of treatments with Esquive (alone or in combination) where T1 and T10 showed higher MPDC levels for Touriga Nacional comparatively to Cabernet Sauvignon was also verified for T11 and was confirmed by the high correlation showed by these three variables (Figure 4A). It is also noteworthy that all the treatments with chemical formulations, namely Tessior + LC2017 (T4, T5, T6) and Tessior applied alone (T7, T8, T9) showed high correlation amongst themselves, all being positively correlated with PC1. The only treatment with a BCA (Esquive +LC2017) showing a high correlation with the chemical-based ones was T3. The other *Trichoderma* treatments (Esquive + LC2017 and Esquive applied alone) showed mixed behavior, while the treatments T1, T10, and T11 were highly correlated with each other, but negatively correlated with T2 and almost uncorrelated with T12. For 2020 (Figure 4B), a similar behavior was observed for most of the treatments. Almost all the treatments showed a high correlation with precipitation, except for treatment 4, which revealed again a high correlation with precipitation during the previous year and for treatment 13 again. In this case, treatments 3 and 9 exhibited a similar correlation with both precipitation and temperature. Therefore, for both years of the assay, almost all isolates showed a higher correlation with precipitation rather than temperature, except for treatment T13 apparently more strongly related to temperature. This treatment also showed a negative correlation with T8, and very low to almost no correlation to the remaining treatments, which could be related to the fact that T13 corresponds to the inoculated control of isolate Bt105. During this year, treatments T2 and T4 showed a negative correlation. These are again treatments with different modes of action since T2 is a treatment of Esquive + LC 2017 inoculated with isolate LA-SOL3 and T4 is a treatment of Tessior + LC2017 inoculated with isolate Bt105. As was observed for the 2019 analysis, this negative correlation may be attributed to T2 being apparently slightly more effective in Touriga Nacional (MPDC = 58.6%) than in Cabernet Sauvignon (MPDC = 53.8%) while the opposite result is found for T4, with an MPDC of 67.7% in Cabernet Sauvignon opposed to 52.5% in Touriga Nacional. On this 2020 analysis, it is also noteworthy that all the treatments inoculated with isolate LA-SOL3 (T2, T5, T8, T11, T14) appear to have a high correlation amongst themselves, showing a low correlation with the remaining treatments. The exception was T5 which also showed a high correlation with the referred treatments, but nonetheless, this correlation shows that not only the variety and the type of product applied but also the isolate inoculated during this assay, may have some influence on the efficacy of the treatments tested.

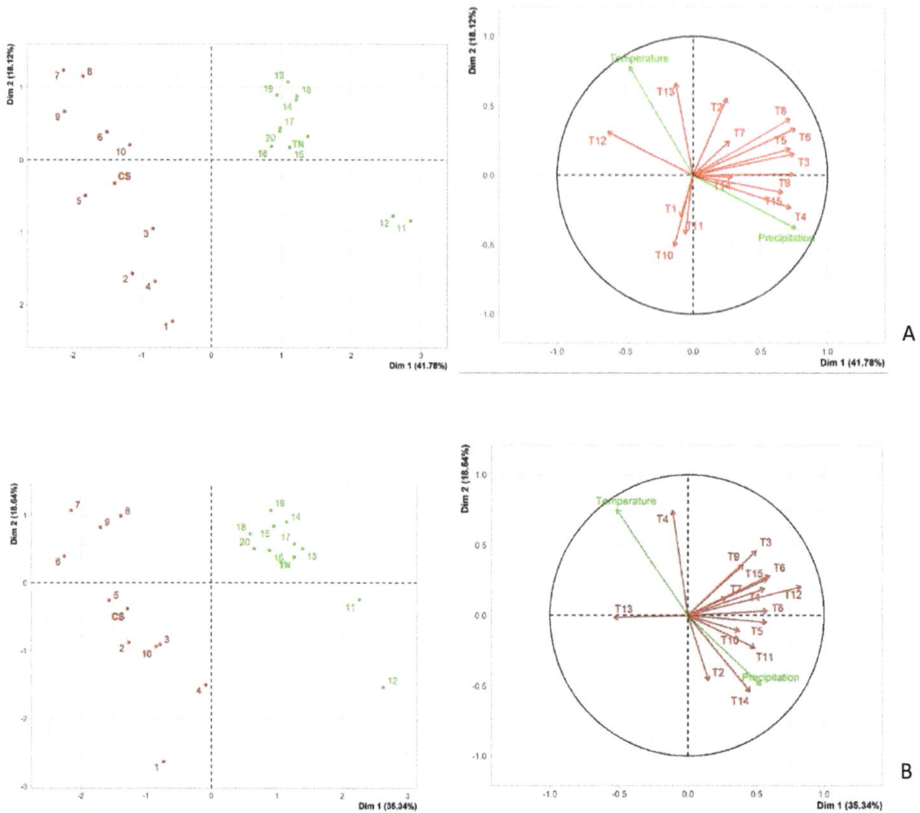

Figure 4. Principal component analysis (PCA) for the interactions of all treatments, 15 variables including all the product/isolate combinations (T represents the treatment numbers which are stated in Table 2), as well as inoculated controls on both variables and the meteorological variables (Figure 2). Red marks represent the observations for Cabernet Sauvignon (CS) and green marks represent the Touriga Nacional (TN) observations. (**A**)-PCA analysis for 2019; (**B**)-PCA analysis for 2020.

3. Discussion

The present study reports the first field assessment on the efficacy of using management strategies, integrating biological and chemical pruning wound protectants with a copper-based site-targeted formulation, against Botryosphaeria dieback pathogens. Higher demand for more sustainable management practices and increasing restrictions on pesticide use has led to an expansion of IPM programs, involving organic products, BCAs, improved cultural practices, and responsible pesticide use [2,26]. To date, and to the best of our knowledge, no field assays have been performed using the two selected pruning wound protectants against *Lasiodiplodia* spp., in combination with copper-based protection with an elicitor effect. Our results show that the efficacy of the wound protectants, when applied alone or in combination with the copper-based product, varied between both years and according to the cultivar under study.

3.1. Efficacy of Pruning Wound Protection Products Applied Alone

For 2019, the highest MPDC values observed for Cabernet Sauvignon were found for the treatment with Tessior, while the lowest values were recorded for the application of Esquive alone. Nevertheless, the same pattern was not observed in 2020 when, in general, both treatments were able to reduce MPR to a similar extent. Pitt et al. [12] reported similar results in Australia, where liquid and paste formulations showed better efficacy against *Diplodia seriata* and *Diplodia mutila* than *Trichoderma*-based products. More recently, Martínez-Diz et al. [41] also described a higher efficacy of Tessior compared to Esquive, on pruning wounds infected with *D. seriata* and *Phaeomoniella chlamydospora*. A slightly different trend could be observed for cultivar Touriga Nacional in 2019 with the highest MPDC values for grapevines treated with Esquive alone. In 2020, this cultivar showed similar efficacy of the Esquive treatment comparing with the Tessior treatment, the only exception being the MDPC values found for isolate CBS124060 which were significantly lower than those reported for Tessior. In fact, all the 2020 treatments induced a slightly higher efficacy than observed in 2019, except for the Esquive treatment applied alone. Similarly, chemical fungicides, namely benomyl, were less effective than *Trichoderma* spp. treatments on wounds inoculated with *Diaporthe ampelina, D. seriata, E. lata, Neofusicoccum australe, Neofusicoccum parvum, L. theobromae* and *P. chlamydospora* [1,59]. This specific *T. atroviride* strain I-1237 has shown efficacy in reducing GTD incidence and severity in preliminary assays carried out in both Portuguese [55] and French [54] vineyards. Nevertheless, Martínez-Diz et al. [41] did not recently find significant differences between pathogen re-isolation from pruning wounds treated with *T. atroviride* strain I-1237 and inoculated non-treated controls.

The variable effectiveness of Trichoderma-based treatments has been previously reported by authors [34,41,51,60,61] and could be attributed to several reasons. The main advantage of using Trichoderma-based products as pruning wound protectants is the long-term protection conferred by the fungus growing in the wood. Therefore, the success of the protection provided by these products is dependent on the establishment of *Trichoderma* spp. within the wound. The influence of grapevine cultivar to wound colonization by *Trichoderma* spp. has also been highlighted by Mutawila et al. [52], probably related to the plant defense responses that differ between cultivars. These different defense responses to *T. atroviride* have been recently reported by Leal et al. [62], where the authors observed that *T. atroviride* may act as a priming stimulus for Tempranillo plantlets, while no stimulus could be verified for Chardonnay. This is in agreement with our PCA analysis which showed a negative correlation among treatments with *T. atroviride* (Esquive), likely due to a difference of efficacy amongst cultivars, which reinforces the importance of the cultivar in wound colonization by *Trichoderma* spp. It is known that not only cultivar but also meteorological conditions, such as temperature and precipitation, as well as the time of application influence the establishment and persistence of *Trichoderma* spp. [63]. Both cultivar and meteorological conditions factors could explain the differences of *T. atroviride* strain I-1237 colonization in our study. According to Esquive manufacturer, this specific *T. atroviride* strain I-1237 can grow at temperatures above 5 °C. In our study, the average temperatures recorded during pruning and treatments performed were 12.2 °C and 14 °C in 2019 and 2020, respectively. This may explain the apparent higher success on the colonization of the pruning wounds, especially verified for all the treatments with Esquive on Cabernet Sauvignon. Martínez-Diz et al. [41], while using the same commercial formulation of *T. atroviride* strain I-1237 (Esquive), found significantly lower colonization in Spanish vineyard trials. Nevertheless, this study was conducted on a different cultivar (Godello) and under different meteorological conditions. In this latter study, temperatures recorded during pruning and treatment application were significantly lower than those recorded during our study (6.5 °C and 9.8 °C in Spain to 12.2 °C and 14.0 °C in Portugal). The PCA analysis performed during this study also showed that most of the treatments where *T. atroviride* strain I-1237 was used were more correlated with precipitation than with temperature, being the only exceptions of treatments 12 and 13 in 2019. Nevertheless, in

2020, there were higher precipitation levels during the months of April and May, and on the PCA analysis performed for this year, all treatments with *T. atroviride* showed a high correlation with precipitation. This seems to reinforce the hypothesis that cultivar and meteorological conditions, especially rainfall, may apparently impact both the establishment and persistence of *T. atroviride* strain I-1237 in pruning wounds. As expected, for both treatments with Esquive, the highest percentages of colonization by *T. atroviride* strain I-1237 led to the highest mean percentage of disease control values. Further research is still needed to prove this hypothesis in GTDs which is characterized by extremely high complexity.

During our study, the application of Tessior (pyraclostrobin + boscalid + liquid polymer) was able to provide some reduction in infection by *Lasiodiplodia* spp. The highest mean percentages of disease control were obtained for Cabernet Sauvignon during 2019, with values as high as 78.3%, while during the next growing season, MPDC values only reached values as high as 55.5%. For Touriga Nacional, as referred to earlier, Tessior application was apparently not as effective as that of *T. atroviride*. The maximum level of MPDC was 45.2% for 2020, and MPDC values as low as 13.3% could be found during the previous year. Application of pyraclostrobin alone was effective in reducing infection by Botryosphaeria dieback fungi under field conditions [35,42], and a mixture of pyraclostrobin with metiram in the nursery when diluted in the soaking water prior to grafting [64]. Prior to this work, only preliminary studies and one in-depth study have been conducted by applying Tessior to pruning wounds, but all of them for controlling *Diplodia* spp. and *P. chlamydospora*. Preliminary field studies were conducted in Greece [65,66], Germany [64,67], and Spain [64] where Tessior was effective in reducing infection by the referred fungi. Recently, a more extensive study was conducted in Spanish vineyards by Martínez-Diz et al. [41] where Tessior showed a high MPDC for both *D. seriata* and *P. chlamydospora* compared to several other commercially available fungicides and BCAs. No previous studies have been conducted using Tessior as a pruning wound protectant against *Lasiodiplodia* spp., and although on the referred study, a botryosphaeriaceous fungi was used, *D. seriata* is considered a less aggressive species in comparison to *Lasiodiplodia* spp. [7]. In our study, we used two isolates of *L. theobromae* and one isolate of *L. mediterranea*. The establishment of the pathogens within grapevines depends on several factors, including not only meteorological variables and cultural practices, but also pathogen intrinsic properties such as aggressiveness [42]. Van Niekerk et al. [68] attributed the higher pathogen infection level on inoculated wounds to higher percentages of rainfall in the Stellenbosch area. This is in agreement with the PCA analysis of most of the treatments in our field study, which were strongly correlated with precipitation. The only treatment that for both years had a higher correlation with temperature was T13 corresponding to the inoculated control of isolate Bt105 (*L. theobromae*). It has been suggested that the expression of virulence factors in *L. theobromae* can be modulated by temperature [8,69]. This isolate was collected in Portugal, and it was previously considered to be highly virulent against both cultivars under study [70], suggesting that it may be better adapted to the local conditions than the other isolates under study. Moreover, our PCA analysis showed a high correlation between all the treatments in which plants were inoculated with isolate LA-SOL3 and between plants inoculated with isolate CBS124060 in 2020. Consequently, the difference in MPDC values found herein, comparatively to previous studies, may be due to the meteorological variables but also related to the pathogen used, and the difference in aggressiveness between them. Another factor that may be noteworthy is the difference in susceptibility between the two cultivars used. In fact, on average higher MPDC values were obtained for Cabernet Sauvignon during both growing seasons of the field assay, compared to the values found for Touriga Nacional. This tendency is verified, not only for the treatments with Tessior, but for most of the treatments. Previous studies conducted on the same cultivars using the same isolates, proved their high aggressiveness towards grapevine in both field and greenhouse assays [70], suggesting that the difference in efficacy in reducing infection by *Lasiodiplodia* spp. may also depend on cultivar susceptibility. Moreover, Sofia et al. [71] showed that from a set of four different cultivars, Touriga Nacional was one of the most susceptible to *P. chlamydospora*, suggesting that this particular cultivar may be highly susceptible to GTDs in general. Therefore, and given the differences found between both growing seasons and both cultivars, further research is recommended to evaluate each of the components of these products to understand how their efficacy may be affected by factors such as time of application, pathogen species aggressiveness and cultivar susceptibility, pruning wound size, and vineyard terroir.

3.2. Efficacy of a Strategy Combining the Application of Pruning Wound Protectants with LC2017

The great diversity of species currently associated with Botryosphaeria dieback, but also with all GTDs, combined with the intrinsic differences in hundreds of cultivars planted worldwide under different terroir conditions, makes the research for an effective management strategy to reduce the impact of GTDs an extremely difficult challenge. To this extent, in this work, we tried to integrate currently used products for pruning wound protection with the application of a site-targeted product based on a copper (II) compound (copper sulfate) and synthetic nanostructured particles of hydroxyapatite (HA) which have already shown interesting drug delivery properties in planta [46–48,72]. In our work, the application of this new site-targeted copper (II) formulation (LC2017), does not seem to greatly increase the efficacy of the pruning wound protection strategy. In some cases, the combination of the two strategies actually increased the mean percentage of disease control, but in others, the opposite was also verified. For example, in 2019, and for the inoculation of isolate CBS124060 on Cabernet Sauvignon, an MPDC of 70.8% was obtained for the treatment with Esquive + LC2017 while only 16.7% was recorded for the treatment with Esquive alone. During the next year (2020) the same value of 60% was found for both treatments. In 2019, on Cabernet Sauvignon, the application of Tessior + LC2017 caused also an MPDC of 81.5% for isolate Bt105, while only 37% was found for the treatment with Tessior alone. For the two other isolates under study, LA-SOL3 showed an increase in MPDC when Tessior was applied alone, while the isolate CBS124060 showed an opposite trend. Although the highest concentration of LC2017 used in the in vitro assay for the inhibition of mycelial growth showed fungistatic effect for all the isolates and *T. atroviride* strain I-1237, there is no apparent pattern on the influence of the LC2017 application as part of an IPM for any of the cultivars, or specific isolate. Moreover, the fungistatic effect verified for *T. atroviride* strain I-1237 did not seem to impact the colonization of pruning wounds by this BCA, since the highest values of re-isolation percentage could be found for the treatments combining Esquive and LC2017 except for Touriga Nacional during 2019. Moreover, no relevant correlation between the treatments using LC2017 was observed for both years by the PCA analysis, indicating a low influence of LC2017 application on the MPDC values found in this study.

The copper formulation used (LC2017) has already shown both fungistatic and fungicidal effects in vitro and *in planta* against *Phaeoacremonium minimum* [46]. Mondello et al. [48] have also reported that LC2017 showed the same in vitro fungistatic effect on *D. seriata* and *N. parvum* as it was verified for our *L. theobromae* isolates. The same authors described as well, the ability of this formulation to activate GTD-related plant defense reactions [48]. Di Marco et al. [73] also tested a copper formulation against *P. chlamydospora* and *P. minimum*, which revealed the ability to reduce conidial germination in in vitro assays but was not able to reduce *P. chlamydospora* colonization on young potted grapevines. Amposah et al. [36] tested a copper hydroxide formulation against Botryosphaeria dieback fungi, and were not able to obtain any significant control both in in vitro and in plant tests. More recently, Mondello et al. [48] performed greenhouse in planta assays that did not show significant differences in stem necrosis length of plants treated with LC2017 and inoculated with *D. seriata* and *N. parvum*. This agrees with our results since the same copper formulation used does not appear to provide a significant impact when combined with pruning wound protectants in controlling these specific Botryosphaeria dieback fungi. Therefore, further research is needed on testing all the components of this formulation, since HA applied alone has shown a non-fungitoxic and stimulant activity on *P. minimum* [48] and *Botrytis cinerea* [71]. However, the efficacy of copper (II) products has been previously related not only as a fungicide but also as an elicitor of some plant defense responses. Aziz et al. [74] observed this eliciting effect for CuSO4 sprayed on leaves, and Battiston et al. [48] have also shown that this formulation of CUSPHy was able to strongly induce several plant defense genes. Mondello et al. [48] also showed that LC2017 (HA + Cu (II)) had the same elicitation potential as BTH (*S*-methyl benzo (1,2,3) thiadizole-7-carbohthioate), a commonly marketed elicitor (BION®, Syngenta, France) by inducing genes related with chitinase and glucanase synthesis and also genes related to the biosynthesis pathways of several phenolic compounds.

Therefore, new methods to efficiently deliver fungicides to specific targeted areas of the plant are considered to have great potential to improve GTD control, and further research is needed to investigate not only the impact that these formulations may have on several GTD pathogens but also on the plant microbiome. Further long-term field trials should also be undertaken to investigate the effectiveness that a prolonged management control combining pruning wound protection and site-targeted fungicides might provide in controlling GTDs.

During this work, to ensure a proper establishment of infection with the fungi under study, approximately 2000 spores of each of the *L. theobromae* and *L. mediterranea* isolates were used to challenge each pruning wound. This represents a high inoculum pressure compared to the levels that pruning wounds are usually exposed to under natural field conditions. This suggests that due to the higher inoculum applied to the pruning wounds, the real efficacy of all the treatments tested against *Lasiodiplodia* spp. may have been underestimated [28].

In conclusion, this study demonstrated the field potential of pruning wound protection formulations to control *Lasiodiplodia* spp., applied alone or in combination with a novel site-targeted copper (II) formulation. The efficacy of the studied products differed between the two cultivars used and between the two growing seasons of the duration of the assay. Nevertheless, some measure of control was achieved for all treatments studied, despite the application of the copper-based product LC2017 not showing clear evidence of improving the control obtained by the pruning wound protection products applied alone. Still, this new site target copper-based product may prove to be a viable way of reducing the amount of copper applied for diseases management on vineyards, especially with further European Union restrictions on copper use in agriculture. Furthermore, the combination of effective pruning wound protection management by using a combination of a BCA with a more sustainable copper-based product has already proven to have interesting results not only as an elicitor by strongly inducing plant defense genes but also in the control of *P. minimum*, as well as *Plasmopora viticola*, may lead, in the long term, to healthier grapevines, which may reduce the expression of GTD symptoms in vineyards. However, this hypothesis still needs to be clarified by conducting studies using the same management strategy over a longer period of time, targeting not only other grapevine cultivars but also other Botryosphaeriaceae species. Thus, good pruning practices and wound protection combined with sustainable management strategies, such as biostimulants and host resistance inducers, can reduce the impact of Botryosphaeria dieback, not only by controlling pathogens already common in a certain wine region but also by making it difficult for the potential establishment of new species.

4. Materials and Methods

4.1. In Vitro Assays

4.1.1. Mycelial Growth Inhibition Assay

Prior to field application of the LC2017 (CuSPHy + HA) product on the field, mycelial growth assays were conducted using three *Lasiodiplodia* spp. isolates, Bt105, LA-SOL3, and CBS124060 (Table 3). An assay was also performed using the *T. atroviride* strain I-1237 (Esquive®, product developed by Agrauxine S.A. and commercialized by Idai Nature S. L.), to test the compatibility between both products. To obtain the *T. atroviride* strain I-1237, a solution was made by directly suspending the Esquive® product, on 100 mL of sterile distilled water. A 1.5 mL aliquot was transferred to a 90 mm Petri dish containing 20 mL of Potato-Dextro-Agar (PDA, Difco, Sparks, MD, USA), and spread onto the surface using a sterile plastic loop. Petri dishes were then incubated at 25 °C for 7 days in absolute darkness. For the mycelial growth assays, a stock solution of product LC2017 was made by suspending the product at the recommended field concentration (250 L/ha) to be applied immediately after pruning, in 1000 mL of sterile distilled water (SDW). Six different concentrations were made in SDW and added to 50 °C molten PDA, and 20 mL was poured into each 90 mm Petri dish, with six replicate plates allowed for each combination of LC2017 concentration and isolate (both *Lasiodiplodia* spp. and *T. atroviride* strain I-1237). The test range of LC2017 product concentration ranged from 0.025 to 12.5 mL L^{-1} and the six concentrations tested were evenly distributed across that range. Four hours after preparing the plates, 3 mm diameter discs were cut from the actively growing margin of one-week-old colonies all the isolates and placed on the center of each plate. Control plates contained only PDA. Plates were incubated at 25 °C for 48 h, in complete darkness, after which the two perpendicular diameters of the colonies were measured using a digital caliper. Mycelial growth inhibition (GI) was calculated according to Battiston et al. [46]: GI = [(DC − DO)/DC] × 100, where DC is the diameter of mycelial growth in the control plates and DO is the diameter of mycelial growth in treated plates. To establish if the effect of LC2017 on the tested fungi was only fungistatic, inhibited fungal disks were reinoculated onto fresh PDA plates and their growth revival was observed after 48 h.

Table 3. *Lasiodiplodia* spp. isolates used for pruning wound inoculation.

Species	Isolates	Geographic Origin
L. theobromae	Bt105	Alentejo, Portugal
	LA-SOL3	Sol Sol, Piura, Peru
L. mediterranea	CBS 124060	Sicily, Italy

4.1.2. Dual Culture Antagonism Assay

Dual culture antagonism assays were also performed to evaluate the antagonistic capability of the *T. atroviride* strain I-1237 (Esquive) against the three *Lasiodiplodia* spp. isolates targeted for study (Bt105, LA-SOL3 and CBS124060), using dual culture assays [49,75]. *Trichoderma atroviride* strain I-1237 cultures used for this assay were obtained using the same method described for the mycelial growth inhibition assays. Mycelium plugs with 5 mm diameter of *T. atroviride* strain I-1237 and each *Lasiodiplodia* sp. isolate were cut from the actively growing margin of 3-day-old colonies, growing on PDA. Plugs were placed on opposite edges of 90 mm Petri dishes containing 15 mL of PDA. Plates were then incubated for 5 days in the dark at 22 °C. Each *Lasiodiplodia* sp. isolate was grown individually under the same conditions as control plates. Each combination *Trichoderma/Lasiodiplodia* spp. was replicated four times and the assay was performed twice. The percentage of mycelium growth inhibition was calculated using the formula, percent inhibition (PI) = [(B − A)/B] × 100 [42], where A is the radius of pathogen mycelium growth on the dual culture plates, and B is the radius of *Lasiodiplodia* spp. growth on the control plates.

4.2. Field Assays

4.2.1. Experimental Field

The assay was conducted between 2019 and 2020, on an experimental vineyard located at Instituto Superior de Agronomia (ISA) (38°42′33.5″ N 9°11′15.8″ W) Lisbon, Portugal, planted in 1998. Two cultivars were used for this assay, Touriga Nacional (TN) and Cabernet Sauvignon (CS), both grafted onto 140 Ruggeri. Vines were trained as bilateral cordons. Traditional cultural practices in the vineyard were kept throughout the whole assay, and disease management followed an IPM. The products applied were selected with the care of not containing any substance that could casually interfere with the assay.

4.2.2. Fungal Isolates Used and Inoculum Preparation

Two *L. theobromae* and one *L. mediterranea* isolates were used for this assay (Table 3). *Lasiodiplodia theobromae* isolates Bt105 and LA-SOL3, were collected in Portugal and Peru, respectively, and were stored at the culture collection of Instituto Superior de Agronomia. Both were isolated from grapevine wood showing symptoms of cankers and wood necrosis. The *L. mediterranea* isolate used is from the CBS culture collection from the Westerdijk Fungal Biodiversity Institute, in Utrecht, Netherlands, with the accession CBS 124060. Although *L. mediterranea* is currently not reported in Portugal, it has been previously reported in other European countries and so, one isolate of this species was also included in this study, not only for comparison but also to investigate the efficacy of the studied products towards this species. Isolates were maintained in PDA and transferred to Petri dishes with PDA to promote colony growth. Cultures were incubated at 25 °C in complete darkness for 8 days. After incubation, cultures were plated onto 6 mm Petri dishes containing 2% water agar with autoclaved pine needles (*Pinus pinea*) and incubated at 25 °C under fluorescent light for a 12 h photoperiod, to pycnidia sporulation [76–78]. On the day of the inoculation, conidia were harvested by collecting pycnidia formed on the pine needles to a 1.5 mL Eppendorf tube containing sterile distilled water (SDW), crushing them with the help of a pestle, followed by shaking the tube in a vortex for one minute. Spore suspensions obtained were filtered through cheesecloth and the concentration was adjusted to 1×10^5 spores/mL with the use of a hemocytometer (Brand, Wertheim, Germany).

4.2.3. Experimental Design and Treatment Plan

Products used in this assay can be found in Table 4. Esquive was applied according to label dosage and manufacturer's instructions and Tessior is a formulation ready for application. LC2017 was applied with the timing and dosage recommended by the manufacturer (Table 4). A total of 15 treatments were set up on grapevines of both cultivars, Touriga Nacional and Cabernet Sauvignon in a completely randomized design, with 10 repetitions per treatment. One shoot was inoculated per grapevine making a total of 150 plants used from each cultivar. The combination of treatments and isolates inoculated can be found in Table 5.

Table 4. Treatments tested for control of *Lasiodiplodia theobromae* and *Lasiodiplodia mediterranea* under field conditions.

Product Name	Manufacturer	Application Time	Application Rate	Active Ingredient
Esquive®	Idai Nature	After pruning	4 kg/ha	*Trichoderma atroviride* strain I-1237 (1×10^8 CFU g^{-1})
Tessior®	BASF Agricultural Solutions Portugal	After pruning	n/a	Pyraclostrobin 0.48% + boscalid 0.95%
LC2017	Natural development Group®	Immediately after harvest	400 L/ha	Hydroxyapatite (HA) loaded with cooper (II) sulphate pentahydrate (CuSPHy + HA)
		After pruning (Winter)	250 L/ha	
		Four leaves developed	250 L/ha	
		Summer pruning	400 L/ha	
		Veraisson	400 L/ha	

Table 5. Treatment plan designed to assess the efficacy of three products against pruning wound infection by *Lasiodiplodia theobromae* and *Lasiodiplodia mediterranea*. Combinations of products used and inoculations spore solution volume.

Treatment	Product	Inoculation	Spore Solution Volume (µL)
1	Esquive + LC2017	*L. theobromae* (Bt105)	20
2	Esquive + LC2017	*L. theobromae* (LA-SOL3)	20
3	Esquive + LC2017	*L. mediterranea* (CBS124060)	20
4	Tessior + LC2017	*L. theobromae* (Bt105)	20
5	Tessior + LC2017	*L. theobromae* (LA-SOL3)	20
6	Tessior + LC2017	*L. mediterranea* (CBS124060)	20
7	Tessior	*L. theobromae* (Bt105)	20
8	Tessior	*L. theobromae* (LA-SOL3)	20
9	Tessior	*L. mediterranea* (CBS124060)	20
10	Esquive	*L. theobromae* (Bt105)	20
11	Esquive	*L. theobromae* (LA-SOL3)	20
12	Esquive	*L. mediterranea* (CBS124060)	20
13	Inoculated non treated Control	*L. theobromae* (Bt105)	20
14	Inoculated non treated Control	*L. theobromae* (LA-SOL3)	20
15	Inoculated non treated Control	*L. mediterranea* (CBS124060)	20

4.2.4. Product Application and Pathogen Inoculation

For both seasons, immediately after harvest, one application of LC2017 was carried out as recommended by the manufacturer (Table 4; Figure 5). Harvest occurred during the month of September for both years of the assay. For both cultivars, one-year-old canes with a similar appearance, namely length, were selected for treatment followed by inoculation and were pruned at 2 cm above the third bud. After pruning, both products were prepared according to the label's rate. Esquive was weighted and mixed with water in a 200 mL spray bottle with a concentration of 4 kg/ha (Figure 5). Tessior (ready to apply solution) and was applied using the equipment specially designed for this product application (Figure 5). Untreated controls were mock-treated with SDW, and the wound protectants were allowed to dry for a few hours. This was followed by application with LC2017, as indicated by the manufacturer (Table 4, Figure 5), using a backpack sprayer. One day after the treatment, inoculation with the selected was performed, by applying 20 µL of the spore suspension (≈2000 spores) on each wound using a micropipette (Figure 5). After inoculation, the pruning wounds were protected for one week using Parafilm M® (Bemis, Sheboygan Falls, WI, USA) to prevent dehydration and promote spore germination. Pruning and artificial inoculation were performed on the 14 and 15 February 2019, and on the 20 and 21 February 2020, during the winter dormancy, taking into consideration that all the procedures were made during favorable meteorological conditions, namely cloudy and humid, but avoiding rain periods. The same precautions were taken for the remaining applications with LC2017, with the consideration of also avoiding days with strong winds to minimize spray drift.

4.2.5. Pathogen Recovery and Identification

For both years, canes were recovered after harvest, during the month of October, and stored in a cold chamber (4 °C) until further processing. For pathogen re-isolation, the bark of each cane was removed, and a sample was collected from about 1 cm below the pruning wound (Figure 5). Four pieces of wood were collected from the border of necrotic internal tissue, surface disinfected with a 7% sodium hypochlorite solution, rinsed in SDW, and plated onto 9 mm Petri dishes containing PDA amended with chloramphenicol (PanReac, AppliChem, Darmstadt, Germany) at 250 mg/L. Plates were incubated at 25 °C, in the dark, and assessed for counting Botryosphaeriaceae and *Trichoderma* spp. colonies (Figure 5). A representative set of *Lasiodiplodia* spp. and *Trichoderma* spp. isolates was selected for identity confirmation. A DNeasy Plant Mini Kit from Qiagen® (Venlo, The Netherlands) was used to extract genomic DNA from 8-day-old cultures grown in PDA and incubated at 25 °C, in the dark, following the manufacturer's instructions. The identity of *Lasiodiplodia theobromae* and *L. mediterranea* was confirmed by sequencing part of the translation elongation factor 1α gene (tef1-α) by using the primers EF1-688F and EF1–1251R [79], while *T. atroviride* strain I-1237 was confirmed by sequencing the internal transcribed spacer region (ITS) using the universal primers ITS5 and ITS4 [80]. Amplified DNA was visualized on agarose gels stained with GreenSafe Premium (Nzytech, Lisbon, Portugal), and was visualized using a UV transilluminator to assess PCR amplification. PCR products were purified using an Illustra ExoProStar Enzymatic PCR and Sequencing Clean-up Kit (GE Life Sciences, Buckinghamshire, UK). PCR products were sequenced both ways at STABVIDA (Lisbon, Portugal) and compared with sequences from GenBank in BLAST searches.

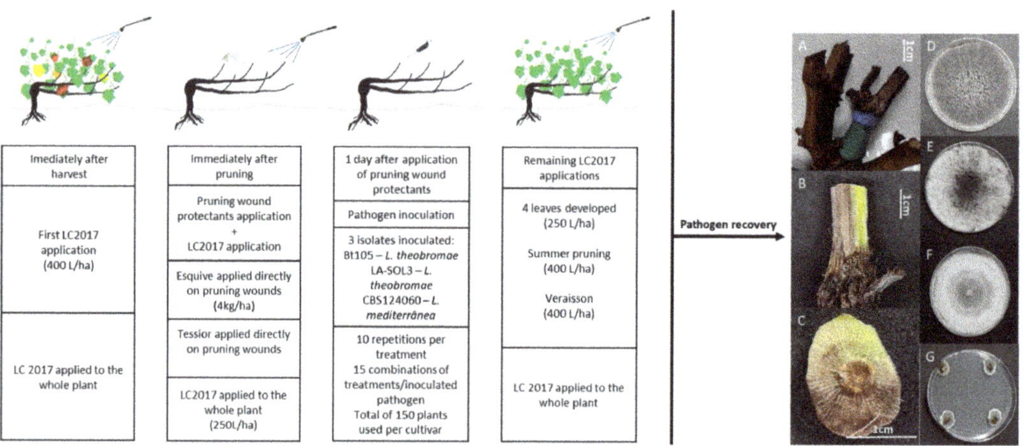

Figure 5. Diagram showing the several steps of the field assay. Information on the products used can be found in Table 4, and a description of the different treatments can be found in Table 5. (**A**) Marked sample collected from the field prior to analysis; (**B**) Sample with the bark removed showing sign of necrosis; (**C**) Sample collected about 1 cm below the pruning wound, to be divided into four pieces and plated onto PDA; (**D**) Isolate Bt105 (*Lasiodiplodia theobromae*) recovered from infected pruning wounds; (**E**) Isolate LA-SOL3 (*Lasiodiplodia theobromae*) recovered from infected pruning wounds; (**F**) Isolate CBS124060 (*Lasiodiplodia mediterranea*) recovered from infected pruning wounds; (**G**) Petri dish containing wood obtained from non-inoculated control plants showing no signs of pathogen growth.

4.2.6. Meteorological Data

For both years of the experiment, daily temperature and rainfall were obtained from the Portuguese Institute for Sea and Atmosphere (IPMA—Instituto do Mar e da Atmosfera). These data were collected on the Lisbon reference meteorological station, which is located approximately 7 km from the vineyard tested.

4.2.7. Statistical Analysis

All statistical analyses were performed using the R program (www.r-project.org (accessed on 30 September 2021)). The experimental data for both in vitro assays (mycelial growth inhibition and dual culture antagonism) were compared using an analysis of variance (ANOVA) followed by a Tukey's test ($p = 0.05$). Prior to this analysis, Levene's test was performed in order to verify the homogeneity of variance. Results of the dual culture antagonism assay were plotted using the R package ggplot2. The efficacy of the wound protectants was calculated as the mean percentage recovery (MPR) of the isolates under study. Normality and homogeneity of variance were tested using Levene's test and when necessary, data were transformed into the arcsine of the square root of the proportion to verify the assumption of homogeneity of variance. For both years and cultivars, an ANOVA was used to compare the differences in the mean percentage of recovery (MPR).

The means were compared using Tukey's test at the 5% significance level (p = 0.05). The mean percentage of disease control was also calculated according to Sosnowski et al. [49,50] and Martínez-Diz et al. [41], using the formula MPDC = 100 × [1 − (MPR treatment/MPR inoculated control)]. To better visualize the results for all the treatments (15 variables, including all the product/isolate combinations, as well as inoculated controls) on both cultivars and their interaction with the meteorological variables (temperature and rainfall), a principal component analysis (PCA) was performed on the results of all the variables. For this analysis, data obtained from the treatments and meteorological variables were considered as two individual data sets or quantitative blocks, and cultivar was considered as a qualitative variable. This analysis was also performed using the R program with the Factoshiny v2.4 package [81].

Author Contributions: P.R., C.R., A.A. and F.F. designed the experiments; P.R. and A.G. implemented the methodology; P.R. formal analysis; P.R. original draft preparation; A.A., F.F. and C.R. review of the original draft; A.A., F.F. and C.R. supervision; A.A. and C.R. funding acquisition. All authors have read and agreed to the published version of the manuscript.

Funding: This research was funded by FEDER funding through COMPETE program (POCI-01-0145-FEDER-016788) and Programa Operacional Regional de Lisboa-POR Lisboa (LISBOA-01-0145-FEDER-016788) and by national funding through FCT within the research project ALIEN (PTDC/AGR-PRO/2183/2014). The authors are thankful to FCT/MCTES for financing CESAM (UIDB/50017/2020 +UIDP/50017/2020), LEAF—Linking Landscape, Environment, Agriculture and Food (UIDB/04129/2020 and UIDP/04129/2020), and the PhD grant of Pedro Reis (SFRH/BD/131766/2017).

Acknowledgments: The authors would like to thank Natural development Group®, for providing the LC2017 product used during this work, Agrauxine S A, for providing the Esquive®, BASF BASF Agricultural Solutions Portugal for providing Tessior®, and to the Portuguese Institute for Sea and Atmosphere (IPMA-Instituto do Mar e da Atmosfera) for providing the meteorological data used on this work.

Conflicts of Interest: The authors declare no conflict of interest. The funders had no role in the design of the study; in the collection, analyses, or interpretation of data; in the writing of the manuscript, or in the decision to publish the results.

References

1. Gramaje, D.; Úrbez-Torres, J.R.; Sosnowski, M.R. Managing grapevine trunk diseases with respects to etiology and epidemiology: Current strategies and future prospects. *Plant Dis.* **2017**, *102*, 12–39. [CrossRef] [PubMed]
2. Bertsch, C.; Ramirez-Suero, M.; Magnin-Robert, M.; Larignon, P.; Chong, J.; Abou-Mansour, E.; Spagnolo, A.; Clément, C.; Fontaine, F. Trunk diseases of grapevine: Complex and still poorly understood. *Plant Pathol.* **2012**, *62*, 243–265. [CrossRef]
3. Kaplan, J.; Travadon, R.; Cooper, M.; Hillis, V.; Lubell, M.; Baumgartner, K. Identifying economic hurdles to early adoption of preventative practices: The case of trunk diseases in California winegrape vineyards. *Wine Econ. Policy* **2016**, *5*, 127–141. [CrossRef]
4. Pitt, W.M.; Trouillas, F.P.; Gubler, W.D.; Savocchia, S.; Sosnowski, M.R. Pathogenicity of Diatrypaceous Fungi on Grapevines in Australia. *Plant Dis.* **2013**, *97*, 749–756. [CrossRef] [PubMed]
5. Pitt, W.M.; Urbez-Torres, J.R.; Trouillas, F.P. Dothiorella and Spencermartinsia, new species and records from grapevines in Australia. *Australas. Plant Pathol.* **2014**, *44*, 43–56. [CrossRef]
6. Rolshausen, P.E.; Akgül, D.S.; Perez, R.; Eskalen, A.; Gispert, C. First report of wood canker caused by *Neoscytalidium dimidiatum* on grapevine in California. *Plant Dis.* **2013**, *97*, 1511. [CrossRef] [PubMed]
7. Úrbez-Torres, J.R. The status of Botryosphaeriaceae species infecting grapevines. *Phytopathol. Mediterr.* **2011**, *50*, S5–S45. [CrossRef]
8. Yang, T.; Groenewald, J.Z.; Cheewangkoon, R.; Jami, F.; Abdollahzadeh, J.; Lombard, L.; Crous, P.W. Families, genera, and species of Botryosphaeriales. *Fungal Biol.* **2017**, *121*, 322–346. [CrossRef] [PubMed]
9. Úrbez-Torres, J.R.; Leavitt, G.M.; Guerrero, J.C.; Guevara, J.; Gubler, W.D. Identification and Pathogenicity of Lasiodiplodia theobromae and Diplodia seriata, the Causal Agents of Bot Canker Disease of Grapevines in Mexico. *Plant Dis.* **2008**, *92*, 519–529. [CrossRef]
10. Úrbez-Torres, J.R.; Gubler, W.D. Pathogenicity of Botryosphaeriaceae Species Isolated from Grapevine Cankers in California. *Plant Dis.* **2009**, *93*, 584–592. [CrossRef]
11. Van Niekerk, J.M.; Crous, P.W.; Groenewald, J.Z.; Fourie, P.H.; Halleen, F. DNA Phylogeny, Morphology and Pathogenicity of Botryosphaeria Species on Grapevines. *Mycologia* **2004**, *96*, 781–798. [CrossRef]

12. Pitt, W.M.; Huang, R.; Steel, C.C.; Savocchia, S. Pathogenicity and epidemiology of Botryosphaeriaceae species isolated from grapevines in Australia. *Australas. Plant Pathol.* **2013**, *42*, 573–582. [CrossRef]

13. Rangel-Montoya, E.A.; Paolinelli, M.; Rolshausen, P.E.; Valenzuela-Solano, C.; Hernandez-Martinez, R. Characterization of Lasiodiplodia species associated with grapevines in Mexico. *Phytopathol. Mediterr.* **2021**, *60*, 237–251. [CrossRef]

14. Linaldeddu, B.T.; Deidda, A.; Scan, B.; Franceschini, A.; Serra, S.; Berraf-Tebbal, A.; Zouaoui Boutiti, M.; Ben Jamâa, M.L.; Phillips, A.J.L. Diversity of Botryosphaeriaceae species associated with grapevine and other woody hosts in Italy, Algeria and Tunisia, with descriptions of *Lasiodiplodia exigua* and *Lasiodiplodia mediterranea* sp. Nov. *Fungal Divers.* **2015**, *71*, 201–214. [CrossRef]

15. Zhang, W.; Groenewald, J.Z.; Lombard, L.; Schumacher, R.K.; Phillips, A.J.L.; Crous, P.W. Evaluating species in *Botryosphaeriale*. *Pers. Mol. Phylogeny Evol. Fungi* **2021**, *46*, 63–115. [CrossRef]

16. Phillips, A.J.L.; Alves, A.; Abdollahzadeh, J.; Slippers, B.; Wingfield, M.J.; Groenewald, J.Z.; Crous, P.W. The Botryosphaeriaceae: Genera and species known from culture. *Stud. Mycol.* **2013**, *76*, 51–167. [CrossRef]

17. Correia, K.C.; Câmara, M.; Barbosa, M.; Sales, R.; Agustí-Brisach, C.; Gramaje, D. Fungal trunk pathogens associated with table grape decline in Northeastern Brazil. *Phytopathol. Mediterr.* **2013**, *52*, 380–387.

18. Urbez-Torres, J.R.; Peduto, F.; Striegler, R.K.; Urrea-Romero, K.E.; Rupe, J.C.; Cartwright, R.D.; Gubler, W.D. Characterization of fungal pathogens associated with grapevine trunk diseases in Arkansas and Missouri. *Fungal Divers.* **2012**, *52*, 169–189. [CrossRef]

19. Yan, J.-Y.; Xie, Y.; Zhang, W.; Wang, Y.; Liu, J.-K.; Hyde, K.D.; Seem, R.C.; Zhang, G.-Z.; Wang, Z.-Y.; Yao, S.-W.; et al. Species of Botryosphaeriaceae involved in grapevine dieback in China. *Fungal Divers.* **2013**, *61*, 221–236. [CrossRef]

20. Taylor, A.; Hardy, G.E.S.J.; Wood, P.; Burgess, T. Identification and pathogenicity of *Botryosphaeria* species associated with grapevine decline in Western Australia. *Australas. Plant. Pathol.* **2005**, *34*, 187–195. [CrossRef]

21. Arkam, M.; Alves, A.; Lopes, A.; Čechová, J.; Pokluda, R.; Eichmeier, A.; Zitouni, A.; Mahamedi, A.E.; Berraf-Tebbal, A. Diversity of Botryosphaeriaceae causing grapevine trunk diseases and their spatial distribution under different climatic conditions in Algeria. *Eur. J. Plant Pathol.* **2021**, 1–20. [CrossRef]

22. Kaiser, W.; Rivero, V.; Valverde, B.E. First report of Diplodia cane dieback of grapevine in Bolivia. *Plant Dis.* **2009**, *93*, 320. [CrossRef] [PubMed]

23. Burruano, S.; Mondello, V.; Conigliaro, G.; Alfonzo, A.; Spagnolo, A.; Mugnai, L. Grapevine decline in Italy caused by *Lasiodiplodia theobromae*. *Phytopathol. Mediterr.* **2008**, *47*, 132–136.

24. Galvez, E.R.; Maldonado, E.; Alves, A. Identification and pathogenicity of Lasiodiplodia theobromae causing dieback of table grapes in Peru. *Eur. J. Plant Pathol.* **2015**, *141*, 477–489. [CrossRef]

25. Rego, C.; Nascimento, T.; Pinto, P.; Oliveira, H. First report of *Lasiodiplodia theobromae* associated with cankers and dieback of grapevine (*Vitis vinifera*) in Portugal. In Proceedings of the 9th Conference of the European Foundation for Plant Pathology, Évora, Portugal, 15–18 November 2010.

26. Aroca, A.; Raposo, R.; Gramaje, D.; Armengol, J.; Martos, S.; Luque, J. First Report of Lasiodiplodia theobromae Associated with Decline of Grapevine Rootstock Mother Plants in Spain. *Plant Dis.* **2008**, *92*, 832. [CrossRef]

27. Akgül, D.S.; Savas, N.G.; Eskalen, A. First Report of Wood Canker Caused by Botryosphaeria dothidea, Diplodia seriata, Neofusicoccum parvum, and Lasiodiplodia theobromae on Grapevine in Turkey. *Plant Dis.* **2014**, *98*, 568. [CrossRef]

28. Mondello, V.; Battiston, E.; Pinto, C.; Coppin, C.; Trotel-Aziz, P.; Clément, C.; Mugnai, L.; Fontaine, F. Grapevine trunk diseases: A review of fifteen years of trials for their control with chemicals and biocontrol agents. *Plant Dis.* **2018**, *102*, 1189–1217. [CrossRef] [PubMed]

29. Reis, P.; Gaspar, A.; Alves, A.; Fontaine, F.; Lourenco, I.; Saramago, J.; Mota, M.; Rego, C. Early Season Symptoms on Stem, Inflorescences and Flowers of Grapevine Associated with Botryosphaeriaceae Species. *Plants* **2020**, *9*, 1427. [CrossRef]

30. Úrbez-Torres, J.R.; Battany, M.; Bettiga, L.J.; Gispert, C.; McGourty, G.; Roncoroni, J.; Smith, R.J.; Verdegaal, P.; Gubler, W.D. Botryosphaeriaceae Species Spore-Trapping Studies in California Vineyards. *Plant Dis.* **2010**, *94*, 717–724. [CrossRef]

31. Van Niekerk, J.M.; Calitz, F.J.; Halleen, F.; Fourie, P.H. Temporal spore dispersal patterns of grapevine trunk pathogens in South Africa. *Eur. J. Plant Pathol.* **2010**, *127*, 375–390. [CrossRef]

32. Elena, G.; Luque, J. Pruning debris of grapevine as a potential inoculum source of Diplodia seriata, causal agent of Botryosphaeria dieback. *Eur. J. Plant Pathol.* **2016**, *144*, 803–810. [CrossRef]

33. Sosnowski, M.R.; McCarthy, G. Economic impact of grapevine trunk disease management in Sauvignon Blanc vineyards of New Zealand. *Wine Vitic. J.* **2017**, *32*, 42–48.

34. Pitt, W.M.; Sosnowski, M.R.; Huang, R.; Qiu, Y.; Steel, C.C.; Savocchia, S. Evaluation of Fungicides for the Management of Botryosphaeria Canker of Grapevines. *Plant Dis.* **2012**, *96*, 1303–1308. [CrossRef] [PubMed]

35. Díaz, G.A.; Latorre, B.A. Efficacy of paste and liquid fungicide formulations to protect pruning wounds against pathogens associated with grapevine trunk diseases in Chile. *Crop. Prot.* **2013**, *46*, 106–112. [CrossRef]

36. Amposah, N.T.; Jones, E.E.; Ridgeway, H.J.; Jaspers, M.V. Evaluation of fungicides for the management of Botryosphaeria dieback diseases of grapevines. *Pest. Manag. Sci.* **2012**, *68*, 676–683. [CrossRef] [PubMed]

37. Sosnowski, M.R.; Mundi, M.C. Pruning wound protection strategies for simultaneous control of Eutypa and Botryosphaeira dieback in New Zealand. *Plant. Dis.* **2019**, *103*, 519–525. [CrossRef]

38. Rego, C.; Reis, P.; Dias, A.; Correia, R. Field evaluation of fungicides against Botryosphaeria dieback canker and Phomopsis cane and leaf spot. *Phytopathol. Mediterr.* **2014**, *53*, 581–582.

39. Bester, W.; Crous, P.W.; Fourie, P.H.; Crous, P.W.; Fourie, P.H. Evaluation of fungicides as potential grapevine pruning wound protectants against Botryosphaeria species. *Australas. Plant Path.* **2007**, *36*, 73–77. [CrossRef]

40. Mutawila, C.; Halleen, F.; Mostert, L. Development of benzimidazole resistant Trichoderma strains for the integration of chemical and biocontrol methods of grapevine pruning wound protection. *BioControl* **2015**, *60*, 387–399. [CrossRef]

41. Martínez-Diz, M.D.P.; Díaz-Losada, E.; Díaz-Fernández, Á.; Bouzas-Cid, Y.; Gramaje, D. Protection of grapevine pruning wounds against Phaeomoniella chlamydospora and Diplodia seriata by commercial biological and chemical methods. *Crop. Prot.* **2021**, *143*, 105465. [CrossRef]

42. Rolshausen, P.E.; Úrbez-Torres, J.R.; Rooney-Lathman, S.; Eskalen, A.; Smith, R.J.; Gubler, W.D. Evaluation of pruning wound susceptibility and protection against fungi associated with grapevine trunk diseases. *Am. J. Enol. Vitic.* **2010**, *61*, 113–119.

43. Epstein, L.; Sukhwinder, K.; VanderGheynst, J.S. Botryosphaeria-related dieback and control investigated in non-coastal California grapevines. *Calif. Agric.* **2008**, *62*, 161–166. [CrossRef]

44. Wu, H.; Spagnolo, A.; Marivingt-Mounir, C.; Clément, C.; Fontaine, F.; Chollet, J.F. Evaluating the combined effect of systemic phenylpyrrole fungicide and the plant growth-promoting rhizobacteria *Paraburkholderia pthytofirmans* (strain PsJN:gfp2x) against the grapevine trunk pathogen *Neofusicoccum parvum*. *Pest. Manag. Sci.* **2020**, *76*, 3838–3848. [CrossRef] [PubMed]

45. Fischer, J.; Beckers, S.J.; Yiamsawas, D.; Thines, E.; Landfester, K.; Wurm, F.R. Targeted drug delivery in plants: Enzyme-responsive lignin nanocarriers for the curative treatment of the worldwide grapevine trunk disease Esca. *Adv. Sci.* **2019**, *6*, 1802315. [CrossRef] [PubMed]

46. Battiston, E.; Compant, S.; Antonielli, L.; Mondello, V.; Clément, C.; Simoni, A.; Di Marco, S.; Mugnai, L.; Fontaine, F. In planta Activity of Novel Copper(II)-Based Formulations to Inhibit the Esca-Associated Fungus Phaeoacremonium minimum in Grapevine Propagation Material. *Front. Plant Sci.* **2021**, *12*, 649694. [CrossRef] [PubMed]

47. Battiston, E.; Salvatici, M.C.; Lavacchi, A.; Gatti, A.; Di Marco, S.; Mugnai, L. Functionalisation of a nano-structured hydroxyapatite with copper (II) compounds as pesticide: In situ TEM and ESEM observations of treated *Vitis vinifera* L. leaves. *Pest. Man. Sci.* **2018**, *74*, 1903–1915. [CrossRef] [PubMed]

48. Mondello, V.; Fernandez, O.; Guise, J.-F.; Trotel-Aziz, P.; Fontaine, F. In planta Activity of the Novel Copper Product HA + Cu(II) Based on a Biocompatible Drug Delivery System on Vine Physiology and Trials for the Control of Botryosphaeria Dieback. *Front. Plant Sci.* **2021**, *12*, 693995. [CrossRef]

49. Úrbez-Torres, J.R.; Tomaselli, E.; Pollard-Flamand, J.; Boulé, J.; Gerin, D.; Pollastro, S. Characterization of *Trichoderma* isolates from southern Italy, and their potential biocontrol activity against grapevine trunk disease fungi. *Phytopathol. Mediterr.* **2020**, *59*, 425–439.

50. Berbegal, M.; Ramón-Albalat, A.; León, M.; Armengol, J. Evaluation of long-term protection from nursery to vineyard provided byTrichoderma atrovirideSC1 against fungal grapevine trunk pathogens. *Pest Manag. Sci.* **2020**, *76*, 967–977. [CrossRef]

51. Mutawila, C.; Fourie, P.H.; Halleen, F.; Mostert, L. Grapevine cultivar variation to pruning wound protection by *Trichoderma* species against trunk pathogens. *Phytopathol. Mediterr.* **2011**, *50*, S264–S276.

52. Mutawila, C.; Halleen, F.; Mostert, L. Optimisation of time of application of *Trichoderma* biocontrol agents for protection of grapevine pruning wounds. *Aust. J. Grape Wine Res.* **2016**, *22*, 279–287. [CrossRef]

53. Mounier, E.; Boulisset, F.; Cortés, F.; Cadiou, M.; Dubournet, P.; Pajot, E. Esquive®WP limits development of grapevine trunk diseases and safeguards the production potential of vineyards. In *Biocontrol of Major Grapevine Diseases: Leading Research*; CABI: São Paulo, Brazil, 2016; pp. 160–170.

54. Mounier, E.; Cortes, F.; Cadious, M.; Pajot, E. The benefits of *Trichoderma atroviride* strain I-1237 for the protection of grapevines against trunk diseases: From the nursery to the vineyard. *Phytopathol. Mediterr.* **2014**, *53*, 591–592.

55. Reis, P.; Letousey, P.; Rego, C. *Trichoderma atroviride* strain I-1237 protects pruning wounds against grapevine wood pathogens. *Phytopathol. Mediterr.* **2017**, *56*, 580.

56. Rusin, C.; Cavalcanti, F.R.; De Lima, P.C.G.; Faria, C.M.D.R.; Almança, M.A.K.; Botelho, R.V. Control of the fungi Lasiodiplodia theobromae, the causal agent of dieback, in cv. syrah grapevines. *Acta Sci. Agron.* **2020**, *43*, e44785. [CrossRef]

57. Sosnowski, M.; Creaser, M.; Wicks, T.; Lardner, R.; Scott, E.S. Protection of grapevine pruning wounds from infection byEutypa lata. *Aust. J. Grape Wine Res.* **2008**, *14*, 134–142. [CrossRef]

58. Sosnowski, M.R.; Ayres, M.; Wicks, T.; McCarthy, M.; Scott, E. In search of resistance to grapevine trunk diseases. *Wine Vitic.* **2013**, *28*, 55–58.

59. Kotze, C.; van Niekerk, J.; Mostert, L.; Halleen, F.; Fourie, P.H. Evaluation of biocontrol agents for pruning wound protection against trunk pathogens infection. *Phytopathol. Mediterr.* **2011**, *56*, 536.

60. John, S.; Wicks, T.J.; Hunt, J.S.; Lorimer, M.F.; Oakey, H.; Scott, E.S. Protection of grapevine pruning wounds from infection by *Eutypa lata* using *Trichoderma harzianum* and *Fusarium lateritium*. *Australas. Plant. Pathol.* **2005**, *34*, 569–575. [CrossRef]

61. Halleen, F.; Fourie, P.; Lombard, P. Protection of Grapevine Pruning Wounds against Eutypa lata by Biological and Chemical Methods. *S. Afr. J. Enol. Vitic.* **2016**, *31*, 125–132. [CrossRef]

62. Leal, C.; Richet, N.; Guise, J.F.; Gramaje, D.; Armengol, J.; Fontaine, F.; Trotel-Aziz, P. Cultivar contributes to the beneficial effects of *Bacillus subtilis* PTA-271 and *Trichoderma atroviride* SC1 to protect grapevine against *Neofusicoccum parvum*. *Front. Microbiol.* **2021**, *12*, 726132. [CrossRef]

63. Pertot, I.; Caffi, T.; Rossi, V.; Mugnai, L.; Hoffmann, C.; Grando, M.S.; Gary, C.; Lafond, D.; Duso, C.; Thiery, D.; et al. A critical review of plant protection tools for reducing pesticide use on grapevine and new perspectives for the implementation of IPM in viticulture. *Crop. Protect.* **2017**, *97*, 70–84. [CrossRef]

64. Rego, C.; Nascimento, T.; Cabral, A.; Silva, M.J.; Oliveira, H. Control of grapevine wood fungi in commercial nurseries. *Phytopathol. Mediterr.* **2009**, *48*, 128–135.

65. Kühn, A.; Zappata, A.; Gold, R.E.; Zito, R.; Kortekamp, A. Susceptibility of grape pruning wounds to grapevine trunk diseases and effectiveness of a new BASF wound protectant. *Phytopathol. Mediterr.* **2017**, *56*, 536.

66. Samaras, A.; Ntasioy, P.; Testempasis, S.; Theocharis, S.; Koundouras, S.; Karaoglanidis, G. Evaluation of the fungicide Tessior (boscalid and pyraclostrobin) for control of grapevine trunk diseases in Greece. *Phytopathol. Mediterr.* **2019**, *58*, 421.

67. Lengyel, S.; Gold, R.E.; Fischer, J.; Yemelin, A.; Thines, E.; Kühn, A. Early detection project- detection and quantification of *Phaeomoniella chlamydospora* and *Botryosphaeria* spp. in *Vitis vinifera* wood samples. *Phytopathol. Mediterr.* **2019**, *58*, 406–407.

68. van Niekerk, J.M.; Halleen, F.; Crous, P.W.; Fourie, P.H. The distribution and symptomatology of grapevine trunk pathogens are influenced by climate. *Phytopathol. Mediterr.* **2011**, *50*, S98–S111.

69. Félix, C.; Duarte, A.S.; Vitorino, R.; Guerreiro, A.C.L.; Domingues, P.; Correia, A.C.M.; Alves, A.; Esteves, A.C. Temperature Modulates the Secretome of the Phytopathogenic Fungus Lasiodiplodia theobromae. *Front. Plant Sci.* **2016**, *7*, 1096. [CrossRef] [PubMed]

70. Reis, P.; Gaspar, A.; Alves, A.; Fontaine, F.; Rego, C. Susceptibility of different grapevine cultivars to infection by *Lasiodiplodia theobromae* and *Lasiodiplodia mediterranea. Plant. Dis.* Under review.

71. Sofia, J.; Mota, M.; Gonçalves, M.T.; Rego, C. Response of four Portuguese grapevine cultivars to infection by Phaeomoniella chlamydospora. *Phytopatol. Mediterr.* **2019**, *57*, 506–518.

72. Battiston, E.; Antonielli, L.; Di Marco, S.; Fontaine, F.; Mugnai, L. Innovative Delivery of Cu(II) Ions by a Nanostructured Hydroxyapatite: Potential Application in Planta to Enhance the Sustainable Control of Plasmopara viticola. *Phytopathology* **2019**, *109*, 748–759. [CrossRef] [PubMed]

73. Di Marco, S.; Osti, F.; Mugnai, L. First studies on the potential of a copper formulation for the control of leaf stripe disease within Esca complex in grapevine. *Phytopathol. Mediterr.* **2011**, *50*, 300–309.

74. Aziz, A.; Trotel-Aziz, P.; Dhuicq, L.; Jeandet, P.; Couderchet, M.; Vernet, G. Chitosan Oligomers and Copper Sulfate Induce Grapevine Defense Reactions and Resistance to Gray Mold and Downy Mildew. *Phytopathology* **2006**, *96*, 1188–1194. [CrossRef]

75. Rahman, M.A.; Begum, M.F.; Alam, M.F. Screening of *Trichoderma* isolates as a biological control agent against *Ceratocystis paradoxa* causing pineapple disease of sugar cane. *Microbiology* **2009**, *37*, 277–285.

76. Crous, P.W.; Slippers, B.; Wingfield, M.J.; Rheeder, J.P.; Marasas, W.F.; Philips, A.J.; Alves, A.; Burgess, T.; Barber, P.; Groenewald, J.Z. Phylogenetic lineages in the Botryosphaeriaceae. *Stud. Mycol.* **2006**, *55*, 235–253. [CrossRef] [PubMed]

77. Phillips, A. Botryosphaeria species associated with diseases of grapevines in Portugal. *Phytopathol. Mediterr.* **2002**, *41*, 3–18.

78. Úrbez-Torres, J.R.; Leavitt, G.M.; Voegel, T.; Gubler, W.D. Identification and distribution of *Botryosphaeria* species associated with grapevine cankers in California. *Plant. Dis.* **2006**, *90*, 1490–1503. [CrossRef] [PubMed]

79. Alves, A.; Crous, P.W.; Correia, A.; Phillips, A.J.L. Morphological and molecular data reveal cryptic speciation in *Lasiodiplodia theobromae. Fungal Divers.* **2008**, *28*, 1–13.

80. White, T.J.; Bruns, T.; Lee, S.; Taylor, J. Amplified and direct sequencing of fungal ribosomal RNA genes for phylogenetics. In *PCR Protocols: A Guide to Methods and Applications*; Innis, M.A., Gelfand, D.H., Sninsky, J.J., White, T.J., Eds.; Academic: San Diego, CA, USA, 1990; pp. 315–322.

81. Vaissie, P.; Monge, A.; Husson, F. Perform Factorial Analysis from 'FactoMineR' with a Shiny Application. 2021. Available online: https://CRAN.R-project.org/package=Factoshiny (accessed on 1 November 2021).

Article

Inhibitory Activity of Shrimp Waste Extracts on Fungal and Oomycete Plant Pathogens

Soumia El boumlasy [1,2,†], Federico La Spada [2,†], Nunzio Tuccitto [3,4], Giovanni Marletta [3,4], Carlos Luz Mínguez [5], Giuseppe Meca [5], Ermes Ivan Rovetto [2], Antonella Pane [2], Abderrahmane Debdoubi [1] and Santa Olga Cacciola [2,*]

[1] Laboratory of Materials-Catalysis, Chemistry Department, Faculty of Science, University Abdelmalek Essaadi, Tetouan B.P. 2117, Morocco; soumiaelboumlasy@gmail.com (S.E.b.); debdoubi@hotmail.com (A.D.)
[2] Department of Agriculture, Food and Environment, University of Catania, 95123 Catania, Italy; federico.laspada@unict.it (F.L.S.); ermes.rovetto@hotmail.com (E.I.R.); apane@unict.it (A.P.)
[3] Consorzio per lo Sviluppo dei Sistemi a Grande Interfase, CSGI, Viale A. Doria 6, 95125 Catania, Italy; n.tuccitto@unict.it (N.T.); gmarletta@unict.it (G.M.)
[4] Department of Chemical Sciences, Università degli Studi di Catania, Viale A. Doria 6, 95125 Catania, Italy
[5] Laboratory of Food Chemistry and Toxicology, Faculty of Pharmacy, University of Valencia, Av. Vicent Andrés Estellés s/n, 46100 Burjassot, Spain; carlos.luz@uv.es (C.L.M.); giuseppe.meca@uv.es (G.M.)
* Correspondence: olgacacciola@unict.it; Tel.: +39-095-7147371
† The authors contributed equally to this work.

Citation: El boumlasy, S.; La Spada, F.; Tuccitto, N.; Marletta, G.; Mínguez, C.L.; Meca, G.; Rovetto, E.I.; Pane, A.; Debdoubi, A.; Cacciola, S.O. Inhibitory Activity of Shrimp Waste Extracts on Fungal and Oomycete Plant Pathogens. *Plants* **2021**, *10*, 2452. https://doi.org/10.3390/plants10112452

Academic Editors: Carlos Agustí-Brisach and Eugenio Llorens

Received: 12 October 2021
Accepted: 10 November 2021
Published: 13 November 2021

Publisher's Note: MDPI stays neutral with regard to jurisdictional claims in published maps and institutional affiliations.

Abstract: (1) Background: This study was aimed at determining the *in vitro* inhibitory effect of new natural substances obtained by minimal processing from shrimp wastes on fungi and oomycetes in the genera *Alternaria*, *Colletotrichum*, *Fusarium*, *Penicillium*, *Plenodomus* and *Phytophthora*; the effectiveness of the substance with the highest *in vitro* activity in preventing citrus and apple fruit rot incited by *P. digitatum* and *P. expansum*, respectively, was also evaluated. (2) Methods: The four tested substances, water-extract, EtOAc-extract, MetOH-extract and nitric-extract, were analyzed by HPLC-ESI-MS-TOF; *in vitro* preliminary tests were carried out to determine the minimal inhibitory/fungicidal concentrations (MIC and MFC, respectively) of the raw dry powder, EtOAc-extract, MetOH-extract and nitric-extract for each pathogen. (3) Results: in the agar-diffusion-assay, nitric-extract showed an inhibitory effect on all pathogens, at all concentrations tested (100, 75, 50 and 25%); the maximum activity was on *Plenodomus tracheiphilus*, *C. gloeosporioides* and *Ph. nicotianae*; the diameters of inhibition halos were directly proportional to the extract concentration; values of MIC and MFC of this extract for all pathogens ranged from 2 to 3.5%; the highest concentrations (50 to 100%) tested *in vivo* were effective in preventing citrus and apple fruit molds. (4) Conclusions: This study contributes to the search for natural and ecofriendly substances for the control of pre- and post-harvest plant pathogens.

Keywords: metabolites; phenolic compounds; inhibitory effect; citrus; apple; HPLC-ESI-MS-TOF; post-harvest diseases; mal secco disease; MIC; MFC

1. Introduction

Plant pathogenic fungi are responsible for many serious diseases that affect agricultural productions both pre- and post-harvest. In this respect, the losses of products along the post-harvest chains (i.e., warehousing, transport and final distribution) determine strong impactful consequences, especially in agriculture-based-economy countries [1–3]. To minimize production losses and maintain crop sustainability, several strategies based on the application of different means, such as physical, chemical and biological, have been adopted over time [4,5]. Currently, one of the most consolidated and effective means for controlling fungal diseases is represented by chemical synthetic fungicides [4,6]. However, their use negatively affects both human health and the preservation of the environment. Moreover, the restricted number of active ingredients which are allowed for post-harvest treatments increases the risk of selection of fungicide resistant plant pathogens, with the consequent dramatic reduction of the efficacy of synthetic fungicides [7]. For these reasons, during past years, their application has been strictly limited by several governmental institutions worldwide [8,9].

In order to satisfy the growing request for high-quality and, at the same time, safe and eco-friendly products, throughout the past two decades, the research field strongly focused on the investigation of the potentialities of alternative means to synthetic fungicides to control plant diseases; these include antagonistic microorganisms or derivatives thereof, natural biostimulants [7,9,10], as well as natural antimicrobial compounds [11,12].

With the perspective of reducing environmental pollution and related consequences for human health, nowadays, the scientific research is also strongly focused on valorizing wastes, especially those largely generated by processing industries [13]. Within this framework, the shrimp market has stood out for considerable development, especially during the past few years. In this respect, it has been estimated that in 2020, the production of shrimp reached a total of 5.03 million tons around the globe, with an amount of waste ranging between 40–50% per ton of fresh product [14–16]. Therefore, the wastes generated by shrimp processing industries in food production are clearly undergoing a dramatic increase [17]. Shrimp wastes generated for production of human food are represented by heads, intestines, tails and shells [17], which are usually disposed by throwing into garbage heaps [18], ocean dumping, incineration and land filling [19]. Therefore, an inevitable increase in generated wastes could be determined by their non-use [20].

Shrimp are, overall, considered a high-value aquaculture product [17], not only because of the nutritional properties of the meat used for human consumption, but also for the composition of their wastes; in fact, their major constituents are proteins (35–50%), chitin (15–25%), calcium and phosphorus (10–15%), and other substances (such as amino acids, vitamins, carotenoids, astaxanthin, polyunsaturated fatty acids and other enzymes) [15,21–23]. For this reason, nowadays, the valorization of shrimp wastes is a consolidated practice.

Shrimp wastes as such have been used for feeding in veterinary practice and aquaculture [17] as well as in compost fertilizer [24,25]. Dried shrimp wastes are also used in animal feeding in mixtures with other agricultural raw materials; however, since drying processes are usually carried out directly along the beaches, these practices of the use of shrimp wastes favor additional pollution, especially in coastal areas [17]. A further strategy for the use of shrimp wastes includes both the extraction of bioactive molecules or the secondary chemically-mediated transformation of some parts of these into other bioactive compounds; one of these is the chitosan, the large-scale production of which is commonly carried out by alkaline deacetylation of the chitin extracted from shrimp shells [26]. Chitosan has several useful applications in various fields, including medicine, cosmetics, agriculture, paper and textile industries, biotechnologies and bioremediation of the environment (water treatment) [15,27]; however, the acid/alkaline-mediated industrial processes for its production from shrimp wastes have serious environmental consequences [17,18,26].

The aforementioned products arising from shrimp wastes represent, therefore, a precious asset in several fields of application; however, it is an accepted fact that their processing generates highly impactful new wastes, which in turn contribute to environmental pollution and, consequently, negatively affect human health.

The investigation of the potentialities of new products arising from a minimal and sustainable processing of shrimp wastes stands, therefore, as an essential challenge for scientific research. Considering that plant pathology is strongly focused on finding eco-friendly strategies for controlling plant pathogens and related diseases, the present study evaluated the effectiveness of new substances obtained by the minimal processing of shrimp wastes in the *in vitro* and *in vivo* control of major fungal and oomycete pathogens of the genera *Alternaria*, *Colletotrichum*, *Fusarium*, *Penicillium*, *Plenodomus* and *Phytophthora*.

2. Results

In this study, wastes from the shrimp species Parapenaeus longirostris were processed to obtain four substances: (i) "Water-extract", (ii) "EtOAc-extract", (iii) "MetOH-extract" and (iv) "Nitric-extract". All these extracts were analyzed, to determine their composition in metabolites and phenolic compounds, by HPLC-ESI-MS/TOF. Then, the antifungal activity of the "dry-powder", "EtOAc-extract", "MetOH-extract" and "Nitric-extract" was preliminarily tested *in vitro* by an agar diffusion test toward several fungal and oomycete pathogens. "Dry-powder", "EtOAc-extract" and "MetOH-extract" did not demonstrate any inhibitory effect in the mycelial growth of all pathogens under study (data not shown); therefore, they were not further tested. "Nitric-extract" was the only extract that negatively affected the mycelial growth of all pathogens; the diameter of the inhibition halos consequently observed at each concentration was, therefore, recorded at the end of the incubation period (see Figure 1a–o). The most effective substance resulting from the *in vitro* test was further investigated to determine its efficiency in terms of minimal inhibitory concentration (MIC) and minimal fungicidal concentration (MFC). The *in vivo* effectiveness of the selected substance in the control of post-harvest infections of fruits by *Penicillium* spp. was finally tested.

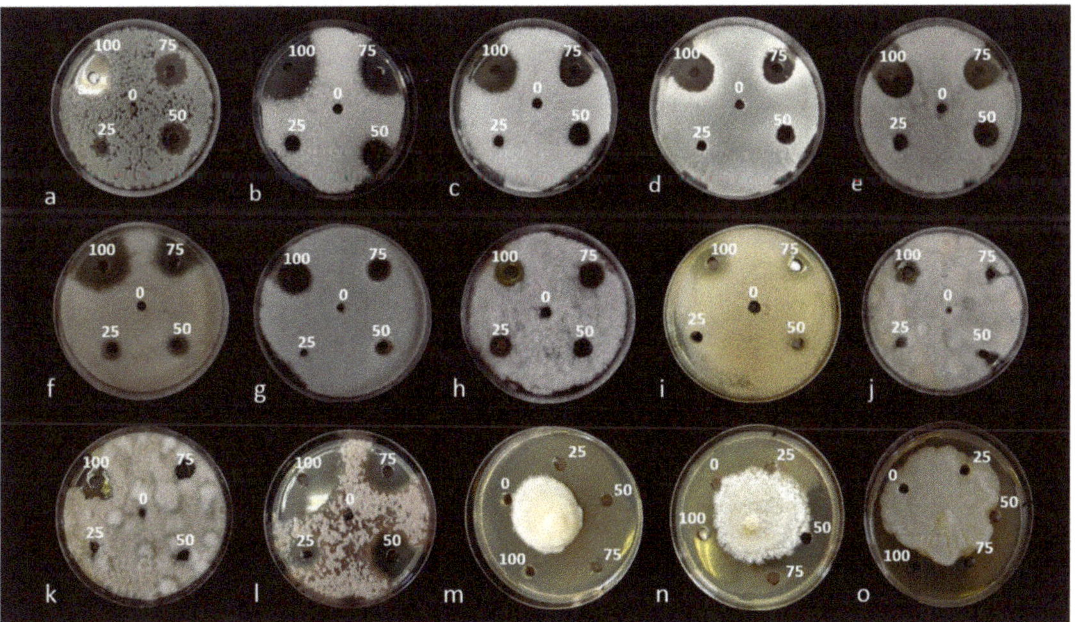

Figure 1. Agar diffusion test. Inhibition halos determined by the Nitric-extract at different concentrations (0, 25, 50, 75, 100%), after 3 days of incubation at 25 °C on PDA: (**a**) *Penicillium digitatum* P1PP0; (**b**) *P. commune* CECT 20767; (**c**) *P. expansum* CECT 2278; (**d**) *P. italicum* CECT 20909; (**e**) *Colletotrichum acutatum* UW14; (**f**) *C. karsti* CAM; (**g**) *C. gloeosporioides* C2; (**h**) *Fusarium proliferatum* CBS 145950; (**i**) *F. sacchari* CBS 145949; (**j**) *Alternaria arborescens* 803; (**k**) *A. alternata* 646; (**l**) *Plenodomus tracheiphilus* Pt2. Inhibition halos at different concentrations after 15 days of incubation at 25 °C on PDA: (**m**) *Phytophthora nicotianae* T2.C-M1A; (**n**) *Ph. nicotianae* T3-B-K1A; (**o**) *Ph. citrophthora* Ax1Ar.

2.1. Metabolites and Phenolic Compounds Detected in Test Substances by HPLC-ESI-MS/TOF

The metabolites detected by HPLC-ESI-MS/TOF in the analyzed substances are presented, as a heat map, in Figure 2. Colors are based on the relative abundance (logarithmic scale) of the metabolites detected, where red represents high abundance and green represents low abundance. Overall, among all substances examined, the analysis evidenced the presence of a total of 54 metabolites already known in the literature. In particular, the "Water-extract" showed 50 metabolites, which is the highest number recovered; "EtOAc-extract" and "Nitric-extract" contained 36 and 35 metabolites, respectively; finally, only 25 metabolites were detected in the "MetOH-extract". Some marked differences were observed among the substances; in particular, a higher abundance of free amino acids, such as phenyalanine, proline, serine, tyrosine and valine, was evidenced in the "Water-extract" and "MetOH-extract" over the "EtOAc-extract" and "Nitric-extract". Really high relative abundances in some metabolites were also observed; in particular, 2-Hydroxyisocaproic acid, 3-(4-Hydroxyphenyl) propionic acid and 4-aminobenzoic acid in "MetOH-extract", docosahexaenoic acid in "EtOAc-extract" and the phenylalanine in the "Water-extract" and "MetOH-extract".

The most important phenolic acids detected by HPLC-ESI-MS/TOF in the substances analyzed are presented in Table 1. Their abundance is expressed in mg/kg of each substance. The most abundant phenolic compound detected in all the analyzed substances was benzoic acid, whose amount ranged from a minimum of 0.87 mg/kg in "Nitric-extract" to a maximum of 3.57 mg/kg in "EtOAc-extract". In order of abundance, vanillin (0.21–2.04 mg/kg) and syringic acid (0.16–1.21 mg/kg), which had the highest concentrations of "MetOH-extract", were detected. The p-coumaric (4-hydroxycinnamic acid) acid was another phenolic compound recovered in all the substances; its abundance ranged from a minimum of 0.27 mg/kg in the "Water-extract" to a maximum of 0.88 mg/kg in the sample "Nitric-extract". The "Nitric-extract" also reported the highest concentration of 1-2-Dihydroxybenzene (0.86 mg/kg). Few phenolic compounds were detected just in one substance; among these, the 3-(4-hydroxy-3-methoxyphenyl) propionic acid and ellagic acid were detected only in "Nitric-extract", while sinapic acid only in the "Water-extract".

Table 1. Concentration of phenolic compounds detected in the tested substances (mean value ± standard deviation).

Phenolic Compounds (mg/kg)	Test Substances			
	Water-Extract	EtOAc-Extract	MetOH-Extract	Nitric-Extract
1-2-Dihydroxybenzene	Nd	0.48 ± 0.03	Nd	0.86 ± 0.02
3-(4-hydroxy-3-methoxyphenyl)	Nd	Nd	Nd	0.25 ± 0.03
Benzoic acid	2.48 ± 0.02	3.57 ± 0.04	3.08 ± 0.05	0.87 ± 0.02
Caffeic acid	0.16 ± 0.02	0.17 ± 0.03	0.58 ± 0.01	0.16 ± 0.04
Ellagic acid	Nd	Nd	Nd	0.20 ± 0.01
Gallic acid	Nd	Nd	Nd	Nd
Hydroxicinnamic acid	0.05 ± 0.03	0.42 ± 0.01	0.27 ± 0.02	Nd
P-Coumaric acid	0.27 ± 0.02	0.34 ± 0.02	0.83 ± 0.04	0.88 ± 0.01
Sinapic acid	0.10 ± 0.03	Nd	Nd	Nd
Syringic acid	0.16 ± 0.02	Nd	1.21 ± 0.03	0.22 ± 0.01
Vanillic acid	0.27 ± 0.01	Nd	Nd	0.56 ± 0.03
Vanillin	0.21 ± 0.02	0.37 ± 0.03	2.04 ± 0.03	0.36 ± 0.01

2.2. In Vitro Preliminary Tests

Results from the *in vitro* preliminary tests evidenced an inhibitory effect on the growth of the pathogens examined only for the waste shrimp extracted with nitric acid, named "Nitric-extract". Additionally, none of the control solutions (each solvent used for the preparation of the respective extract) inhibited mycelial growth. In the agar diffusion test, "Nitric-extract" at concentrations of 100, 75 and 50% showed an inhibitory effect on all strains of fungal and oomycete pathogens, while at concentration of 25%, an inhibitory effect was still observed only on *Ph. nicotianae* T2.C-M1A, *F. sacchari* CBS 145949, *A. alternata* 646, *P. digitatum* P1PP0, *P. commune* CECT 20767, *C. gloeosporioides* C2, *F. proliferatum* CBS 145950, *Pl. tracheiphilus* Pt2 and *Ph. nicotianae* T3-B-K1A, in order of significance (Table 2 and Figure 2). The diameter of inhibition halos was directly proportional to the concentration of the extract (Table 2). Significant differences in the inhibitory effects of the extracts were noticed among fungal and oomycete species as well as between species of the same genus and even between strains of the same species (Table 2). At the maximum dose, which is 100% of the extract concentration, the highest inhibitory effect was on *Pl. tracheiphilus* Pt2; at 75% concentration, the highest inhibitory activity was on *Pl. tracheiphilus* Pt2 and *Ph. nicotianae* T2.C-M1A; at 50%, on *Ph. nicotianae* T2.C-M1A; and at the lowest dose (25% extract concentration), on *Ph. nicotianae* T2.C-M1A as well as on three typically post-harvest pathogens, i.e., *F. sacchari* CBS 145949, *A. alternata* 646 and *P. digitatum* P1PP0.

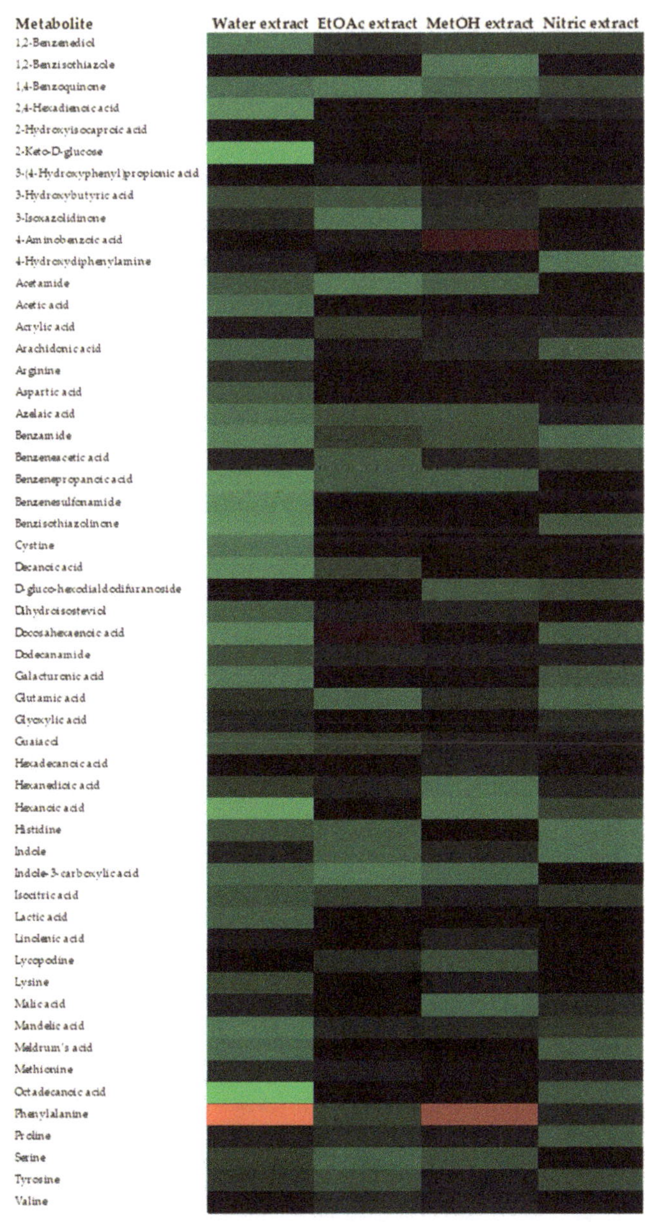

Figure 2. Heat map representing the relative abundances of metabolites detected in different shrimp extracts.

Table 2. Inhibitory effect of different concentrations (from 25 to 100%) of shrimp nitric-extract on the mycelium growth of 12 fungal and three oomycete plant pathogens, determined with the agar diffusion test by measuring the diameter of the inhibition halo around the wells. The incubation period was three days for fungi and 15 days for oomycetes.

	25% Nitric-Extract (Mean ± SD)	50% Nitric-Extract (Mean ± SD)	75% Nitric-Extract (Mean ± SD)	100% Nitric-Extract (Mean ± SD)
Penicillium digitatum P1PP0	14.00 ± 2.65 **c** [1]; *(ab)* [2]	20.00 ± 1.00 **b**; *(bc)*	25.00 ± 1.00 **a**; *(bcd)*	26.00 ± 1.00 **a**; *(de)*
P. commune CECT 20767	12.00 ± 1.73 **d**; *(bc)*	23.00 ± 0.00 **c**; *(ab)*	30.00 ± 1.00 **b**; *(bc)*	34.00 ± 1.73 **a**; *(b)*
P. expansum CECT 2278	0.00 ± 0.00 **d**; *(e)*	13.00 ± 1.73 **c**; *(ef)*	22.00 ± 0.00 **b**; *(cdef)*	27.00 ± 1.00 **a**; *(cde)*
P. italicum CECT 20909	0.00 ± 0.00 **d**; *(e)*	12.00 ± 1.73 **c**; *(f)*	20.00 ± 1.73 **b**; *(cdef)*	25.00 ± 0.00 **a**; *(def)*
Colletotrichum acutatum UW14	0.00 ± 0.00 **d**; *(e)*	15.00 ± 1.00 **c**; *(def)*	20.00 ± 0.00 **b**; *(cdef)*	22.00 ± 0.00 **a**; *(efg)*
C. karsti CAM	0.00 ± 0.00 **c**; *(e)*	11.00 ± 1.00 **b**; *(f)*	13.00 ± 1.00 **b**; *(f)*	19.00 ± 2.65 **a**; *(gh)*
C. gloeosporioides C2	12.00 ± 1.73 **c**; *(bc)*	15.00 ± 1.00 **c**; *(def)*	23.00 ± 1.73 **b**; *(cdef)*	32.00 ± 3.46 **a**; *(bc)*
Fusarium proliferatum CBS 145950	12.00 ± 1.00 **b**; *(bc)*	13.00 ± 2.65 **b**; *(ef)*	15.00 ± 1.00 **ab**; *(def)*	18.00 ± 1.73 **a**; *(gh)*
F. sacchari CBS 145949	15.00 ± 1.00 **c**; *(ab)*	17.00 ± 1.73 **c**; *(cde)*	21.00 ± 1.73 **b**; *(cdef)*	27.00 ± 0.00 **a**; *(cde)*
Alternaria arborescens 803	0.00 ± 0.00 **c**; *(e)*	12.00 ± 1.73 **b**; *(f)*	18.00 ± 1.73 **a**; *(def)*	20.00 ± 1.73 **a**; *(fgh)*
A. alternata 646	14.00 ± 1.00 **c**; *(ab)*	19.00 ± 1.00 **b**; *(bcd)*	24.00 ± 1.00 **a**; *(bcde)*	25.00 ± 1.73 **a**; *(def)*
Plenodomus tracheiphilus Pt2	10.00 ± 2.00 **d**; *(c)*	15.00 ± 1.73 **c**; *(def)*	34.00 ± 1.00 **b**; *(a)*	43.00 ± 1.00 **a**; *(a)*
Ph. nicotianae T2.C-M1A	16.00 ± 1.73 **d**; *(a)*	26.00 ± 1.00 **c**; *(a)*	34.00 ± 0.00 **b**; *(ab)*	30.00 ± 2.65 **a**; *(bcd)*
Ph. nicotianae T3-B-K1A	4.00 ± 1.73 **d**; *(d)*	12.00 ± 2.00 **c**; *(f)*	16.00 ± 1.00 **b**; *(def)*	20.00 ± 1.73 **a**; *(fgh)*
Ph. citrophthora Ax1Ar	0.00 ± 0.00 **c**; *(e)*	12.00 ± 0.00 **b**; *(f)*	14.00 ± 1.73 **ab**; *(ef)*	16.00 ± 1.00 **a**; *(h)*

[1] In a horizontal direction, for each pathogen, values with different bold letters are statistically different according to Tukey's honestly significant difference (HSD) test ($p \leq 0.05$). [2] In the vertical direction, for the concentrations 25% Nitric-extract, 50% Nitric-extract, 75% Nitric-extract, 100% Nitric-extract, values with different letters (in *italic* and within brackets) are statistically different according to Tukey's honestly significant difference (HSD) test ($p \leq 0.05$).

2.3. Determination of MIC and MFC

To further test the inhibitory activity of "Nitric-extract" on the growth of pathogens, the minimum inhibitory concentration (MIC) and the minimum fungicidal concentration (MFC) were determined, and results are summarized in Table 3. The values of both MIC and MFC for all pathogens were in the range 2–3.5%. In more detail, the highest values of MIC (3.5%) were recorded for *P. expansum* CECT 2278 and *F. sacchari* CBS 145949, while the lowest (2%) were recorded for *C. gloeosporioides* C2, *Ph. nicotianae* T3-B-K1A and *Pl. tracheiphilus* Pt 2. Values of MFC were the same as MIC for the majority of the strains. Only for strains *P. commune* CECT 20767, *A. alternata* 646, *Ph. nicotianae* T3-B-K1A and *Ph. citrophthora* Ax1Ar, MFC was higher than MIC, indicating that for these four strains, MIC exerted only a fungistatic effect.

Table 3. Minimum inhibitory concentration (MIC) and minimum fungicidal concentration (MFC) determined by the nitric-extract.

	Nitric-Extract (%)	
Pathogen (Species, Strains)	**MIC**	**MFC**
Penicillium digitatum P1PP0	3.0	3.0
P. commune CECT 20767	2.5	3.0
P. expansum CECT 2278	3.5	3.5
P. italicum CECT 20909	3.0	3.0
Colletotrichum acutatum UW14	2.5	2.5
C. karsti CAM	3.0	3.0
C. gloeosporioides C2	2.0	2.0
Fusarium proliferatum CBS 145950	3.0	3.0
F. sacchari CBS 145949	3.5	3.5
Alternaria arborescens 803	2.5	2.5
A. alternata 646	2.5	3.0
Plenodomus tracheiphilus Pt2	2.0	2.0
Phytophthora nicotianae T2.C-M1A	2.5	2.5
P. nicotianae T3-B-K1A	2.0	2.5
P. citrophthora Ax1Ar	2.5	3.0

2.4. In Vivo Antifungal Activity

The antifungal activity of "Nitric-extract" was finally tested *in vivo* on citrus (oranges and lemons) and apple fruits artificially infected by *P. digitatum* and *P. expansum*, respectively. Results are summarized below.

2.4.1. Antifungal Activity on Oranges

Three days post inoculation with *P. digitatum* P1PP0 of oranges, all concentrations of "Nitric-extract" significantly reduced rot severity compared to the water control (treatment ID01) (Figure 3). However, except for "Nitric-extract" applied as such (ID02), each of the other concentrations was not statistically different from the respective control.

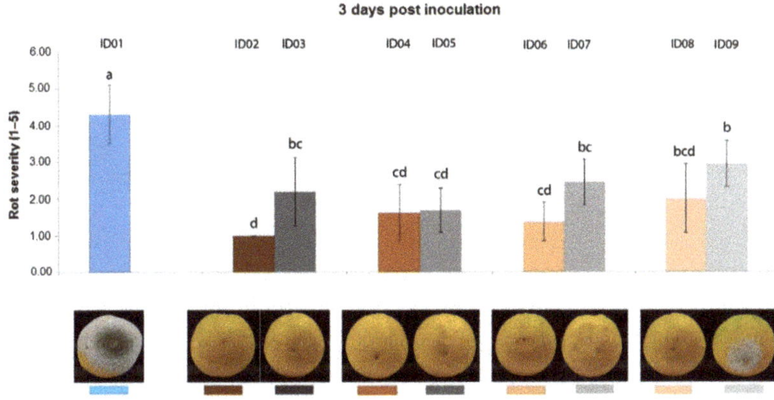

Figure 3. Rot severity caused by *Penicillium digitatum* strain P1PP0 in orange (*Citrus* × *sinensis*) fruits cv. Valencia treated with water (ID01) or nitric-extract as such (ID02), 75% Nitric-extract (ID04), 50% Nitric-extract (ID06), 25% Nitric-extract (ID08) and respective controls (NaNO$_3$ 0.17 g/mL—ID03; NaNO$_3$ 0.17 g/mL diluted in sterile distilled water (sdw) at 75%—ID05; NaNO$_3$ 0.17 g/mL diluted in sdw at 50%—ID07; NaNO$_3$ 0.17 g/mL diluted in sdw at 25%—ID09) 3 days after inoculation. Values sharing the same letters are not statistically different according to Tukey's HSD (honestly significant difference) test ($p \leq 0.05$). Bars represent SD.

Five days after inoculation (Figure 4), all concentrations of "Nitric-extract" still demonstrated values of rot severity significantly lower than the water control (treatment ID01); however, in this case, only treatment with "Nitric-extract" at 25% (ID08) significantly differed from the respective control (ID09), although this difference was not statistically significant in comparison to the other control treatments (ID03, ID05, ID07).

2.4.2. Antifungal Activity on Lemons

Three days after inoculation of *P. digitatum* P1PP0 in lemons (Figure 5), all tested concentrations of "Nitric-extract" significantly reduced rot severity compared to the control. Additionally, among these, "Nitric-extract" as such (ID02) and "Nitric-extract" at 75% (ID02) significantly reduced rot severity compared to all other control solution (treatments ID03, ID05, ID07 and ID09). "Nitric-extract" as such (ID02) and "Nitric-extract" 75% (ID04) were also the only treatments that, five days after inoculation, still maintained significant effectiveness in the reduction of rot severity in lemons (Figure 6).

2.4.3. Antifungal Activity on Apples

Results from the trial carried out on apple fruits inoculated with *P. expansum* CECT 2278 evidenced that, three days post inoculation (Figure 7), "Nitric-extract" as such (ID02), at 75% (ID04) and at 50% significantly reduced rot severity in comparison with any other treatment and controls. Five days post inoculation, only "Nitric-extract" as such (ID02) still significantly reduced rot severity (Figure 8).

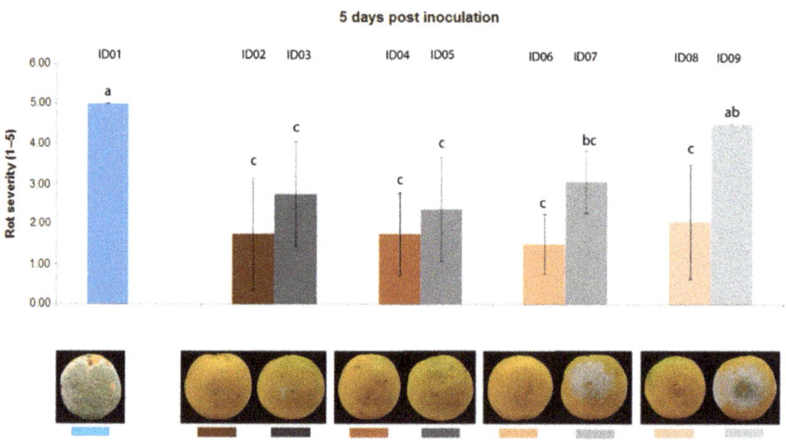

Figure 4. Rot severity caused by *Penicillium digitatum* strain P1PP0 in orange (*Citrus × sinensis*) fruits cv. Valencia treated with water (ID01) or nitric-extract as such (ID02), 75% Nitric-extract (ID04), 50% Nitric-extract (ID06), 25% Nitric-extract (ID08) and respective controls (NaNO$_3$ 0.17 g/mL—ID03; NaNO$_3$ 0.17 g/mL diluted in sterile distilled water (sdw) at 75%—ID05; NaNO$_3$ 0.17 g/mL diluted in sdw at 50%—ID07; NaNO$_3$ 0.17 g/mL diluted in sdw at 25%—ID09) 5 days after inoculation. Values sharing the same letters are not statistically different according to Tukey's HSD test ($p \leq 0.05$). Bars represent SD.

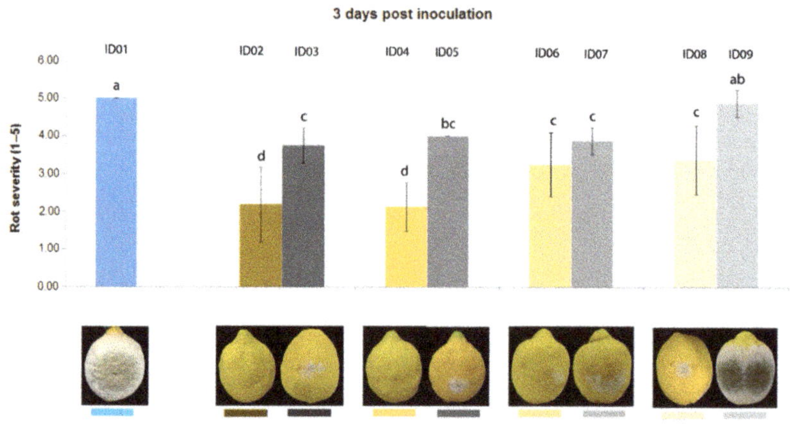

Figure 5. Rot severity caused by *Penicillium digitatum* strain P1PP0 in lemon (*Citrus × limon*) fruits cv. Femminello Siracusano treated with water (ID01) or Nitric-extract as such (ID02), 75% Nitric-extract (ID04), 50% Nitric-extract (ID06), 25% Nitric-extract (ID08) and respective controls (NaNO$_3$ 0.17 g/mL—ID03; NaNO$_3$ 0.17 g/mL diluted in sterile distilled water (sdw) at 75%—ID05; NaNO$_3$ 0.17 g/mL diluted in sdw at 50%—ID07; NaNO$_3$ 0.17 g/mL diluted in sdw at 25%—ID09) 3 days after inoculation. Values sharing the same letters are not statistically different according to Tukey's HSD test ($p \leq 0.05$). Bars represent SD.

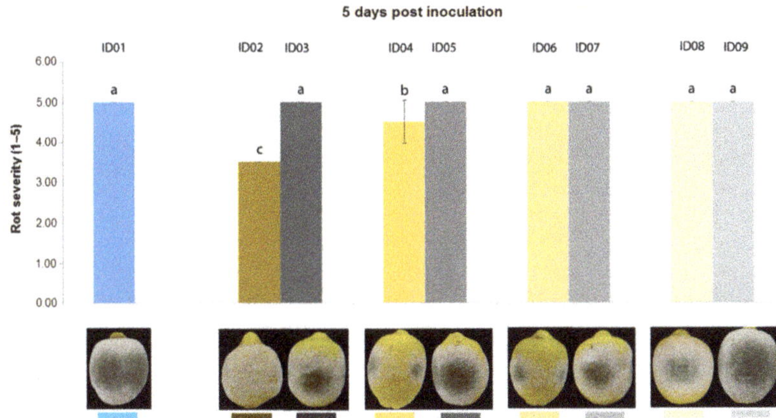

Figure 6. Rot severity caused by *Penicillium digitatum* strain P1PP0 in lemon (*Citrus* × *limon*) fruits cv. Femminello Siracusano fruits treated with water (ID01) or Nitric-extract as such (ID02), 75% Nitric-extract (ID04), 50% Nitric-extract (ID06), 25% Nitric-extract (ID08) and respective controls (NaNO$_3$ 0.17 g/mL—ID03; NaNO$_3$ 0.17 g/mL diluted in sterile distilled water (sdw) at 75%—ID05; NaNO$_3$ 0.17 g/mL diluted in sdw at 50%—ID07; NaNO$_3$ 0.17 g/mL diluted in sdw at 25%—ID09) 5 days after inoculation. Values sharing the same letters are not statistically different according to Tukey's HSD test ($p \leq 0.05$). Bars represent SD.

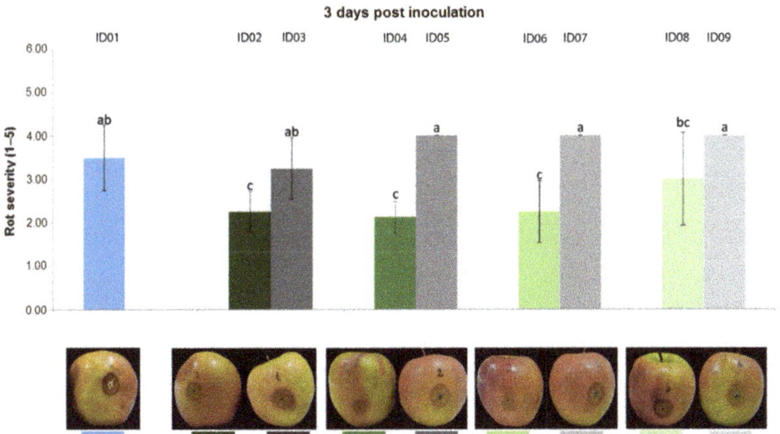

Figure 7. Rot severity caused by *Penicillium expansum* strain CECT 2278 in apple (*Malus domestica*) fruits cv. Braeburn treated with water (ID01) or Nitric-extract as such (ID02), 75% Nitric-extract (ID04), 50% Nitric-extract (ID06), 25% Nitric-extract (ID08) and respective controls (NaNO$_3$ 0.17 g/mL—ID03; NaNO$_3$ 0.17 g/mL diluted in sterile distilled water (sdw) at 75%—ID05; NaNO$_3$ 0.17 g/mL diluted in sdw at 50%—ID07; NaNO$_3$ 0.17 g/mL diluted in sdw at 25%—ID09) 3 days after inoculation. Values sharing the same letters are not statistically different according to Tukey's HSD test ($p \leq 0.05$). Bars represent SD.

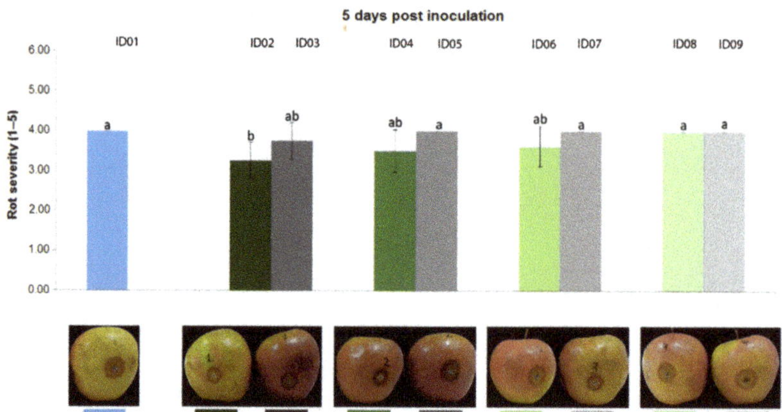

Figure 8. Rot severity caused by *Penicillium expansum* strain CECT 2278 in apple (*Malus domestica*) fruits cv. Braeburn treated with water (ID01) or Nitric-extract as such (ID02), 75% Nitric-extract (ID04), 50% Nitric-extract (ID06), 25% Nitric-extract (ID08) and respective controls (NaNO$_3$ 0.17 g/mL—ID03; NaNO$_3$ 0.17 g/mL diluted in sterile distilled water (sdw) at 75%—ID05; NaNO$_3$ 0.17 g/mL diluted in sdw at 50%—ID07; NaNO$_3$ 0.17 g/mL diluted in sdw at 25%—ID09) 5 days after inoculation. Values sharing the same letters are not statistically different according to Tukey's HSD test ($p \leq 0.05$). Bars represent SD.

3. Discussion

This study evaluated, for the first time, the potentialities of minimally processed shrimp wastes in the *in vitro* inhibitory activity on fungal and oomycete plant pathogens, and their effectiveness in controlling post-harvest rots caused by *Penicillium* spp. in citrus and apple fruits. To this aim, wastes from the shrimp species *Parapenaeus longirostris* were dried and grounded to result in a "dry-powder", which was further processed leading to four different extracts "Water-extract", "EtOAc-extract", "MetOH-extract" and "Nitric-extract". Acid hydrolysis is mandatory for the mineralization of calcium-containing shrimp waste, and hydrolysis is commonly performed by hydrochloric, acetic, phosphoric, sulfuric, nitric and lactic acids. Nitric acid was selected as, among the above-mentioned acids, it has the slowest reaction kinetics [28], which allows for better digestion control. All these substances, "Water extract", "EtOAc-extract", "MetOH-extract" and "Nitric-extract, were analyzed to determine their composition in metabolites and phenolic compounds. Then, the "dry-powder", "EtOAc-extract", "MetOH-extract" and "Nitric-extract" were also preliminarily tested *in vitro*, in order to select the substance with the highest mycelial growth inhibitory activity. "Nitric-extract" was the most effective substance and was further investigated to determine its antifungal properties (in terms of MIC and MFC) and *in vivo* antifungal activity.

Results from the chemical analysis showed that all substances extracted from the shrimp waste were miscellaneous mixtures of a conspicuous number of metabolites and phenolic compounds. Interestingly, a high relative abundance of the 2-Hydroxyisocaproic, 3-(4-Hydroxyphenyl) propionic and 4-Aminobenzoic acids in "MetOH-extract", and of docosahexaenoic acid in "EtOAc-extract" were reported. Various studies reported fungicidal activity for these molecules when tested as pure substances; 2-hydroxyisocaproic acid was effective against *Candida* and *Aspergillus* species [29]; 3-(4-Hydroxyphenyl) propionic acid contains the hydroxyl group, which has been reported as one of the substance responsible for the antifungal activity of *Lactobacillus paracasei* [30]. Moreover, the para-aminobenzoic acid showed antibiotic activity toward *Staphylococcus aureus* [31]; a *Pseudomonas aeruginosa*-bioconverted oil extract of docosahexaenoic acid was effective against the mycelial growth of several plant pathogens, including *Botrytis cinerea*, *Colletotrichum capsici*, *Fusarium oxys-*

porum, F. solani, Phytophthora capsici, Rhizoctonia solani and *Sclerotinia sclerotiorum* [32]. However, in the present study, two extracts, "MetOH-extract" and "EtOAc-extract", containing a higher amount of the above-mentioned acids, showed no inhibitory activity on mycelial growth.

An additional interesting metabolite present in all substances was phenylalanine, which was also detected in high amount in "Water-extract" and "MetOH-extract". A recent study [33] reported that post-harvest treatments of mango, avocado and citrus fruits with phenylalanine induced resistance against infections caused by *Colletotrichum gloeosporioides, Lasiodiplodia theobromae* and *P. digitatum*, respectively, although *in vitro* tests carried out in the same study evidenced no inhibitory effects toward the same pathogens. Therefore, although lacking of fungicidal action, the "Water-extract" and "MetOH-extract", which showed a high amount of phenylalanine, could provide strong resistance induction properties to control post-harvest disease. It goes without saying that, since phenylalanine was also detected in "EtOAc-extract" and "Nitric-extract", these samples could also have resistance induction properties, as demonstrated for other extracts of natural origin [34]. This possibility assumes a particular significance of the extract "Nitric-extract", which was the only substance tested that demonstrated clear and strong *in vitro* antifungal activity as well as significant *in vivo* control of infective processes. Additional studies are, therefore, ongoing, to verify possible resistance induction properties of all the minimally processed shrimp wastes produced in this study. Quite interestingly, although the exoskeleton of shellfish is the main raw material for the extraction of chitosan, whose inhibitory activity on post-harvest fruit rots is well documented [35], this biopolymer was not present in the extracts examined in this study. As a consequence, it can be inferred that other substances are responsible for the antimycotic activity showed by the "Nitric-extract".

With reference to composition in phenolic compounds, analyses evidenced the presence, in all tested substances, of molecules whose antimicrobial activity is supported by a wide range of literature [2,36–44]. Some of these compounds have been also applied as eco-friendly alternatives to synthetic fungicides [1,45]. Among the phenolic compounds, the molecules that recurred in all analyzed substances were the benzoic, caffeic and p-coumaric acids and the vanillin. Benzoic and caffeic acids have important preservative properties that determine the inhibition of fungal growth [43,46]. Vanillin (4-hydroxy-3-methoxybenzaldehyde) is considered one of the most important additives used in the food industry; it is characterized by effective inhibitory activity toward a wide range of microorganisms, thus causing a delay in the growth of yeasts and fungi [36,40]. The p-coumaric acid (4-hydroxycinnamic acid), which, in "Nitric-extract", had the highest concentration, is the main phenolic acid contained in the peel of sweet oranges [44], and is well known for its efficacy in negatively affecting the growth of post-harvest pathogens, such as *Monilinia fructicola, Botrytis cinerea* and *Alternaria alternata* [2]. Interestingly, "Nitric-extract" also reported the highest concentration of catechol (1-2-dihydroxybenzene) and the exclusive presence of dihydroferulic (3-(4-hydroxy-3-methoxyphenyl) propionic acid) and ellagic acids. Catechol shows significant activity in the control of *Fusarium oxysporum* and *Penicillium italicum* [38]. Dihydroferulic acid significantly inhibits the *in vitro* growth of *Saccharomyces cerevisiae, Aspergillus fumigatus* and *A. flavus* [39]. Moreover, ellagic acid, which possesses well-documented antibacterial activity [37], shows extraordinary antifungal effects toward *Botrytis cinerea* [41], as well as a significant growth inhibition of several fungal species belonging to the genera *Trichophyton* and *Candida* [42]. Finally, phenolic compounds are hypothesized to be, at least in part, responsible for the strong broad-spectrum antifungal activity shown by a pomegranate peel extract [34].

Overall, unlike the "Water-extract", "EtOAc-extract" and "MetOH-extract", "Nitric-extract" results were characterized by p-coumaric acid and catechol, both present at high concentrations, and by the exclusive presence of the acids dihydroferulic and ellagic; these molecules could be, therefore, responsible for the antifungal activity of this extract. Synergetic action of some of the molecules detected in "Nitric-extract" also cannot be excluded. This effect has already been observed for the active components of extracts from different natural matrices. This is the case, for example, of pomegranate, whose high biological value is recognized as being the result of the synergistic chemical action of the total phytoconstituents of the fruit rather than of single extracted components [47–49].

The quantity and quality of the molecules that were active (individually or in synergy) in determining the *in vitro* antifungal activity of the tested substances could also be related to the extraction process. By comparing the compositions of the three extracts, namely, "EtOAc-extract", "MetOH-extract" and "Nitric-extract", the three applied extraction processes had different efficiencies. The choice of the best solvent for the extraction of precise bioactive components from a specific matrix is a crucial aspect for reaching the expected qualitative and quantitative yield of the desired molecules in the final extract [34]. Examples of this aspect are provided by studies carried out on pomegranate extracts; Al-Zoreky [50] observed that the 80% methanolic extract was richer in polyphenols compared to hot water and diethyl ether extracts and, therefore, led to higher antimicrobial activity against pathogenetic bacteria. Tayel et al. [51] found that, regardless the concentration of specific bioactive components, a methanolic pomegranate peel extract was more effective than ethanol and water extracts in controlling *Penicillium digitatum*. In view of these aspects, it is quite surprising that, among the extracts, only "Nitric-extract" provided *in vitro* antifungal efficacy and, at the same time, neither "EtOAc-extract" nor "MetOH-extract" resulted in an inhibitory effect on mycelial growth.

Results from the *in vitro* preliminary test together with those from MIC and MFC tests overall demonstrated that the pathogens mostly affected by "Nitric-extract" were *Pl. tracheiphilus* Pt 2, *C. gloeosporioides* C2 and *Ph. Nicotianae*—both tested isolates. *Plenodomus tracheiphilus* is the causal agent of 'mal secco, one of the most destructive diseases affecting lemon trees [52]. Because of the vascular propagation of the pathogen in all aerial parts of the infected plant, the management of the disease is complicated [53]. It is commonly carried out by the pruning of diseased twigs, withered shoots and suckers, followed by the spraying of the canopy with copper-based fungicides, which can reduce the occurrence of new *Pl. tracheiphilus*-infections. However, many copper-based treatments are not cost effective in commercial lemon groves, and also represent a significant source of environmental pollution [53]. Another copper-susceptible pathogen is *C. gloeosporioides*, the causative agent of anthracnoses in several fruits and vegetables [54] as well as of twig and shoot dieback in citruses [55]. *Phytophthora nicotianae* is very likely the most widespread and destructive *Phytophthora* species worldwide, affecting a very wide host range of more than 255 plant species [8,56,57]. Control strategies may be different depending on the specific situation, although the pathogen is markedly sensitive to Metalaxyl and Fosetyl Al, fungicides which are commonly used for controlling plant diseases affecting roots, collars and stems [56]. Results from this study pose "Nitric-extract" as a promising alternative to the use of conventional fungicides in controlling not only *Pl. tracheiphilus*, *C. gloeosporioides* and *Ph. nicotianae*, but all pathogens tested in the present study. To this aim, further investigations are needed to evaluate the phytotoxicity, if any, of the extract, its attitude to systemic translocation, which is of particular relevance in the case of tracheomycoses, such as 'mal secco' caused by *Pl. tracheiphilus*, as well as the most effective method of application, e.g., by drenching, spraying or incorporation into fruit coatings, which also depends on the type of disease.

As a preliminary step towards the application of "Nitric-extract" to control plant diseases, its effectiveness was tested *in vivo* against molds caused by *Penicillium* species in orange, lemon and apple fruits, which are the most economically important post-harvest diseases affecting these fruits [58,59]. Post-harvest molds of citrus and apple fruits are traditionally controlled by the application of highly effective chemicals, such as imidazole and bendimidazole (thiabendazole) fungicides [60,61]. More recently, as a consequence of the selection of imidazole- and bendimidazole-resistant strains of *Penicillium*, several other synthetic fungicides, including azoxystrobin, fludioxonil, cyprodinil and pyrimethanil, have been proposed as alternatives for the chemical control of these post-harvest fruit diseases [7,60–63]. Like imidazoles and benzimidazoles, all these fungicides are effective at relatively low doses but are characterized by a high acute toxicity [64–69].

There is boundless literature evaluating the efficacy of alternative strategies to the use of conventional synthetic fungicides for the control of postharvest molds of *Penicillium* species [70–78]. A novelty in the present study is the *in vivo* control of *Penicillium* spp. using a natural substance that is derived from minimum waste treatment.

Overall, treatments with "Nitric-extract" at the highest concentrations were the most effective in positively affecting the reduction of rot severity in all tested fruits. Additionally, an interesting weak positive effect was also observed in all control treatments, including $NaNO_3$ in water solution (ID03, ID05, ID07 and ID09), although, *in vitro*, they were not effective in inhibiting the mycelial growth of all pathogens included in this study. As already observed for other inorganic salts [74], it cannot be excluded that the *in vivo* effectiveness of $NaNO_3$ was not the consequence of direct antifungal activity, but the possible result of the triggering of defense mechanisms in fruits. Further tests are ongoing to verify this hypothesis.

The results from the treatments with "Nitric-extract" demonstrated that three days post-treatment, "Nitric-extract" as such determined a significant reduction of rot severity over any other treatment in all fruits (oranges, lemons and apples). Additionally, "Nitric-extract" at 75% significantly reduced rot severity in lemon and apples over controls; "Nitric-extract" at 50% had a significant effect over controls only on apples; the concentration of 25% was as effective as the controls in all fruits. Five days post-treatment, "Nitric-extract" as such still maintained significant effects in reduction of rots only in lemons and apples; "Nitric-extract" at 75% demonstrated significant reduction of rot severity only in lemons; finally, "Nitric-extract" at 50% and at 25% were as effective as the controls in all tested fruits. Overall, the results showed an interesting performance of "Nitric-extract" in controlling postharvest mold caused by *Penicillium* spp., although the effective dose was much higher than that of traditional synthetic fungicides [7], and, as with other eco-friendly alternatives to synthetic fungicides [7,74], its use may not provide complete protection. A successful strategy for improving its efficacy or reducing fungicide residues from post-harvest fruit treatments could include the use of "Nitric-extract" in a mixture with conventional fungicides applied at a concentration lower than the standard dose, or by incorporating it in a fruit coating.

This study is part of a research program aimed at exploring the antifungal activity of extracts obtained from minimally processed shrimp wastes and their possible application in agriculture. The antifungal activity shown *in vitro* against a wide range of fungal and oomycete pathogens by the nitric extract appears promising and could be exploited in the context of new strategies for the management of plant diseases caused by these pathogens. *In vivo* preliminary results suggest a possible use of nitric extract for post-harvest treatments against citrus and apple molds caused by *Penicillium* species. To this aim, and to optimize the efficacy of treatments, next steps will be to define the methods and times of application. In this study, nitric extracts were applied to fruits 24 h after inoculation with the pathogen, indicating curative efficacy. However, an additional aspect that would merit further investigation is whether nitric extract, like other natural substances, is able to elicit plant defense mechanisms against infections by pathogens. In this case, the treatment of fruits with this extract might also have preventive efficacy against infections by molds. Regarding this, it cannot be ruled out that the other shrimp waste extracts, which, in preliminary *in vitro* tests did not show inhibitory activity on the mycelium growth, may also be effective *in vivo* acting as resistance elicitors. Last but not least, a prerequisite for the use of nitric extracts of shrimp waste to prevent post-harvest molds is to evaluate if the treatment leaves unpleasant odors on the fruits. A sensory analysis using an electron nose is planned to clarify this aspect. Although the effective dose of nitric shrimp waste extract is far higher than the label dose of synthetic fungicides used to control post-harvest fruit diseases, this extract, as a natural substance, could be an interesting alternative to traditional post-harvest chemical treatments, as it is more eco-friendly and far less toxic than synthetic fungicides.

4. Materials and Methods

4.1. Preparation of Shrimp Waste Substances

Around 5 kg of shrimp waste (cephalothorax, head and carapace) of the species *Parapenaeus longirostris*, common name deep-water rose shrimp, was collected in a local fish market in Catania (Italy), in February 2021. Shrimp waste was kept on ice until processed in the laboratory and firstly, washed with distilled water; then, dried in an oven at 30 °C for a week. The dried sample was powdered and homogenized. Then, 10 g of shrimp waste powder was packed in plastic food bags labelled "dry-powder" and stored at −20 °C until further use; 10 g of shrimp waste powder was processed to extraction (20 min long sonication by means of ArgoLab DU-100) with (i) 50 mL of ethyl acetate (EtOAc from Aldrich) or (ii) 50 mL of methanol (MetOH from Aldrich). Then, the supernatant was carefully collected and transferred into a clean beaker. The powder was then re-subjected to the described procedure for a total of three times. The 150 mL of supernatant were firstly filtrated and then evaporated under vacuum at the temperature of 40 °C until a crude extract was obtained. The crude extracts, representing the "EtOAc-extract" and "MetOH-extract", were stored at −20 °C until further use. (iii) To obtain "Nitric-extract", 20 mL of nitric acid (HNO3, 65% from Aldrich) was added to 5 g of shrimp waste powder; the mixture was then stirred for 1 h at 150 rpm and then the acid was neutralized by adding 80 mL of NaOH (0.10 g/mL). pH was verified to be around pH 5. To obtain the "water-extract", 25 mL of water with 1% of acetic acid were added to 5 g of shrimp waste powder; the mixture was homogenized by vortexing and ultrasonication. The liquid extracts were then filtrated and stored at −20 °C until further use.

4.2. Analysis of Metabolites Present in Shrimp Waste Samples by HPLC-ESI-MS/TOF

The differential analysis of the metabolites contained in the four substances tested was carried out by HPLC-ESI-MS-TOF. Before the analysis, each sample was subjected to specific pretreatments. In particular, "EtOAc-extract" and "MetOH-extract" were dissolved in a methanol solution at 1% of acetic acid. Finally, "Water-extract" and "Nitric-extract" were mixed to acidified water. Each sample was finally filtered with 0.22 μm filter and then analyzed using an UPLC (1290 Infinity LC, Agilent Technologies, Santa Clara, CA, USA) coupled with a quadrupole time of flight mass spectrometer (Agilent 6546 LC/Q-TOF) operating in positive and negative ionization mode. Chromatographic separation was performed with an Agilent Zorbax RRHD SB-C18, 2.1 mm × 50 mm, 1.8 μm column. Mobile phase A was composed of Milli-Q water and acetonitrile was used for mobile phase B (both phases were acidified with 0.1% formic acid), with gradient elution, as follows: 0 min, 2% B; 22 min 95% B; 25 min, 5% B. The column was equilibrated for 3 min before every analysis. The flow rate was 0.4 mL/min, and 5 μL of sample was injected. Dual AJS ESI source conditions were as follows: gas temperature: 325 °C; gas flow: 10 L/min; nebulizer pressure: 40 psig; sheath gas temperature: 295 °C; sheath gas flow: 12 L/min; capillary voltage: 4000 V; nozzle voltage: 500 V; Fragmentor: 120 V; skimmer: 70 V; product ion scan range: 100–1500 Da; MS scan rate: 5 spectra/s; MS/MS scan rate: 3 spectra/s; maximum precursors per cycle: 2; and collision energy: 10, 20, 40 eV. The analysis of the metabolites was carried out in triplicate. Untargeted LC/Q-TOF based metabolomics approach was used to identify the metabolic profiling of shrimp waste extracts. Integration, data elaboration and identification of metabolites were managed using MassHunter Qualitative Analysis software B.08.00 and library PCDL Manager B.08.00.

4.3. Fungal and Oomycete Strains, Culture Conditions and Propagules Production

Fungal and oomycete strains were included in this study. Most of them had been previously characterized [7,8,55,79,80]. The complete list of strains tested in this study is as follows: four *Penicillium* spp. (*P. digitatum* P1PP0, *P. commune* CECT 20767, *P. expansum*

CECT 2278 and *P. italicum* CECT 20909); three *Phytophthora* spp. (*Ph. nicotianae* strains T3-B-K1A and T2.C-M1A, *Ph. citrophthora* strain Ax1Ar); *Plenodomus tracheiphilus* strain Pt2; two *Alternaria* species (*A. alternata* strain 646, and *A. arborescens* strain 803); three *Colletotrichum* species (*C. acutatum* strains UW14, *C. karsti* strain CAM and *C. gloeosporioides* strain C2); two *Fusarium* species (*F. proliferatum* strain CBS 145950 and *F. sacchari* strain 145949). All strains were from the collection of the laboratory of Molecular Plant Pathology of the Di3A (University of Catania, Catania, Italy).

4.4. In Vitro Preliminary Screening for Selecting the Most Effective Extract

The antifungal activity of the "dry-powder", "EtOAc-extract", "MetOH-extract" and "Nitric-extract" were preliminarily checked in order to select, among them, the most promising one to be used in further tests.

For testing the effectiveness of the "dry-powder" in affecting mycelial growth, 16 g of shrimp waste powder were homogenized with 1 L of autoclaved PDA and poured in 90 mm Petri dishes. For each pathogen, a mycelial plug (diameter 3 mm) from a 7-day-old culture grown on PDA at 25 °C was transferred in the center of a "dry-powder"—amended PDA plate; control cultures of each pathogen, obtained by subcultures in "dry-powder"—non amended PDA plates, were included in the test. The plates were incubated at room temperature (20 ± 2 °C) for three days (for fungal pathogens) or for 15 days (for oomycete pathogens). At the end of the incubation period, no negative effects were observed in mycelial growth compared with controls for any of the pathogens. The "dry-powder" was not further tested.

The effect of "EtOAc-extract", "MetOH-extract" and "Nitric-extract" on the mycelial growth of the pathogens was tested at different concentrations. To this purpose, "EtOAc-extract" and "MetOH-extract" were separately diluted in 1% dimethyl sulfoxide to obtain, for each substance, four solutions at the following concentrations 10, 25, 50 and 100 mg/mL; "Nitric-extract" was diluted in water to obtain the following concentrations: 25, 50, 75 and 100%.

"EtOAc-extract", "MetOH-extract", "Nitric-extract" and each fungal pathogen were tested separately in a 90 mm PDA plate as it follows: 500 µL of a suspension of conidia of the fungal pathogen (concentration 10^4 conidia/mL) were homogeneously spread on the surface of a PDA plate; by using a cork borer, five wells (diameter 3 mm, each) were then realized on the PDA plate; then, 60 µL of each concentration of the substance were pipetted into the respective well; the plates were finally incubated at 25 °C for three days. For the oomycete pathogens (*Phytophthora* spp.), the influence of "EtOAc-extract", "MetOH-extract" and "Nitric-extract" was tested separately as follows: for each *Phytophthora* strain, a mycelial plug (diameter 3 mm) from a 7-day-old culture grown on PDA at 25 °C was transferred in the center of a PDA plate and surrounded by 5 wells at a distance of 3 cm from the plug; then, 60 µL of each concentration of the substance tested were pipetted into the respective well. The plates were then incubated at 25 °C for 15 days.

In all the experiments, the possible mycelial growth inhibitory activity induced by each solvent used for the preparation of the respective extract was verified by *in vitro* tests performed as described above. For all pathogens and substances at each concentration, all the tests were performed in triplicate.

4.5. Minimum Inhibitory Concentration (MIC) and Minimum Fungicidal Concentration (MFC) of Nitric-Extract

The minimum inhibitory concentration (MIC) and the minimum fungicidal concentration (MFC) are dilution end points of a substance which completely inhibits the growth or kills the fungi tested; both are widely used in routine tests of substances with antimicrobial activity [9,81]. The minimum inhibitory concentration (MIC), defined as the lowest concentration of the test substance that inhibits visible growth, was determined with a microdilution method. For each pathogen, in a 2.0 mL tube, 400 µL of "Nitric-extract" at specific concentrations were added to 400 µL of sterile PDB and to 200 µL of spores suspension (concentration 10^4 spores/mL) to obtain 10 serial dilutions (1 mL each) of the substance tested (final concentrations 0.5, 1.0, 1.5, 2.0, 2.5, 3.0, 3.5, 4.0, 4.5, 5.0%). Then, the tubes were incubated at 25 °C for 3 days.

After the incubation period, the MIC was the lowest concentration where no cloudiness was visible in the tubes, which means that no pathogen growth was observed. The determination of MFC was an additional step of the MIC test. The MFC is defined as the lowest concentration of a substance required to kill a fungal pathogen corresponding to no visible subculture growth on an unamended culture medium in environmental conditions favorable to the growth. In the present study, the evaluation of the MFC was carried out by transferring 10 μL from each of the wells where solution cloudiness was not observed into PDA medium. The inoculated plates were incubated at 25 °C for 3 days. The MFC for each pathogen was represented by the plated concentration that did not lead to any mycelial growth after the incubation period.

4.6. Evaluating the In Vivo Antifungal Activity of Nitric-Extract in Preventing Fruit Rots

The antifungal activity of the "Nitric-extract" was evaluated *in vivo* against infections caused by *P. digitatum* and *P. expansum* on citrus (oranges and lemons) and apple fruits, respectively.

4.6.1. Nitric-Extract Dilutions

For the test, "Nitric-extract" was tested in all fruits (orange, lemon and apple) as such (ID02) or as three serial dilutions in sterilized distilled water (sdw) (concentrations; 75%—ID04; 50%—ID06; 25%—ID08). In addition, ID03, ID05, ID07 and ID09 were the respective controls.

In this experiment, four control groups were considered: (i) water (ID01); (ii) a solution of nitric acid (HNO3, 65%) and sodium hydroxide (NaOH, 0.1 g/mL) at the ratio 1:4—they are the solvent and base used for the preparation of "Nitric-extract", respectively, which leads to a water solution of $NaNO_3$ at the concentration 0.002 mol/mL (0.17 g/mL); (iii) $NaNO_3$ 0.17 g/mL diluted in sdw at 75%—ID05; (iv) $NaNO_3$ 0.17 g/mL diluted in sdw at 50%—ID07; $NaNO_3$ 0.17 g/mL diluted in sdw at 25%—ID09 (Table 4).

Table 4. List of treatments of the *in vivo* tests with shrimp powder extract (Nitric-extract). Control IDs were obtained by adding sterile distilled water (sdw) to $NaNO_3$ (0.17 g/mL).

ID of Treatment	Tested Substance
ID01	WATER
ID02	100% Nitric-extract ($NaNO_3$ 0.17 g/mL)
ID03	$NaNO_3$ 0.17 g/mL
ID04	75% Nitric-extract
ID05	$NaNO_3$ 0.17 g/mL diluted in sdw at 75%
ID06	50% Nitric-extract diluted in sdw
ID07	$NaNO_3$ 0.17 g/mL diluted in sdw at 50%
ID08	25% Nitric-extract
ID09	$NaNO_3$ 0.17 g/mL diluted in sdw at 25%

4.6.2. Fruits

All fruits used in this test came from organic crops. Citrus fruits were mature oranges (*Citrus × sinensis*) cv. Valencia and lemons (*Citrus × limon*) cv. Femminello Siracusano, while apples (*Malus domestica*) were of the cv. Braeburn. Before the tests, all fruits were preliminarily surface-disinfected by dipping in 1% NaClO (NaClO 0.5% for apples) for 2 min, rinsing under tap water and air-drying at room temperature (20 ± 2 °C).

4.6.3. Fungal Pathogens and Inoculum Preparation

The strains used in the trial were *P. digitatum* strain P1PP0 and *P. expansum* strain CECT 2278. For each strain, the inoculum was represented by a conidial suspension at the concentration 10^6 conidia/mL.

4.6.4. Inoculation

Surface-disinfected fruits (oranges, lemons and apples) were wounded with a 2 mm-diameter plastic tip at four points along the equatorial surface; then, 10 µL of conidial suspension (*P. digitatum* strain P1PP0 for citrus and *P. expansum* strain CECT2278 for apples) was pipetted into each wound. Inoculated fruits were incubated in a plastic container at 20 °C and 80% RH (relative humidity) for 24 h. For all fruit (oranges, lemons and apples), the treatment with "Nitric-extract" as such or as a dilution was carried out as follows: after the incubation period, at each inoculation point, 20 µL of the substance was placed into the wound; overall, 3 fruits per treatment were used. An additional control group, represented by 3 fruits wounded as above, received 20 µL of sterile distilled water (sdw) per wound. The experiment was repeated another two times, with similar results. Analysis of variance did not reveal any differences among the experiments (F not significant); therefore, only the results of a single experiment are reported here.

4.6.5. Evaluation of the Efficacy of the Nitric-Extract in Preventing Fruit Rot

The antifungal activity of "Nitric-extract", as such, or as a dilution, was recorded at 3 and 5 days after inoculation and expressed as rot severity, rated according to empirical scales, from 1 to 5. This scale was different according to the fruit. For citrus fruits, the scale 25% was as follows: 1. absence of symptoms or signs of the pathogen; 2. slight presence of rot; 3. clear presence of rot and slight appearance of mycelium; 4. rot and clear presence of white mycelium; 5. clear presence of soft rot, white mycelium and sporulation. For apple fruits the scale was as follows: 1. absence of symptoms or signs of the pathogen; 2. slight presence of rot; 3. clear presence of rot and slight appearance of mycelium; 4. presence of rot, white mycelium and slight appearance of sporulation; 5. clear presence of soft rot, white mycelium and sporulation.

All data were subjected to one-way analysis of variance (ANOVA) using the R software (https://www.r-project.org/) (accessed on 9 November 2021). In order to normalize the distributions, data were transformed in square-root values, but untransformed values are reported in the respective graphs. Tukey's HSD (honestly significant difference) post-hoc test was applied to evidence significant statistical differences ($p \leq 0.05$).

Author Contributions: Conceptualization, S.O.C., N.T., G.M. (Giuseppe Meca), G.M. (Giovanni Marletta), A.P. and A.D.; methodology, S.O.C., N.T., G.M., S.E.b., F.L.S., E.I.R. and C.L.M.; software, F.L.S. and C.L.M.; validation, S.O.C., N.T., G.M. (Giuseppe Meca), G.M. (Giovanni Marletta), A.P. and A.D.; formal analysis, N.T., S.E.b. and C.L.M.; investigation, S.E.b., E.I.R., C.L.M. and F.L.S.; resources, S.O.C., A.P., N.T., G.M. (Giuseppe Meca) and G.M. (Giovanni Marletta); data curation, S.E.b. and F.L.S.; writing—original draft preparation, S.E.b. and F.L.S.; writing—review and editing, S.O.C., N.T., G.M. (Giuseppe Meca), A.P., G.M. (Giovanni Marletta) and A.D.; visualization, S.O.C.; supervision, A.D. and S.O.C. All authors have read and agreed to the published version of the manuscript.

Funding: This study was funded by the University of Catania, Italy "Investigation of phytopathological problems of the main Sicilian productive contexts and eco-sustainable defense strategies (MED-IT-ECO)" PiaCeRi-PIAno di inCEntivi per la Ricerca di Ateneo 2020–22 linea 2" (5A722192155) and the project "Smart and innovative packaging, postharvest rot management and shipping of organic citrus fruit (BiOrangePack)" under Partnership for Research and Innovation in the Mediterranea Area (PRIMA)-H2020 (E69C20000130001). F.L.S. was supported by a fellowship funded by the BiOrangePack project. This study is part of E.I.R. activity of a Ph.D. student in "Agricultural, Food and Environmental Science", University of Catania; XXXVI Cycle. E.I.R. is supported by a grant funded by the National Operative Program (PON) "Research and Innovation" 2014–2020, line "Innovative doctorates with industrial characterization") (CCI 2014IT16M2OP005).

Institutional Review Board Statement: Not applicable.

Informed Consent Statement: Not applicable.

Data Availability Statement: Raw data can be shared upon reasonable request.

Acknowledgments: We are grateful to A. Davies for the English revision of the text.

Conflicts of Interest: The authors declare no conflict of interest. The funders had no role in the design of the study; in the collection, analyses, or interpretation of data; in the writing of the manuscript, or in the decision to publish the results.

References

1. Korukluoglu, M.; Sahan, Y.; Yigit, A. Antifungal Properties of Olive Leaf Extracts and Their Phenolic Compounds. *J. Food Saf.* **2008**, *28*, 76–87. [CrossRef]
2. Hernández, A.; Ruiz-Moyano, S.; Galván, A.I.; Merchán, A.V.; Pérez Nevado, F.; Aranda, E.; Serradilla, M.J.; de Guía Córdoba, M.; Martín, A. Anti-fungal activity of phenolic sweet orange peel extract for controlling fungi responsible for post-harvest fruit decay. *Fungal Biol.* **2021**, *125*, 143–152. [CrossRef] [PubMed]
3. Wei, L.; Zhang, J.; Tan, W.; Wang, G.; Li, Q.; Dong, F.; Guo, Z. Antifungal activity of double Schiff bases of chitosan derivatives bearing active halogeno-benzenes. *Int. J. Biol. Macromol.* **2021**, *179*, 292–298. [CrossRef]
4. Meng, D.; Garba, B.; Ren, Y.; Yao, M.; Xia, X.; Li, M.; Wang, Y. Antifungal activity of chitosan against *Aspergillus ochraceus* and its possible mechanisms of action. *Int. J. Biol. Macromol.* **2020**, *158*, 1063–1070. [CrossRef] [PubMed]
5. Abade, A.; Ferreira, P.A.; de Barros Vidal, F. Plant Diseases recognition on images using Convolutional Neural Networks: A Systematic Review. *Comput. Electron. Agric.* **2021**, *185*, 106125. [CrossRef]
6. Huang, F.-C.; Molnár, P.; Schwab, W. Cloning and functional characterization of carotenoid cleavage dioxygenase 4 genes. *J. Exp. Bot.* **2009**, *60*, 3011–3022. [CrossRef] [PubMed]
7. La Spada, F.; Aloi, F.; Coniglione, M.; Pane, A.; Cacciola, S.O. Natural Biostimulants Elicit Plant Immune System in an Integrated Management Strategy of the Postharvest Green Mold of Orange Fruits Incited by *Penicillium digitatum*. *Front. Plant Sci.* **2021**, *12*, 684722. [CrossRef]
8. La Spada, F.; Stracquadanio, C.; Riolo, M.; Pane, A.; Cacciola, S.O. *Trichoderma* Counteracts the Challenge of *Phytophthora nicotianae* Infections on Tomato by Modulating Plant Defense Mechanisms and the Expression of Crinkler, Necrosis-Inducing *Phytophthora* Protein 1, and Cellulose-Binding Elicitor Lectin Pathogenic Effectors. *Front. Plant Sci.* **2020**, *11*, 583539. [CrossRef]
9. Stracquadanio, C.; Quiles, J.M.; Meca, G.; Cacciola, S.O. Antifungal Activity of Bioactive Metabolites Produced by *Trichoderma asperellum* and *Trichoderma atroviride* in Liquid Medium. *JoF* **2020**, *6*, 263. [CrossRef]
10. du Jardin, P. Plant biostimulants: Definition, concept, main categories and regulation. *Sci. Hortic.* **2015**, *196*, 3–14. [CrossRef]
11. Wang, L.; Wu, H.; Qin, G.; Meng, X. Chitosan disrupts *Penicillium expansum* and controls postharvest blue mold of jujube fruit. *Food Control* **2014**, *41*, 56–62. [CrossRef]
12. Yang, R.; Miao, J.; Shen, Y.; Cai, N.; Wan, C.; Zou, L.; Chen, C.; Chen, J. Antifungal effect of cinnamaldehyde, eugenol and carvacrol nanoemulsion against *Penicillium digitatum* and application in postharvest preservation of citrus Fruit. *LWT* **2021**, *141*, 110924. [CrossRef]
13. Girotto, F.; Alibardi, L.; Cossu, R. Food waste generation and industrial uses: A review. *Waste Manag.* **2015**, *45*, 32–41. [CrossRef]
14. Prameela, K.; Venkatesh, K.; Immandi, S.B.; Kasturi, A.P.K.; Rama Krishna, C.; Murali Mohan, C. Next generation nutraceutical from shrimp waste: The convergence of applications with extraction methods. *Food Chem.* **2017**, *237*, 121–132. [CrossRef]
15. Nirmal, N.P.; Santivarangkna, C.; Rajput, M.S.; Benjakul, S. Trends in shrimp processing waste utilization: An industrial prospective. *Trends Food Sci. Technol.* **2020**, *103*, 20–35. [CrossRef]
16. Taser, B.; Ozkan, H.; Adiguzel, A.; Orak, T.; Baltaci, M.O.; Taskin, M. Preparation of chitosan from waste shrimp shells fermented with *Paenibacillus Jamilae* BAT1. *Int. J. Biol. Macromol.* **2021**, *183*, 1191–1199. [CrossRef] [PubMed]
17. Kandra, P.; Challa, M.M.; Kalangi Padma Jyothi, H. Efficient use of shrimp waste: Present and future trends. *Appl. Microbiol. Biotechnol.* **2012**, *93*, 17–29. [CrossRef] [PubMed]
18. Zhou, Y.; Guo, N.; Wang, Z.; Zhao, T.; Sun, J.; Mao, X. Evaluation of a clean fermentation-organic acid method for processing Sshrimp waste from six major cultivated shrimp Sspecies in China. *J. Clean. Prod.* **2021**, *294*, 126135. [CrossRef]
19. Kumar, A.; Kumar, D.; George, N.; Sharma, P.; Gupta, N. A process for complete biodegradation of shrimp waste by a novel marine isolate *Paenibacillus* sp. AD with simultaneous production of chitinase and chitin oligosaccharides. *Int. J. Biol. Macromol.* **2018**, *109*, 263–272. [CrossRef]
20. Goossens, Y.; Schmidt, T.G.; Kuntscher, M. Evaluation of Food Waste Prevention Measures—The Use of Fish Products in the Food Service Sector. *Sustainability* **2020**, *12*, 6613. [CrossRef]
21. Cheong, J.Y.; Muskhazli, M.; Nor Azwady, A.A.; Ahmad, S.A.; Adli, A.A. Three dimensional optimisation for the enhancement of astaxanthin recovery from shrimp shell wastes by *Aeromonas hydrophila*. *Biocatal. Agric. Biotechnol.* **2020**, *27*, 101649. [CrossRef]
22. Roy, V.C.; Getachew, A.T.; Cho, Y.-J.; Park, J.-S.; Chun, B.-S. Recovery and bio-potentialities of astaxanthin-rich oil from shrimp (*Peneanus monodon*) waste and mackerel (*Scomberomous niphonius*) skin using concurrent supercritical CO_2 extraction. *J. Supercrit. Fluids* **2020**, *159*, 104773. [CrossRef]
23. Shen, Q.; Song, G.; Wang, H.; Zhang, Y.; Cui, Y.; Xie, H.; Xue, J.; Wang, H. Isolation and lipidomics characterization of fatty acids and phospholipids in shrimp waste through GC/FID and HILIC-QTrap/MS. *J. Food Compos. Anal.* **2021**, *95*, 103668. [CrossRef]
24. Pattanaik, S.S.; Sawant, P.B.; Xavier, K.A.M.; Dube, K.; Srivastava, P.P.; Dhanabalan, V.; Chadha, N.K. Characterization of carotenoprotein from different shrimp shell waste for possible use as supplementary nutritive feed ingredient in animal diets. *Aquaculture* **2020**, *515*, 734594. [CrossRef]

25. Santos, V.P.; Marques, N.S.S.; Maia, P.C.S.V.; Lima, M.A.B.d.; Franco, L.d.O.; Campos-Takaki, G.M.d. Seafood Waste as Attractive Source of Chitin and Chitosan Production and Their Applications. *Int. J. Mol. Sci.* **2020**, *21*, 4290. [CrossRef]

26. Eddya, M.; Tbib, B.; EL-Hami, K. A comparison of chitosan properties after extraction from shrimp shells by diluted and concentrated acids. *Heliyon* **2020**, *6*, e03486. [CrossRef] [PubMed]

27. Zhang, Y.; Wang, Y.; Zhang, Z.; Cui, W.; Zhang, X.; Wang, S. Removing copper and cadmium from water and sediment by magnetic microspheres—$MnFe_2O_4$/chitosan prepared by waste shrimp shells. *J. Environ. Chem. Eng.* **2021**, *9*, 104647. [CrossRef]

28. Zhang, Z.; Li, C.; Wang, Q.; Zhao, Z.K. Efficient hydrolysis of chitosan in ionic liquids. *Carbohydr. Polym.* **2009**, *78*, 685–689. [CrossRef]

29. Sakko, M.; Moore, C.; Novak-Frazer, L.; Rautemaa, V.; Sorsa, T.; Hietala, P.; Järvinen, A.; Bowyer, P.; Tjäderhane, L.; Rautemaa, R. 2-hydroxyisocaproic acid is fungicidal for *Candida* and *Aspergillus* species. *Mycoses* **2014**, *57*, 214–221. [CrossRef]

30. Honoré, A.H.; Aunsbjerg, S.D.; Ebrahimi, P.; Thorsen, M.; Benfeldt, C.; Knøchel, S.; Skov, T. Metabolic footprinting for investigation of antifungal properties of *Lactobacillus paracasei*. *Anal. Bioanal. Chem.* **2016**, *408*, 83–96. [CrossRef]

31. Krátký, M.; Konečná, K.; Janoušek, J.; Brablíková, M.; Janďourek, O.; Trejtnar, F.; Stolaříková, J.; Vinšová, J. 4-Aminobenzoic Acid Derivatives: Converting Folate Precursor to Antimicrobial and Cytotoxic Agents. *Biomolecules* **2019**, *10*, 9. [CrossRef]

32. Bajpai, V.K.; Kim, H.R.; Hou, C.T.; Kang, S.C. Microbial conversion and in vitro and in vivo antifungal assessment of bioconverted docosahexaenoic acid (bDHA) used against agricultural plant pathogenic fungi. *J. Ind. Microbiol. Biotechnol.* **2009**, *36*, 695–704. [CrossRef]

33. Kumar Patel, M.; Maurer, D.; Feygenberg, O.; Ovadia, A.; Elad, Y.; Oren-Shamir, M.; Alkan, N. Phenylalanine: A Promising Inducer of Fruit Resistance to Postharvest Pathogens. *Foods* **2020**, *9*, 646. [CrossRef]

34. Belgacem, I.; Pangallo, S.; Abdelfattah, A.; Romeo, F.V.; Cacciola, S.O.; Li Destri Nicosia, M.G.; Ballistreri, G.; Schena, L. Transcriptomic Analysis of Orange Fruit Treated with Pomegranate Peel Extract (PGE). *Plants* **2019**, *8*, 101. [CrossRef] [PubMed]

35. Romanazzi, G.; Feliziani, E.; Sivakumar, D. Chitosan, a Biopolymer with Triple Action on Postharvest Decay of Fruit and Vegetables: Eliciting, Antimicrobial and Film-Forming Properties. *Front. Microbiol.* **2018**, *9*, 2745. [CrossRef] [PubMed]

36. Fitzgerald, D.J.; Stratford, M.; Gasson, M.J.; Narbad, A. Structure−Function Analysis of the Vanillin Molecule and Its Antifungal Properties. *J. Agric. Food Chem.* **2005**, *53*, 1769–1775. [CrossRef] [PubMed]

37. Lu, H.-F.; Tsou, M.-F.; Lin, J.-G.; Ho, C.-C.; Liu, J.-Y.; Chuang, J.-Y.; Chung, J.-G. Ellagic Acid Inhibits Growth and Arylamine N-Acetyltransferase Activity and Gene Expression in Staphylococcus Aureus. *In Vivo* **2005**, *19*, 195–199.

38. Kocaçalışkan, I.; Talan, I.; Terzi, I. Antimicrobial Activity of Catechol and Pyrogallol as Allelochemicals. *Zeitschrift für Naturforschung C* **2006**, *61*, 639–642. [CrossRef]

39. Beck, J.J.; Kim, J.H.; Campbell, B.C.; Chou, S.-C. Fungicidal Activities of Dihydroferulic Acid Alkyl Ester Analogues. *J. Nat. Prod.* **2007**, *70*, 779–782. [CrossRef]

40. Ngarmsak, M. Antifungal Activity of Vanillin on Fresh-Cut Tropical Fruits. *Acta Hortic.* **2007**, 409–416. [CrossRef]

41. Tao, S.; Zhang, S.; Tsao, R.; Charles, M.T.; Yang, R.; Khanizadeh, S. *In vitro* antifungal activity and mode of action of selected polyphenolic antioxidants on *Botrytis Cinerea*. *Arch. Phytopathol. Plant Prot.* **2010**, *43*, 1564–1578. [CrossRef]

42. Li, B.; Lai, T.; Qin, G.; Tian, S. Ambient pH Stress Inhibits Spore Germination of *Penicillium expansum* by Impairing Protein Synthesis and Folding: A Proteomic-Based Study. *J. Proteome Res.* **2010**, *9*, 298–307. [CrossRef] [PubMed]

43. Jeong, M.-H.; Lee, Y.-S.; Cho, J.-Y.; Ahn, Y.-S.; Moon, J.-H.; Hyun, H.-N.; Cha, G.-S.; Kim, K.-Y. Isolation and characterization of metabolites from *Bacillus Licheniformis* MH48 with antifungal activity against plant pathogens. *Microb. Pathog.* **2017**, *110*, 645–653. [CrossRef] [PubMed]

44. Liu, X.; Ji, D.; Cui, X.; Zhang, Z.; Li, B.; Xu, Y.; Chen, T.; Tian, S. *p*-Coumaric acid induces antioxidant capacity and defense responses of sweet cherry fruit to fungal pathogens. *Postharvest Biol. Technol.* **2020**, *169*, 111297. [CrossRef]

45. Zamuz, S.; Munekata, P.E.S.; Dzuvor, C.K.O.; Zhang, W.; Sant'Ana, A.S.; Lorenzo, J.M. The role of phenolic compounds against *Listeria monocytogenes* in food. A review. *Trends Food Sci. Technol.* **2021**, *110*, 385–392. [CrossRef]

46. Khan, F.; Bamunuarachchi, N.I.; Tabassum, N.; Kim, Y.-M. Caffeic Acid and Its Derivatives: Antimicrobial Drugs toward Microbial Pathogens. *J. Agric. Food Chem.* **2021**, *69*, 2979–3004. [CrossRef]

47. Seeram, N.; Adams, L.; Henning, S.; Niu, Y.; Zhang, Y.; Nair, M.; Heber, D. In vitro antiproliferative, apoptotic and antioxidant activities of punicalagin, ellagic acid and a total pomegranate tannin extract are enhanced in combination with other polyphenols as found in pomegranate juice. *J. Nutr. Biochem.* **2005**, *16*, 360–367. [CrossRef] [PubMed]

48. Singh, R.P.; Chidambara Murthy, K.N.; Jayaprakasha, G.K. Studies on the Antioxidant Activity of Pomegranate (*Punica granatum*) Peel and Seed Extracts Using in Vitro Models. *J. Agric. Food Chem.* **2002**, *50*, 81–86. [CrossRef]

49. Vora, A.; Londhe, V.; Pandita, N. Herbosomes enhance the *in vivo* antioxidant activity and bioavailability of punicalagins from standardized pomegranate extract. *J. Funct. Foods* **2015**, *12*, 540–548. [CrossRef]

50. Al-Zoreky, N.S. Antimicrobial activity of pomegranate (*Punica granatum* L.) fruit peels. *Int. J. Food Microbiol.* **2009**, *134*, 244–248. [CrossRef]

51. Tayel, A.A.; El-Baz, A.F.; Salem, M.F.; El-Hadary, M.H. Potential applications of pomegranate peel extract for the control of citrus green mould. *J. Plant Dis. Prot.* **2009**, *116*, 252–256. [CrossRef]

52. Barash, I.; Pupkin, G.; Koren, L.; Ben-Hayyim, G.; Strobel, G.A. A low molecular weight phytotoxin produced by *Phoma tracheiphila*, the cause of mal secco disease in citrus. *Physiol. Plant Pathol.* **1981**, *19*, 17–29. [CrossRef]

53. Migheli, Q.; Cacciola, S.O.; Balmas, V.; Pane, A.; Ezra, D.; di San Lio, G.M. Mal Secco Disease Caused by *Phoma tracheiphila*: A Potential Threat to Lemon Production Worldwide. *Plant Dis.* **2009**, *93*, 852–867. [CrossRef] [PubMed]

54. Kumari, P.; Singh, R.; Punia, R. Studies on Collection, Isolation, Purification and Maintenance of Culture of *Colletotrichum gloeosporioides*. *Int. J. Agric. Innov. Res.* **2017**, *6*, 351–353.

55. Riolo, M.; Aloi, F.; Pane, A.; Cara, M.; Cacciola, S.O. Twig and Shoot Dieback of Citrus, a New Disease Caused by *Colletotrichum* Species. *Cells* **2021**, *10*, 449. [CrossRef] [PubMed]

56. Mammella, M.A.; Martin, F.N.; Cacciola, S.O.; Coffey, M.D.; Faedda, R.; Schena, L. Analyses of the Population Structure in a Global Collection of *Phytophthora nicotianae* Isolates Inferred from Mitochondrial and Nuclear DNA Sequences. *Phytopathology* **2013**, *103*, 610–622. [CrossRef]

57. Biasi, A.; Martin, F.N.; Cacciola, S.O.; di San Lio, G.M.; Grünwald, N.J.; Schena, L. Genetic Analysis of *Phytophthora nicotianae* Populations from Different Hosts Using Microsatellite Markers. *Phytopathology* **2016**, *106*, 1006–1014. [CrossRef]

58. Ismail, M.; Zhang, J. Post-Harvest Citrus Diseases and Their Control. *Outlook Pest Manag.* **2004**, *15*, 29–35. [CrossRef]

59. Errampalli, D. Penicillium expansum (Blue Mold). In *Postharvest Decay*; Elsevier: Amsterdam, The Netherlands, 2014; pp. 189–231. ISBN 978-0-12-411552-1.

60. Eckert, J.W. The Chemical Control of Postharvest Diseases: Deciduous Fruits, Berries, Vegetables and Root/Tuber Crops. *Ann. Rev. Phytopathol.* **1988**, *26*, 433–469. [CrossRef]

61. Davé, B.; Sales, M.; Walia, M. Resistance of different strains of *Penicillium digitatum* to imazalil treatment in California citrus packinghouses. *Proc. Fla. State Hortic. Soc.* **1989**, *102*, 178–179.

62. Kanetis, L.; Förster, H.; Adaskaveg, J.E. Comparative Efficacy of the New Postharvest Fungicides Azoxystrobin, Fludioxonil, and Pyrimethanil for Managing Citrus Green Mold. *Plant Dis.* **2007**, *91*, 1502–1511. [CrossRef] [PubMed]

63. Errampalli, D.; Crnko, N. Control of blue mold caused by *Penicillium expansum* on apples "Empire" with fludioxonil and cyprodinil. *Can. J. Plant Pathol.* **2004**, *26*, 70–75. [CrossRef]

64. Conclusion regarding the peer review of the pesticide risk assessment of the active substance fluoxastrobin. *EFSA Sci. Rep.* **2007**, *102*, 1–84. [CrossRef]

65. Conclusion regarding the peer review of the pesticide risk assessment of the active substance cyprodinil. *EFSA Sci. Rep.* **2005**, *51*, 1–78. [CrossRef]

66. Conclusion regarding the peer review of the pesticide risk assessment of the active substance pyrimethanil. *EFSA Sci. Rep.* **2006**, *61*, 1–70. [CrossRef]

67. Conclusion on the peer review of the pesticide risk assessment of the active substance thiabendazole. *EFSA J.* **2014**, *12*, 1–57. [CrossRef]

68. Conclusion on the peer review of the pesticide risk Assessment of the active substance azoxystrobin. *EFSA J.* **2010**, *8*, 1–110. [CrossRef]

69. Conclusion on the peer review of the pesticide risk Assessment of the active substance imazalil. *EFSA J.* **2010**, *8*, 1–110. [CrossRef]

70. Borrás, A.D.; Aguilar, R.V. Biological control of *Penicillium digitatum* by *Trichoderma viride* on postharvest citrus fruits. *Int. J. Food Microbiol.* **1990**, *11*, 179–183. [CrossRef]

71. Del Río, J.A.; Arcas, M.C.; Benavente-García, O.; Ortuño, A. Citrus polymethoxylated flavones can confer resistance against *Phytophthora citrophthora*, *Penicillium digitatum*, and *Geotrichum* species. *J. Agric. Food Chem.* **1998**, *46*, 4423–4428. [CrossRef]

72. Holmes, G.J.; Eckert, J.W. Sensitivity of *Penicillium digitatum* and *P. italicum* to Postharvest Citrus Fungicides in California. *Phytopathology* **1999**, *89*, 716–721. [CrossRef]

73. Youssef, K.; Ligorio, A.; Sanzani, S.M.; Nigro, F.; Ippolito, A. Control of storage diseases of citrus by pre- and postharvest application of salts. *Postharvest Biol. Technol.* **2012**, *72*, 57–63. [CrossRef]

74. Youssef, K.; Sanzani, S.M.; Ligorio, A.; Ippolito, A.; Terry, L.A. Sodium carbonate and bicarbonate treatments induce resistance to postharvest green mould on citrus fruit. *Postharvest Biol. Technol.* **2014**, *87*, 61–69. [CrossRef]

75. Abo-Elnaga, H.I.G. Control of green mould of citrus by using *Trichoderma harzianum*, humic acid or garlic. *Arch. Phytopathol. Plant Prot.* **2013**, *46*, 1118–1126. [CrossRef]

76. Aloui, H.; Licciardello, F.; Khwaldia, K.; Hamdi, M.; Restuccia, C. Physical properties and antifungal activity of bioactive films containing *Wickerhamomyces anomalus* killer yeast and their application for preservation of oranges and control of postharvest green mold caused by *Penicillium digitatum*. *Int. J. Food Microbiol.* **2015**, *200*, 22–30. [CrossRef] [PubMed]

77. Fallanaj, F.; Ippolito, A.; Ligorio, A.; Garganese, F.; Zavanella, C.; Sanzani, S.M. Electrolyzed sodium bicarbonate inhibits *Penicillium digitatum* and induces defence responses against green mould in citrus fruit. *Postharvest Biol. Technol.* **2016**, *115*, 18–29. [CrossRef]

78. Pangallo, S.; Li Destri Nicosia, M.G.; Raphael, G.; Levin, E.; Ballistreri, G.; Cacciola, S.O.; Rapisarda, P.; Droby, S.; Schena, L. Elicitation of resistance responses in grapefruit and lemon fruits treated with a pomegranate peel extract. *Plant Pathol.* **2017**, *66*, 633–640. [CrossRef]

79. Riolo, M.; Aloi, F.; La Spada, F.; Sciandrello, S.; Moricca, S.; Santilli, E.; Pane, A.; Cacciola, S.O. Diversity of *Phytophthora* Communities across Different Types of Mediterranean Vegetation in a Nature Reserve Area. *Forests* **2020**, *11*, 853. [CrossRef]

80. Aloi, F.; Riolo, M.; Parlascino, R.; Pane, A.; Cacciola, S.O. Bot Gummosis of Lemon (*Citrus × Limon*) Caused by *Neofusicoccum parvum*. *JoF* **2021**, *7*, 294. [CrossRef]
81. Wiegand, I.; Hilpert, K.; Hancock, R.E.W. Agar and broth dilution methods to determine the minimal inhibitory concentration (MIC) of antimicrobial substances. *Nat. Protoc.* **2008**, *3*, 163–175. [CrossRef]

Article

Biological Control of Charcoal Rot in Peanut Crop through Strains of *Trichoderma* spp., in Puebla, Mexico

Saira Jazmín Martínez-Salgado [1], Petra Andrade-Hoyos [2], Conrado Parraguirre Lezama [2], Antonio Rivera-Tapia [3], Alfonso Luna-Cruz [4] and Omar Romero-Arenas [2,*]

[1] Facultad de Ciencias Biológicas, Programa Biotecnología, Benemérita Universidad Autónoma de Puebla, Ciudad Universidad Autónoma de Puebla, Ciudad Universitaria, Puebla 72570, Mexico; jazmin_saira@hotmail.com

[2] Centro de Agroecología, Instituto de Ciencias, Benemérita Universidad Autónoma de Puebla, Edificio VAL 1, Km 1.7 Carretera a San Baltazar Tetela, San Pedro Zacachimalpa, Puebla 72960, Mexico; andrad@colpos.mx (P.A.-H.); conrado.parraguirre@correo.buap.mx (C.P.L.)

[3] Centro de Investigaciones en Ciencias Microbiológicas, Instituto de Ciencias, Benemérita Universidad Autónoma de Puebla, Ciudad Universitaria, Puebla 72570, Mexico; jose.riverat@correo.buap.mx

[4] Instituto de Investigaciones Químico-Biológicas, Universidad Michoacana de San Nicolas de Hidalgo, Morelia 27852, Mexico; aluna@colpos.mx

* Correspondence: biol.ora@hotmail.com; Tel.: +52-22-2229-5500 (ext. 3717)

Citation: Martínez-Salgado, S.J.; Andrade-Hoyos, P.; Parraguirre Lezama, C.; Rivera-Tapia, A.; Luna-Cruz, A.; Romero-Arenas, O. Biological Control of Charcoal Rot in Peanut Crop through Strains of *Trichoderma* spp., in Puebla, Mexico. *Plants* **2021**, *10*, 2630. https://doi.org/10.3390/plants10122630

Academic Editor: Carlos Agustí-Brisach

Received: 22 October 2021
Accepted: 15 November 2021
Published: 30 November 2021

Publisher's Note: MDPI stays neutral with regard to jurisdictional claims in published maps and institutional affiliations.

Abstract: Charcoal rot is an emerging disease for peanut crops caused by the fungus *Macrophomina phaseolina*. In Mexico, peanut crop represents an important productive activity for various rural areas; however, charcoal rot affects producers economically. The objectives of this research were: (a) to identify and morphologically characterize the strain "PUE 4.0" associated with charcoal rot of peanut crops from Buenavista de Benito Juárez, belonging to the municipality of Chietla in Puebla, Mexico; (b) determine the in vitro and in vivo antagonist activity of five *Trichoderma* species on *M. phaseolina*, and (c) determine the effect of the incidence of the disease on peanut production in the field. Vegetable tissue samples were collected from peanut crops in Puebla, Mexico with the presence of symptoms of charcoal rot at the stem and root level. The "PUE 4.0" strain presented 100% identity with *M. phaseolina*, the cause of charcoal rot in peanut crops from Buenavista de Benito Juárez. *T. koningiopsis* (T-K11) showed the highest development rate, the best growth speed, and the highest percentage of radial growth inhibition (PIRG) over *M. phaseolina* (71.11%) under in vitro conditions, in addition, *T. koningiopsis* (T-K11) showed higher production (1.60 ± 0.01 t/ha^{-1}) and lower incidence of charcoal rot under field conditions. The lowest production with the highest incidence of the disease occurred in plants inoculated only with *M. phaseolina* (0.67 ± 0.01 t/ha^{-1}) where elongated reddish-brown lesions were observed that covered 40% of the total surface of the main root.

Keywords: charcoal rot; biological control; percentage of radial growth inhibition (PIRG); incidence of the disease; production

1. Introduction

Peanut (*Arachis hypogea* L.) is a self-pollinated annual tropical legume that belongs to Papilionaceae subfamily, native to South America and valued worldwide for its high content of oil, proteins, and minerals such as iron, calcium, phosphorus, magnesium, selenium, and zinc; in addition to vitamins E, B6, riboflavin, thiamine, and niacin [1,2]. China is the world's leading producer of peanuts, accounting for nearly 41.0% of the total output. In 2019, China was the biggest peanut producer with a production of 17.5 million metric tons. India, Nigeria, and the United States followed with about 6.8, 3.0, and 2.5 million metric tons, respectively [3]. The cultivated area in Mexico is currently 47,532 ha with a production of 81,413 tons in 2019 [4]. State of Puebla ranks third in national production with 9.31 tons [5].

Peanut crop can be affected by various diseases caused by fungi, bacteria, and viruses that affect the yield, fungal diseases being the most worrying that generate significant economic losses [6]. Among fungal diseases of peanuts, charcoal rot is a disease caused by *Macrophomina phaseolina*, recently reported in Mexico [7]. However, charcoal rot affects more than 500 economically important plant species, such as cotton (*Gossypium hirsutum*), the chickpea (*Cicer Arientitanium*), the beans (*Phaseolus vulgaris*), the potato (*Solanum tuberosum*), the soybean (*Glycine max*), the corn (*Zea mays*), and the peanuts (*A. hypogaea*) [8,9].

M. phaseolina is a generalist phytopathogenic fungus originating in the soil and in the seed, present throughout the world [10]. It is characterized by hyaline hyphae with thin walls and light brown or dark brown with septa. Microsclerotia form a compact mass of hardened fungal mycelium that darkens with aging [11]. The *M. phaseolina* phytopathogen can infect the roots of the host plant at the seedling stage through multiple germinating hyphae. Once in the roots, the hyphae affect the vascular system, interrupting the transport of water and nutrients to the upper parts of the plants, causing the yellowing and senescence of the leaves. Charcoal rot mainly affects the lower stem and main root, causing premature death of the host plant [12].

There are several effective fungicides available and labeled for use in peanut (*A. hypogaea*) crop to control various fungal diseases, including demethylation inhibitors, growth inhibitors, and succinate dehydrogenase inhibitors (codes 3, 11, and 7 of the Fungicide Resistance Action Committee, respectively) [13]. Faced with the high costs of fungicides and their potentially harmful effects on people and environment, biological control is considered viable practice for development of sustainable agriculture [14,15].

Use of filamentous fungi as biocontrol agents represents an effective alternative for agricultural production systems [16]. Success and use in agroecological practice are due to its action mechanisms such as competition for space, mycoparasitism, antibiosis [17,18], and production of volatile compounds [19]. Management of charcoal rot by means of antagonistic microorganisms, such as *Pseudomonas fluorescens* and *Trichoderma* spp., has been carried out in economically important crops, among which are soybean [20], the sorghum [21], the beans, and sunflower [22]. However, the control of charcoal rot remains a challenge despite the many efforts that have been made about research. Therefore, the objective of this research was to evaluate antagonistic capacity of five *Trichoderma* species against "PUE 4.0" strain, present in peanut crop through in vitro and field tests in rural communities in Buenavista de Benito Juárez, belonging to the municipality of Chietla in Puebla, Mexico.

2. Materials and Methods

2.1. Area of Isolation

Vegetable tissue samples with rot at the stem and root level were collected in a plot of 3144.3 m^2 of peanut crop with a history of high charcoal rot incidence [7] during the summer–fall 2020 production. Agricultural plot corresponds to the Buenavista de Benito Juárez community, belonging to the municipality of Chietla in the state of Puebla-Mexico, with a warm desert climate (Bwh) and average rainfall of 700 mm [23]. Sampling was directed towards individuals with symptoms associated with genus *Macrophomina*; all samples were kept in plastic bags in a cooler until they were transferred to laboratory, to be processed.

The samples were cut into small 5 mm pieces, disinfected with 1% sodium hypochlorite for 3 min, and washed with sterile water. Finally, they were wrapped with sterile paper towels and placed in a laminar flow chamber at 20 °C for 15 min [24]. Subsequently, the samples were placed upright in Petri dishes with potato dextrose agar medium (PDA, Dioxon) modified with chloramphenicol (20 mg/mL^{-1}) and incubated at 28 °C for 5 days. The identification of fungal colonies associated with the genus *Macrophomina* was carried out by the observation of reproductive structures under a microscope and employing taxonomic keys of Barnett and Hunter [25]. For the microscopic observation, thin layer PDA cultures (microculture technique) were used. The mycelial cultures were observed after eight days of incubation at 28 °C employing lactophenol. The microscopic morphology of the fungi was examined under an optic microscope (Carl Zeiss, Jena, Germany) [26].

2.2. DNA Extraction, PCR Amplification, and Sequencing

This procedure was performed with the 2% cetyl trimethylammonium bromide (CTAB) method according to Doyle and Doyle [27] with some modifications [28]. Genomic DNA was suspended in 100 µL of sterile HPLC water and quantified by spectrophotometry in a NanoDrop 2000c (Thermo Scientific, Waltham, MA, USA). To determine the DNA quality, absorbance values between 1.8 and 2.2 at $A_{280/260}$ and $A_{230/260}$ nm were considered acceptable. Finally, the DNA was diluted to 20 ng μL^{-1} and then stored at −20 °C for PCR amplification.

Molecular identification of "PUE 4.0" strain was carried out based on the analysis of internal transcribed spacer (ITS) region sequences using primer pairs ITS5 (5′-GGAAGTAAAAGTCGTAACAAGG-3′)/ITS4(5′-TCCTCCGCTTATTGATATGC-3′) [29]. The reaction mixture was prepared in a final volume of 15 µL with 1× Taq buffer DNA polymerase, 0.18 µM of each dNTP, 0.18 µL of each primer containing 10 pmol, 0.90 U of GoTaq DNA polymerase (Promega, Madison, WI, USA), and 40 ng μL^{-1} DNA. PCR was performed in a Peltier PTC-200 DNA thermal cycler (Bio-Rad, Santa Rosa, CA, USA). The amplicons were verified by electrophoresis in a 1.5% agarose gel (Seakem, Invitrogen, Carlsbad, CA, USA) and stained with 10,000× GelRed (Biotium, Fremont, CA, USA). All PCR products were cleaned with ExoSAP-IT (Affymetrix, Santa Clara, CA, USA), and both strands were individually sequenced using the BigDye Terminator v3.1 Cycle Sequencing Kit (Applied Biosystems, Carlsbad, CA, USA) in a 3130 Genetic Analyzer Sequencer (Applied Biosystems, Carlsbad, CA, USA) at Postgraduate College Facilities, Mexico, according to Juárez-Vázquez [30]. Sequences were assembled and edited using SeqMan (DNAStar, Madison, WI, USA) and compared to sequences established in GenBank™ using the Blast algorithm.

2.3. Pathogenicity Tests

Fifty peanut plants of the "Virginia Champs" variety provided by local farmers were used. Each two-month-old plant was individually planted in a 1 L plastic pot, containing a sterilized mixture of Peatmoss and Agrellite (1:1 *v/v*). The inoculation of "PUE 4.0" strain (*M. phaseolina*) was by the toothpick method three days after sowing [31]. Development of the plants was done under greenhouse conditions (70% relative humidity and 28 °C) until the appearance of symptoms of disease.

Toothpicks were previously sterilized and then placed in Petri dishes with the *M. phaseolina* colony until 100% colonization was reached. Using toothpicks with mycelium, small 2 mm wounds were made on the roots. Sterile toothpicks were used for control group plants, tests were carried out in duplicate. After three weeks, inoculated plants showed symptoms of wilting chlorosis on leaves and discoloration of the vascular ring, from brown to dark brown: characteristic symptoms of charcoal rot, while control plants remained healthy, fulfilling Koch's postulates.

2.4. In Vitro Assessment of Antagonistic Capacity of Trichoderma spp.

For the evaluation of the antagonism test in vitro, the group of strains of *T. harzianum* (T-H3), *T. asperellum* (T-AS1), *T. hamatum* (T-A12), *T. koningiopsis* (T-K11), and an endemic strain of the *T. harzianum* (T-Ah) study region were used, whose sequences are found in the database of the National Center for Biological Information (NCBI) with access numbers MK780094, MK778890, MK791650, and MK791648, respectively. The dual confrontations were carried out with the strain "PUE 4.0" of *M. phaseolina* (MW585378) in a completely randomized experimental design with five treatments and three repetitions in duplicate.

For the evaluation of mycelial development, 5 mm diameter fragments of the *Trichoderma* strains as well as *M. phaseolina* were inoculated in Petri dishes with PDA (Potato and Dextrose Agar) and incubated in dark conditions at 28 °C for 10 days. The diameter of the mycelium was measured every 12 h with a digital vernier (CD-6 Mitutoyo) to estimate the growth speed (cm/d^{-1}), which was calculated with the linear growth function [32] Equation (1).

$$y = mx + b \tag{1}$$

where:

y = is the distance
m = slope
x = is time
b = the constant factor.

Antagonism and percentage of inhibition were evaluated considering the mycelial growth radius of *Trichoderma* spp. and *M. phaseolina* (with their respective controls). PDA discs (5 mm in diameter) with mycelia of *Trichoderma* spp. and *M. phaseolina* were placed at the extremes of Petri plates containing PDA and incubated at 28 °C for 240 h. Then, mycelial growth was scored every 12 h until the first contact between the mycelia of each antagonist with *M. phaseolina* occurred [33].

The percentage of radial growth inhibition (PIRG) was calculated based on the formula of Equation (2).

$$PIRG\% = (R1 - R2)/R1 \times 100 \tag{2}$$

where:

PIRG = Percent inhibition of radial growth.
R1 = Radial growth (mm) of *M. phaseolina* without *Trichoderma* spp.
R2 = Radial growth (mm) of *M. phaseolina* with *Trichoderma* spp.

Invasion of the antagonist or colonization on the surface of the *M. phaseolina* mycelium was taken as the index of antagonism with the scale proposed by Bell [34] (Table 1).

Table 1. *Trichoderma* strain antagonism evaluated in vitro using Bell's scale [34], considering the invasion of the surface.

Class	Class Features
I	*Trichoderma* completely overgrew the *M. phaseolina* and covered the entire medium surface.
II	*Trichoderma* overgrew at least two-thirds of the medium surface.
III	*Trichoderma* and the *M. phaseolina* each colonized approximately one-half of the medium surface and neither organism appeared to dominate the other.
IV	*M. phaseolina* colonized at least two-thirds of the medium surface and appeared to withstand encroachment by *Trichoderma*.
V	*M. phaseolina* completely overgrew the *Trichoderma* and occupied the entire medium surface.

2.5. Field Experiment: Evaluation of Antagonism

A test was carried out under open field conditions in the community of Buenavista de Benito Juárez (18°27′39″ N; 98°37′11″ W), belonging to the municipality of Chietla in the state of Puebla-Mexico. Nine hundred and sixty (960) "Virginia Champs" variety peanut seedlings provided by community producers were used.

Community producers carried out land preparation three months before the establishment of the crop. First, a 50-cm-deep plow was made to reduce compaction and promote soil drainage, then two 30-cm-deep turns of earth were made to promote aeration ground. The transplant date was 7 July 2020, where peanut seedlings of 30 days of emergence were used, and they were sown at a depth of 8 cm. Finally, they were fertilized with 40 kg/ha of phosphorus (P) and 60 kg/ha of potassium (K) at 15 days after transplantation (dft). The population density was four plants per m^2 spaced at 35 cm each and distributed in 26 rows with 60 cm between the rows of the crop, where two rows were considered to form an experimental block in a straight line.

The experimental design consists of 13 randomized complete blocks with 8 repetitions per treatment, occupying 100 plants per treatment, leaving four plants on the banks, which were not considered, giving a total of 800 plants of the 960 seedlings planted in the study community.

Inoculation of *M. phaseolina* (MW585378) was performed 15 days (dft) on the neck of each of the peanut plants (100 seedlings per treatment) with 1 mL of solution at a concentration of 1×10^8 conidia. After 36 h, the plants were inoculated with strains of *T. harzianum* (T-H3), *T. asperellum* (T-AS1), *T. hamatum* (T-A12), *T. koningiopsis* (T-K11), and *T. harzianum* (T-Ah) at the same concentration as the pathogen (1×10^8 conidia mL^{-1}), for each treatment. For chemical treatment, Cercobin® (Thiophanate methyl) was applied, following the manufacturer's recommendations (500 g in 400 L^{-1} of water per ha). Finally, for the control treatment, only sterile water without the presence of fungal activity was applied.

The incidence of the disease expected by *M. phaseolina* was calculated in 100 plants per treatment at the end of four months that the cultivation lasted. For this, the infected portion was measured in relation to the total length of the roots [35] and it was classified on the scale proposed by Bokhari [36] where: $0 \leq 25\%$ severity, 1 = 26 to 50% severity, 2 = 51 to 75% severity, and $3 \geq 76\%$ severity. Additionally, the complementary variables of total fresh weight of each plant, dry weight of peanut pods per plant, number of pods per plant, and weight of 100 peanut grains per treatment were taken; in addition, the yield was calculated according to Zamurrad [37] at the end of experiment.

2.6. Statistical Analysis

Data were analyzed with ANOVA (two ways) in the statistical package SPSS Statistics version 17 for Windows. Growth speed and the development rate were response variables with three repetitions. Experiments were validated in duplicate in a completely randomized statistical design. The data were subjected to the Bartlett homogeneity test, and subsequently, a Tukey–Kramer comparison test of means was performed with a probability level of $p \leq 0.05$.

Radial growth inhibition data (PIRG) were expressed in percentages and transformed with angular arccosine $\sqrt{x} + 1$. The mean values of the variables, which were the total fresh weight of the plant, dry weight of pods per plant, number of pods per plant, weight of 100 peanut grains, and yield were subjected to an analysis of variance with the same statistical program, using the test of Tukey–Kramer to determine the significant differences between the treatments ($p < 0.05$).

3. Results

3.1. Isolation, Characterization, and Identification of the Causative Agent of Charcoal Rot

Representative isolates from 50 different plants developed typical morphological characteristics of *M. phaseolina*. The fungal colonies were initially whitish grayish-dark brown in color on PDA medium (Figure 1a). After 6 days, semi-compressed mycelium was observed on the culture plate with microsclerotia embedded within the hyphae and absorbed into the agar. The aggregation of hyphae formed jet-black microsclerotia with a size of 74×110 μm (Figure 1b).

Amplification of 5.8S rDNA gene region showed a product of 601 bp, which presented 100% identity with *M. phaseolina* (ID: KF951698) in the Gen Bank nucleotide database of the National Center for Biotechnology Information (https://www.ncbi.nlm.nih.gov/, accessed on 15 February 2021). This sequence was deposited in the same database, with the accession number MW585378.

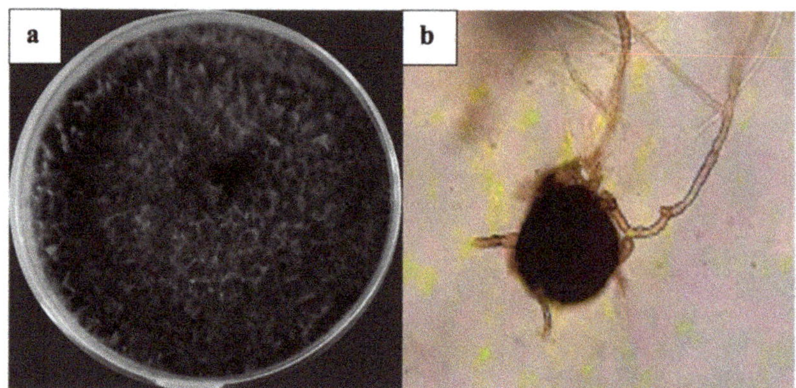

Figure 1. Cultural and morphological characteristics of *M. phaseolina*. (**a**) Colony on the agar plate; (**b**) microsclerotia (100×).

3.2. Pathogenicity Tests

Koch's postulates confirmed that "PUE 4.0" strain showed typical symptoms of charcoal rot 20 days after inoculation. In addition, microsclerotia were observed in the vascular system (Xylem) which caused an upward wilt of stem, as shown in Figure 2b. No symptoms were observed in the control group.

Figure 2. Pathogenicity tests with "PUE 4.0" strain: (**a**) plant of *Arachis hypogea* with charcoal rot and death of foliage at 20 days after inoculation; (**b**) cross-section of peanut root showing rot and the presence of microsclerotia; (**c**) control group without symptoms.

Using PCR of repetitive sequence, it was possible to confirm the identity of the re-isolation (PUE 4.1) of the original strain. This sequence was deposited in the same database, with the accession number MW585379.

3.3. Percentage of Growth Inhibition In Vitro

Areas of interaction were observed between *T. harzianum* (T-H3), *T. asperellum* (T-AS1), *T. hamatum* (T-A12), *T. koningiopsis* (T-K11), and the native strain of *T. harzianum* (T-Ah) against *M. phaseolina* (MW585378), where parasitism greater than 50% was obtained at 240 h.

Development rate and the growth speed had significant differences ($p \leq 0.05$), where *T. koningiopsis* (T-K11) obtained the highest value (Table 2) with 2.18 ± 0.035 mm/hour and 2.23 ± 0.013 cm/d^{-1}, respectively. *M. phaseolina* showed the lowest growth speed (1.67 ± 0.054 cm/d^{-1}).

Table 2. Rate of development, growth speed, and percentage of inhibition of radial growth and antagonism classification on Bell Scale [34].

Name	Development Rate (mm/hour) *	Growth Rate (cm/d^{-1}) *	PICR *	Class Antagonism
T. harzianum (T-H3)	2.17 ± 0.095 [a]	1.32 ± 0.04 [a]		
T. asperellum (T-AS1)	1.86 ± 0.033 [b]	2.16 ± 0.017 [a]		
T. hamatum (T-A12)	1.60 ± 0.05 [c]	2.14 ± 0.01 [a]		
T. koningiopsis (T-K11)	2.18 ± 0.035 [a]	2.23 ± 0.013 [a]		
T. harzianum (Th-Ah)	1.59 ± 0.04 [c]	2.06 ± 0.068 [a]		
M. phaseolina (PUE 4.0)	1.44 ± 0.04 [c]	1.67 ± 0.054 [b]		
M. phaseolina (PUE 4.0) vs. *T. harzianum* (T-H3)			63.55 ± 0.88 [ab]	I
M. phaseolina (PUE 4.0) vs. *T. asperellum* (T-AS1)			53.33 ± 0.76 [bc]	II
M. phaseolina (PUE 4.0) vs. *T. hamatum* (T-A12)			51.55 ± 2.47 [c]	II
M. phaseolina (PUE 4.0) vs. *T. koningiopsis* (T-K11)			71.11 ± 0.44 [a]	I
M. phaseolina (PUE 4.0) vs. *T. harzianum* (Th-Ah)			59.11 ± 4.23 [bc]	II

* Media followed by the same letter do not present significant statutory differences ($p \leq 0.05$) according to Tukey's test. Means followed by the same letter (a, b and c) are not significantly different for $p \leq 0.05$ according to Tukey test.

The percentage of inhibition of radial growth (PIRG) presents significant differences ($p = 0.0001$). The double confrontation between the different *Trichoderma* species showed an inhibition greater than 50% from the tenth day (Table 2). However, the highest percentage of inhibition was obtained with *T. koningiopsis* (T-K11) obtaining 71.11%. Similarly, *T. harzianum* (T-H3) presented the second-best inhibition with 63.55%; both antagonistic strains presented a class I classification (Figure 3) according to the scale established by Bell et al. [34].

3.4. Evaluation of Antagonism in Field

Field results showed that the treatments with antifungal activity were effective in reducing the incidence of charcoal rot in the root, stem, and pods of "Virginia Champs" variety peanut plants under induced infection (Table 3).

The peanut plants that were inoculated with *M. phaseolina* at the time of transplantation presented root rots at 110 days; necrosis and elongated reddish-brown lesions were observed that covered 40% of the total surface of the main root (Figure 4a). In addition, a considerable loss of secondary roots, with abundant dark mycelium that formed black and rounded microsclerotia (Figure 4a) as well as necrosis in peanut pods, affecting their development, was observed (Figure 4b).

Inoculation with *M. phaseolina* caused a reduction in the height of the plant, presenting significant differences ($p = 0.0056$). It was possible to corroborate that *T. koningiopsis* (T-K11) was more effective in promoting the total growth of the plant, which reached a weight of 1417.60 g as well as reduced the incidence of charcoal rot by 76% (Table 3), compared to *T. harzianum* (T-H3), *T. asperellum* (T-AS1), *T. hamatum* (T-A12), and the strain native of *T. harzianum* (T-Ah).

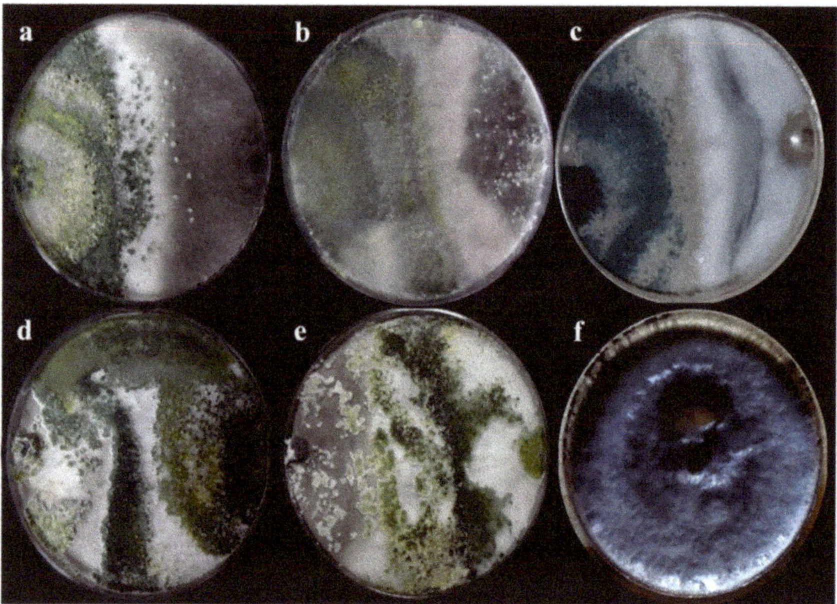

Figure 3. Antagonism of *T. asperellum* (**a**), *T. harzianum* native (**b**), *T. hamatum* (**c**), *T. koningiopsis* (**d**), and *T. harzianum* (**e**) with *M. phaseolina* on the scale by Bell et al. [34] after 132 h in dishes with PDA medium, incubated at 28 °C for 10 days. (**d**,**e**) Class I antagonism; (**a**–**c**) class II antagonism; (**f**) control group of *M. phaseolina*.

Table 3. Antagonistic activity on the incidence of disease, the weight per plant, weight of 100 grains the peanuts, weight, and number of pods.

Treatments	Incidence of Disease	Total Fresh Weight per Plant (g) *	Dry Weight of Pods per Plant (g) *	Number of Pods per Plant *	Weight of 100 Grains (g) *
		M ± SE	M ± SE	M ± SE	M ± SE
M. phaseolina	3	673.20 ± 52.04 [e]	54.80 ± 10.74 [b]	11.00 ± 0.28 [e]	58 ± 0.09 [f]
M. phaseolina vs. *T. hamatum*	2	909.80 ± 62.6 [c]	112.60 ± 14.05 [a]	14.60 ± 0.04 [c]	61 ± 0.02 [e]
M. phaseolina vs. *T. asperellum*	2	970.60 ± 132.71 [bc]	115.60 ± 7.83 [a]	14.88 ± 0.61 [bc]	62 ± 0.04 [d]
M. phaseolina vs. *T. koningiopsis*	1	1417.60 ± 101.61 [a]	124.20 ± 8.60 [a]	16.01 ± 0.71 [a]	64.8 ± 0.01 [a]
M. phaseolina vs. *T. harzianum*	2	1018.60 ± 55.52 [b]	116.20 ± 9.15 [a]	15.35 ± 0.6 [b]	63 ± 0.04 [c]
M. phaseolina vs. *T. harzianum* (native)	2	1007.80 ± 74.28 [bc]	123.00 ± 5.27 [a]	15.13 ± 0.81 [bc]	62.4 ± 0.04 [cd]
M. phaseolina vs. Cercobin®	2	1077.00 ± 112.81 [b]	120.60 ± 18.60 [a]	15.22 ± 0.86 [b]	64 ± 0.03 [b]
Control	0	712.20 ± 105.86 [d]	88.00 ± 7.62 [b]	13.89 ± 0.52 [d]	60.9 ± 0.04 [e]

* Media followed by the same letter do not present significant statutory differences ($p < 0.05$) according to Tukey's test. (n = 100 plants per treatment), M = mean, SE = Standard Error. Means followed by the same letter (a, b, c, d, e and f) are not significantly different for $p \leq 0.05$ according to Tukey test.

Chemical treatment (Cercobin®) presented a 51% reduction of the incidence of charcoal rot without presenting significant differences with *T. harzianum* (T-H3), *T. asperellum* (T-AS1), *T. hamatum* (T-A12), and the strain native of *T. harzianum* (T-Ah) in comparison with plants inoculated only with *M. phaseolina* (Table 3). This can be explained in the following way: the *M. phaseolina* strain is native to the study region and may have acquired resistance towards Cercobin®, which is why the chemical treatment is no longer as effective to control charcoal rot.

Figure 4. Charcoal rot in peanut plants that were inoculated with *M. phaseolina*. (**a**) reddish brown necrosis * present on the surface of the main root with abundant dark mycelium that formed black and rounded microsclerotia on the lateral roots **; (**b**) necrosis in peanut pods.

The potential yield per hectare showed highly significant differences between the treatments ($p = 0.001$), where *T. koningiopsis* (T-K11) showed higher production (1.60 ± 0.01 t/ha^{-1}), presenting 64.04 pods per m^2. The chemical treatment (Cercobin®) was characterized by generating the second highest production, obtaining 60.88 pods per m^2. The lowest production occurred in plants inoculated only with *M. phaseolina* (0.67 ± 0.01 t/ha^{-1}), presenting 44 pods per m^2. The average performance of the yields for the remaining strains of *Trichoderma* spp. oscillate between 1.28 ± 0.01 and 1.38 ± 0.01 ton ha^{-1} (Figure 5). If we consider the average yield per hectare in the region (1.30 t/ha^{-1}), the results reported in the present investigation are superior to the treatments inoculated with biological agents, being able to generate a strategy in the future to control charcoal rot in the rural communities of Buenavista de Benito Juárez, which belongs to the municipality of Chietla in Puebla, Mexico.

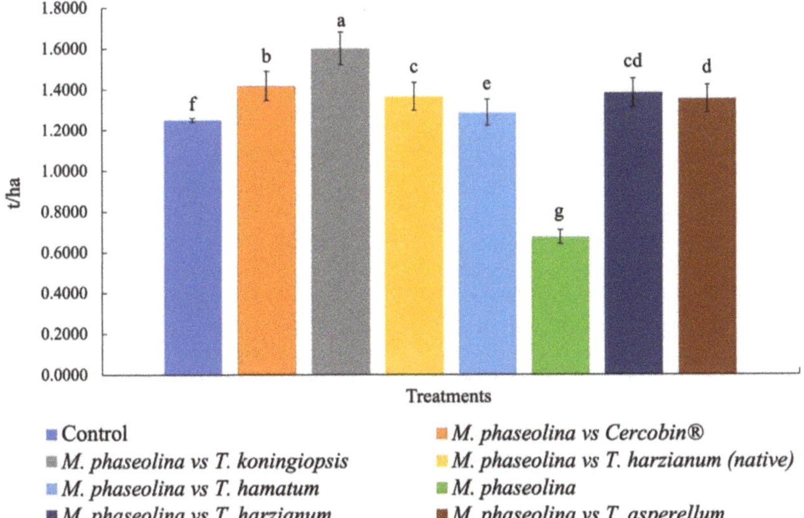

Figure 5. Potential yield (t/ha^{-1}) of peanuts for each treatment in Buenavista de Benito Juárez, which belongs to the municipality of Chietla in Puebla, Mexico. Media followed by the same letter do not present significant statutory differences ($p \leq 0.05$) according to Tukey's test.

4. Discussion

M. phaseolina is considered an important necrotrophic phytopathogen for various crops of agricultural interest, affecting at least 500 plant species in more than 100 families [12,38], where it has been reported to be a causal agent of charcoal rot, affecting the root and stem of peanut crops [9].

According to the morphological characteristics and the amplification of the 5.8S rDNA gene region [39], the identity of "PUE 4.0" strain was confirmed, corresponding to *M. phaseolina*, which coincides with the descriptions reported by Pandey [40] and Márquez [12].

The presence of microsclerotia in the Xylem caused the development of necrotic roots, chlorotic leaves, and premature death of the plants, symptoms that coincide with those reported in strawberry and sunflower crops [41]. This may be due to the presence of phytotoxic metabolites of *M. phaseolina* such as patulin, phaseolinon, and botryodiplodin, which play an important role in the early stages of charcoal rot [42,43]. As far as we know, this is the first report of this pathogen that causes charcoal rot in peanut (*Arachis hypogea* L.) crop in the variety "Virginia Champs" in Puebla, Mexico.

Species of the genus *Trichoderma*, known as green spore fungi, have been widely reported as biological control agents against diseases caused mainly by soil-borne pathogens [33].

In the present study, a parasitism greater than 50% was observed for all *Trichoderma* strains at 240 h against *M. phaseolina*, where there were areas of interaction in the dual culture assays. This may be due to the action capacity of the different *Trichoderma* species, as competition for space and nutrients [18] decreases the growth and development of the strain "PUE 4.0" in vitro.

Cubilla-Ríos [44] evaluated the antagonistic capacity of three *Trichoderma* species: *T. arundinaceum*, *T. brevicompactum*, and *T. harzianum* T34 on two isolates of *M. phaseolina* present in sesame and soybean, and found a greater antagonist activity exerted by *T. harzianum* T34. Likewise, Sreedevi [45] reported that *T. harzianum* reduced growth by 64.4% in *M. phaseolina*, isolated from the root rot of peanuts, results like those obtained in the present investigation with "T-H3" strain from *T. harzianum* (63.55%). According to the literature [46], the properties of *T. harzianum* can be attributed to the production of various volatile substances such as acetaldehyde; isocyanide derivatives, terpenes; derivatives of alpha-pyrone; piperazine, hydrazone derivatives as well as polyketides and alcohols [47]; and responsible for the degradation of the cell wall of fungi. This may be present in the inhibition process against *M. phaseolina* [48].

In Figure 2b,c, an interaction zone surrounding the mycelium of *M. phaseolina* could be observed. It could be rational to infer those volatile substances, enzymes that degrade the cell wall (Chitinase and Glucanase) and a large amount of antibiotics [49,50], exert a greater inhibition on the growth and development of "PUE 4.0" strain. In this sense, the strain of *T. koningiopsis* (T-K11) exerted a greater inhibition of radial growth (PICR) on *M. phaseolina* from the tenth day and a class I classification, according to the scale established by Bell [34]. Ruangwong [51] mentions that *T. koningiopsis* (PSU3-2) contains azetidine, 2-phenylethanol, and ethyl hexadecanoate; compounds that may be associated with antibiosis and suppression of growth on the mycelial of *M. phaseolina* in the present investigation.

The biological control of *M. phaseolina* under field conditions through different isolates of *Trichoderma* reduced the charcoal rot for the peanut crop. In addition, it was possible to observe reddish brown lesions in the total surface of the root in the group of plants inoculated only with "PUE 4.0" strain, characteristic symptoms of *M. phaseolina* mentioned by Ghosh [9] for the peanut crop, and Etebarian [52] for the melon crop.

Of the different microbial antagonists used in the present investigation, *T. koningiopsis* (T-K11) and *T. harzianum* (native) showed more efficient results in reducing charcoal rot and the incidence of the disease. The same finding was found by Hussain [22] and Khan [53], who observed that *T. harzianum* is effective in controlling *M. phaseolina* for mung beans. Likewise, Dubey [54] observed that *T. harzianum* increased the frequency of healthy plants, results similar to those obtained in the present investigation by *T. koningiopsis* (T-K11), which managed to reduce the incidence of charcoal rot by 76% in the field.

If we consider the average of China as the largest peanut producer in the world (1.45 t/ha^{-1}), the average yield obtained in the present investigation with *T. koningiopsis* (T-K11) showed higher production (1.60 $\pm$ 0.01 t/ha^{-1}) [3]. This can be explained by the fact that the genus *Trichoderma* can inhibit the growth of pathogenic fungi and exert positive effects on the absorption of nutrients, plant growth, and yield, in addition to protecting them against biotic and abiotic stress. [55–57]. In a study presented by Osman [58], it was observed that *T. harzianum* alone and combined with yeast improved the yield and quality of peanuts under field conditions. On the other hand, Dania [59], found the highest number of peanut pods per plant (15.67) in a combination of *T. hamatum* and cattle manure, results similar to those obtained in the present investigation by *T. koningiopsis* (16.01) and *T. harzianum* (15.35).

5. Conclusions

It was possible to identify *M. phaseolina* (MW585378 and MW585379) associated with charcoal rot from the crop of peanuts "Virginia Champs" variety, located in the rural communities of Buenavista de Benito Juárez, belonging to the municipality of Chietla in Puebla, Mexico.

The strains *T. koningiopsis* (T-K11) and *T. harzianum* (TH-3) displayed class I of antagonism on the Bell scale. In addition, *T. koningiopsis* (T-K11) displayed the highest rate of development, speed of growth, and percentage of inhibition of PIGR radial growth on *M. phaseolina* (71.11%) under in vitro conditions.

In field conditions, *T. koningiopsis* (T-K11) was more effective in promoting the total growth of the plant reaching a weight of 1417.60 g as well as in reducing the incidence of the disease by 76%.

In the locality of Buenavista de Benito Juárez, belonging to the municipality of Chietla in Puebla, Mexico, *T. koningiopsis* (T-K11) showed higher production (1.60 $\pm$ 0.01 t/ha^{-1}) as well as the chemical treatment (Cercobin®), which obtained the second highest production, obtaining 64 pods per m^2. Lowest production occurred in plants inoculated only with *M. phaseolina* (0.67 $\pm$ 0.01 t/ha) where elongated reddish-brown lesions were observed that covered 40% of the total surface of the main root.

Author Contributions: Conceptualization, O.R.-A., P.A.-H., and S.J.M.-S.; methodology, A.R.-T., S.J.M.-S., and O.R.-A.; software, C.P.L. and O.R.-A.; validation, P.A.-H., A.R.-T., and O.R.-A.; formal analysis, A.L.-C., C.P.L., and O.R.-A.; resources, O.R.-A. and A.R.-T.; original—draft preparation, P.A.-H., S.J.M.-S., and O.R.-A.; writing—review and editing, P.A.-H. and O.R.-A.; visualization, O.R.-A. and A.L.-C.; supervision, P.A.-H.; project administration, O.R.-A.; funding acquisition, O.R.-A. All authors have read and agreed to the published version of the manuscript.

Funding: This research was supported by the program PRODEP 2020 of the Secretaría de educación Pública of Mexico (SEP); the Consejo Nacional de Ciencia y Tecnología (CONACyT) and Benémerita Universidad Autónoma of Puebla.

Institutional Review Board Statement: Not applicable.

Informed Consent Statement: Not applicable.

Data Availability Statement: Informed consent was obtained from all subjects involved in the study.

Acknowledgments: The authors are grateful to Consejo Nacional de Ciencia y Tecnología (CONACyT) and Instituto de Ciencias at Benémerita Universidad Autónoma of Puebla.

Conflicts of Interest: The authors declare no conflict of interest.

References

1. Desmae, H.; Janila, P.; Okori, P.; Pandey, M.K.; Motagi, B.N.; Monyo, E.; Mponda, O.; Okello, D.; Sako, D.; Echeckwu, C.; et al. Genetics, genomics and breeding of groundnut (*Arachis hypogaea* L.). *Plant Breed.* **2018**, *138*, 425–444. [CrossRef]

2.   Ijaz, M.; Perveen, S.; Nawaz, A.; Ul-Allah, S.; Sattar, A.; Sher, A.; Ahmad, S.; Nawaz, F.; Rasheed, I. Eco-friendly alternatives to synthetic fertilizers for maximizing peanut (*Arachis hypogea* L.) production under arid regions in Punjab, Pakistan. *J. Plant Nutr.* **2020**, *43*, 762–772. [CrossRef]

3.   United States Department of Agriculture. Available online: https://www.fas.usda.gov/data (accessed on 8 October 2021).

4.   Food and Agriculture Organization. Available online: https://www.fao.org/faostat/en/#data/QCL (accessed on 8 October 2021).

5.   Servicio de Información Agroalimentaria y Pesquera. Available online: https://www.gob.mx/siap (accessed on 8 October 2021).

6.   Deepthi, K.C.; Reddy, E.N.P. Stem rot disease of groundnut (*Arachis hypogaea* L.) induced by *Sclerotium rolfsii* and its management. *Int. J. Life Sci. Biotechnol. Pharm. Res.* **2013**, *2*, 2250–3137.

7.   Martínez-Salgado, S.J.; Romero-Arenas, O.; Morales-Mora, L.A.; Luna-Cruz, A.; Rivera-Tapia, J.A.; Silva-Rojas, H.V.; Andrade-Hoyos, P. First report of *Macrophomina phaseolina* causing charcoal rot of peanut (*Arachis hypogaea* L.) in Mexico. *Plant Dis.* **2021**. Accepted for publication 18 March 2021. [CrossRef]

8.   Babu, B.K.; Babu, T.K.; Sharma, R. Molecular identification of microbes: I. *Macrophomina phaseolina*. In *Analyzing Microbes*; Arora, D., Das, S., Sukumar, M., Eds.; Springer: Berlin/Heidelberg, Germany, 2013; pp. 93–97. ISBN 978-3-642-34409-1.

9.   Ghosh, T.; Biswas, M.K.; Guin, C.; Roy, P. A review on characterization, therapeutic approaches, and pathogenesis of *Macrophomina phaseolina*. *Plant Cell Biotechnol. Mol. Biol* **2018**, *19*, 72–84.

10.   Basandrai, A.K.; Pandey, A.K.; Somta, P.; Basandrai, D. *Macrophomina phaseolina*–host interface: Insights into an emerging dry root rot pathogen of mungbean and urdbean, and its mitigation strategies. *Plant Pathol.* **2021**, *70*, 1263–1275. [CrossRef]

11.   Lakhran, L.; Ahir, R.R.; Choudhary, M.; Choudhary, S. Isolation, purification, identification, and pathogenicity of *Macrophomina phaseolina* (Tassi) Goid caused dry root rot of chickpea. *J. Pharmacogn. Phytochem.* **2018**, *7*, 3314–3317.

12.   Marquez, N.; Giachero, M.L.; Declerck, S.; Ducasse, D.A. *Macrophomina phaseolina*: General characteristics of pathogenicity and methods of control. *Front. Plant Sci.* **2021**, *12*, 634397. [CrossRef] [PubMed]

13.   Standish, J.R.; Culbreath, A.K.; Branch, W.D.; Brenneman, T.B. Disease, and yield response of a stem-rot-resistant and susceptible peanut cultivar under varying fungicide inputs. *Plant Dis.* **2019**, *103*, 2781–2785. [CrossRef] [PubMed]

14.   Zongo, A.; Konate, A.K.; Koïta, K.; Sawadogo, M.; Sankara, P.; Ntare, B.R.; Desmae, H. Diallel analysis of early leaf spot (*Cercospora arachidicola* Hori) disease resistance in groundnut. *Agronomy* **2019**, *9*, 15. [CrossRef] [PubMed]

15.   Pérez-Torres, E.; Bernal-Cabrera, A.; Milanés-Virelles, P.; Sierra-Reyes, Y.; Leiva-Mora, M.; Marín-Guerra, S.; Monteagudo-Hernández, O. Eficiencia de *Trichoderma harzianum* (cepa a-34) y sus filtrados en el control de tres enfermedades fúngicas foliares en arroz. *Bioagro* **2018**, *3*, 17–26.

16.   Contreras-Cornejo, H.A.; Macías-Rodríguez, L.; del-Val, E.; Larsen, J. Interactions of Trichoderma with plants, insects, and plant pathogen microorganisms: Chemical and molecular bases. In *Co-Evolution of Secondary Metabolites*; Merillon, J.M., Ramawat, K., Eds.; Springer: Cham, Switzerland, 2019; pp. 1–28.

17.   Nawrocka, J.; Szczechy, M.; Malolepsza, U. *Trichoderma atroviride* enhances phenolic synthesis and cucumber protection against *Rhizoctonia solani*. *Plant Protect. Sci.* **2018**, *54*, 17–23. [CrossRef]

18.   Andrade-Hoyos, P.; Silva-Rojas, H.V.; Romero-Arenas, O. Endophytic *Trichoderma* species isolated from *Persea americana* and *Cinnamomum verum* roots reduce symptoms caused by *Phytophthora cinnamomi* in avocado. *Plants* **2020**, *9*, 1220. [CrossRef]

19.   Hernández-Melchor, D.J.; Ferrera-Cerrato, R.; Alarcón, A. *Trichoderma*: Importancia grícola, biotecnológica, y sistemas de fermentación para producir biomasa y enzimas de interés industrial. *Chil. J. Agric. Anim. Sci.* **2019**, *35*, 95–112. [CrossRef]

20.   Khaledi, N.; Taheri, P. Biocontrol mechanisms of *Trichoderma harzianum* against soybean charcoal rot caused by *Macrophomina phaseolina*. *J. Plant Prot. Res.* **2016**, *56*, 21–31. [CrossRef]

21.   Das, I.K.; Indira, S.; Annapurna, A.; Seetharama, N. Biocontrol of charcoal rot in sorghum by Pseudomonas fluorescent associated with the rhizosphere. *Crop. Prot.* **2008**, *27*, 1407–1414. [CrossRef]

22.   Hussain, S.; Ghaffar, A.; Aslam, M. Biological control of *Macrophomina phaseolina* charcoal rot of sunflower and mung bean. *J. Phytopathol.* **1990**, *130*, 157–160. [CrossRef]

23.   García, E. *Modificaciones al Sistema de Clasificación Climática de Köppen*; Serie Libros; Instituto de Geografía UNAM: México, Mexico, 2004; Volume 6.

24.   Morales-Mora, L.A.; Andrade-Hoyos, P.; Valencia-de Ita, M.A.; Romero-Arenas, O.; Silva-Rojas, H.V.; Contreras-Paredes, C.A. Characterization of strawberry associated fungi and in vitro antagonistic effect of *Trichoderma harzianum*. *Rev. Mex. Fitopatol.* **2020**, *38*, 434–449. [CrossRef]

25.   Barnett, H.L.; Hunter, B.B. *Illustrated Genera of Imperfect Fungi*; The American Phytopathological Society: Saint Paul, MN, USA, 1998.

26.   Samson, R.A.; Visagie, C.M.; Houbraken, J.; Hong, S.B.; Hubka, V.; Klaassen, C.H.; Perrone, G.; Seifert, K.A.; Susca, A.; Tanney, J.B.; et al. Phylogeny, identification and nomenclature of the genus *Aspergillus*. *Stud. Mycol.* **2014**, *78*, 141–173. [CrossRef]

27.   Doyle, J.J.; Doyle, J.L. Isolation of plant DNA from fresh tissue. *Focus* **1990**, *12*, 13–15.

28.   Rivera-Jiménez, M.N.; Zavaleta-Mancera, H.A.; Rebollar-Alviter, A.; Aguilar-Rincón, V.H.; García-de los Santos, G.; Vaquera-Huerta, H.; Silva-Rojas, H.V. Phylogenetics and histology provide insight into damping-off infections of 'Poblano' pepper seedlings caused by *Fusarium* wilt in greenhouses. *Mycol. Prog.* **2018**, *17*, 1237–1249. [CrossRef]

29.   White, T.J. *PCR Protocols: A Guide to Methods and Applications*; Academic Press: San Diego, CA, USA, 1990.

30. Juárez-Vázquez, S.B.; Silva-Rojas, H.V.; Rebollar-Alviter, A.; Maidana-Ojeda, M.; Osnaya-González, M.; Fuentes-Aragón, D. Phylogenetic and morphological identification of *Colletotrichum godetiae*, a novel pathogen causing anthracnose on loquat fruit (*Eriobotrya japonica*). *J. Plant. Dis. Prot.* **2019**, *126*, 593–598. [CrossRef]
31. Campbell, M.A.; Li, Z.; Buck, J.W. Development of southern stem canker disease on soybean seedlings in the greenhouse using a modified toothpick inoculation assay. *Crop. Prot.* **2017**, *100*, 57–64. [CrossRef]
32. Zeravakis, G.; Philippoussis, A.; Ioannidou, S.; Diamantopoulou, P. Mycelium growth kinetics and optimal temperature conditions for the cultivation of edible mushroom species on lignocellulosic substrates. *Folia Microbiol.* **2001**, *46*, 231. [CrossRef]
33. Andrade-Hoyos, P.; Luna-Cruz, L.; Osorio-Hernández, E.; Molina-Gayosso, E.; Landero-Valenzuela, N.; Barrales-Cureño, H.J. Antagonismo de *Trichoderma* spp. vs. Hongos Asociados a la Marchitez de Chile. *Rev. Mex. Cienc. Agric.* **2019**, *10*, 1259–1272. [CrossRef]
34. Bell, D.K.; Wells, H.D.; Markham, C.R. In vitro antagonism of *Trichoderma* species against six fungal plant pathogens. *Phytopathology* **1982**, *72*, 379–382. [CrossRef]
35. Perveen, K.; Haseeb, A.; Shukla, P.K. Management of *Sclerotinia sclerotiorum* on *Mentha arvensis* cv. Gomti. *J. Mycol. Plant. Pathol.* **2007**, *37*, 33–36.
36. Bokhari, N.A.; Perveen, K. Antagonistic action of *Trichoderma harzianum* and *Trichoderma viride* against *Fusarium solani* causing root rot of tomato. *Afr. J. Microbiol. Res.* **2012**, *6*, 7193–7197.
37. Zamurrad, M.; Tariq, M.; Shah, F.H.; Subhani, A.; Ijaz, M.; Iqbal, M.S.; Koukab, M. Performance based evaluation of groundnut genotypes under medium rainfall conditions of chakwal. *J. Agric. Food Appl. Sci.* **2013**, *1*, 9–12.
38. Kumar, P.; Gaur, V.K.; Meena, A.K. Screening of different *Macrophomina phaseolina* on susceptible (RMG-62) variety of mungbean. *Inter. J. Pure Appl. Biosci.* **2017**, *5*, 698–702. [CrossRef]
39. Zhang, J.Q.; Zhu, Z.D.; Duan, C.X.; Wang, X.M.; Li, H.J. First report of charcoal rot caused by *Macrophomina phaseolina* on Mungbean in China. *Plant Dis.* **2011**, *95*, 872. [CrossRef]
40. Pandey, A.K.; Burlakoti, R.R.; Rathore, A.; Nair, R.M. Morphological and molecular characterization of *Macrophomina phaseolina* isolated from three legume crops and evaluation of mungbean genotypes for resistance to dry root rot. *Crop. Prot.* **2020**, *127*, 104962. [CrossRef]
41. Hajlaoui, M.R.; Monia, M.H.; Sayeh, M.; Zarrouk, I.; Jemmali, A.; Koikie, S.T. First report of *Macrophomina phaseolina* causing charcoal rot of strawberry in Tunisia. *New Dis. Rep.* **2015**, *32*, 14. [CrossRef]
42. Salvatore, M.M.; Félix, C.; Lima, F.; Ferreira, V.; Naviglio, D.; Salvatore, F.; Duarte, A.S.; Alves, A.; Andolfi, A.; Esteves, A.C. Secondary metabolites produced by *Macrophomina phaseolina* isolated from *Eucalyptus globulus*. *Agriculture* **2020**, *10*, 72. [CrossRef]
43. Khan, S.N. *Macrophomina phaseolina* as causal agent for charcoal rot of sunflower. *Mycopath* **2007**, *5*, 111–118.
44. Cubilla-Ríos, A.A.; Ruíz-Díaz-Mendoza, D.D.; Romero-Rodríguez, M.C.; Flores-Giubi, M.E.; Barúa-Chamorro, J.E. Antibiosis of proteins and metabolites of three species of *Trichoderma* against paraguayan isolates of *Macrophomina phaseolina*. *Agron. Mesoam.* **2019**, *30*, 63–77. [CrossRef]
45. Sreedevi, B.; Charitha-Devi, M.; Saigopal, D.V.R. Isolation and screening of effective *Trichoderma* spp. against the root rot pathogen *Macrophomina phaseolina*. *J. Agric. Technol.* **2011**, *7*, 623–635.
46. Shoaib, A.; Munir, M.; Javaid, A.; Awan, Z.A.; Rafiq, M. Anti-mycotic potential of *Trichoderma* spp. and leaf biomass of *Azadirachta indica* against the charcoal rot pathogen, *Macrophomina phaseolina* (Tassi) Goid in cowpea. *Egypt J. Biol. Pest. Control* **2018**, *28*, 26. [CrossRef]
47. Pitson, S.M.; Seviour, R.J.; McDougall, B.M. B-glucanasa fúngica no celulolítica: Su fisiología y regulación. *Enzym. Microb. Tecnol.* **1993**, *15*, 178–192. [CrossRef]
48. Chang, K.F.; Hwang, S.F.; Wang, H.; Turnbull, G.; Howard, R. Etiology and biological control of Sclerotinia blight of coneflower using Trichoderma species. *Plant Pathol. J.* **2006**, *5*, 15–19. [CrossRef]
49. Khalili, E.; Javed, M.A.; Huyop, F.; Rayatpanah, S.; Jamshidi, S.; Wahab, R.A. Evaluation of *Trichoderma* isolates as potential biological control agent against soybean charcoal rot disease caused by *Macrophomina phaseolina*. *Biotechnol. Biotechnol. Equip.* **2016**, *30*, 479–488. [CrossRef]
50. Woo, S.L.; Scala, F.; Ruocco, M.; Lorito, M. The molecular biology of the interactions between *Trichoderma*, phytopathogenic fungi and plants. *Phytopathology* **2006**, *96*, 181–185. [CrossRef] [PubMed]
51. Ruangwong, O.-U.; Pornsuriya, C.; Pitija, K.; Sunpapao, A. Biocontrol mechanisms of *Trichoderma koningiopsis* PSU3-2 against postharvest anthracnose of Chili Pepper. *J. Fungi* **2021**, *7*, 276. [CrossRef]
52. Etebarian, H.R. Evaluation of *Trichoderma* Isolates for biological control of charcoal stem rot in melon caused by *Macrophomina phaseolina*. *J. Agric. Sci. Technol.* **2006**, *8*, 243–250.
53. Khan, M.R.; Haque, Z.; Rasool, F.; Salati, K.; Khan, U.; Mohiddin, F.A.; Zuhaib, M. Management of root-rot disease complex of mungbean caused by *Macrophomina phaseolina* and *Rhizoctonia solani* through soil application of *Trichoderma* spp. *Crop. Prot.* **2019**, *119*, 24–29. [CrossRef]
54. Dubey, S.C.; Suresh, M.; Singh, B. Evaluation of *Trichoderma* species against *Fusarium oxysporum* f. sp. ciceris for integrated management of chickpea wilt. *Biol. Control* **2007**, *40*, 118–127. [CrossRef]
55. Harman, G.; Howell, C.; Viterbo, A.; Chet, I.; Lorito, M. *Trichoderma* species—Opportunistic, avirulent plant symbionts. *Nat. Rev. Microbiol.* **2004**, *2*, 43–56. [CrossRef]
56. Harman, G.E. Overview of mechanisms and uses of *Trichoderma* spp. *Phytopathology* **2006**, *96*, 190–194. [CrossRef]

57. Tsegaye-Redda, E.T.; Ma, J.; Mei, J.; Li, M.; Wu, B.; Jiang, X. Biological control of soilborne pathogens (*Fusarium oxysporum* f. sp. cucumerinum) of cucumber (*Cucumis sativus*) by *Trichoderma* sp. *J. Life Sci.* **2018**, *12*, 1–12. [CrossRef]
58. Osman, H.A.; Ameen, H.H.; Mohamed, M.; Elkelany, U.S. Efficacy of integrated microorganisms in controlling root-knot nematode *Meloidogyne javanica* infecting peanut plants under field conditions. *Bull. Natl. Res. Cent.* **2020**, *44*, 134. [CrossRef]
59. Dania, V.O.; Eze, S.E. Using *Trichoderma* species in combination with cattle dung as soil amendment improves yield and reduces pre-harvest aflatoxin contamination in groundnut. *J. Agric. Sci.* **2020**, *42*, 449–461. [CrossRef]

Article

Biosolid-Amended Soil Enhances Defense Responses in Tomato Based on Metagenomic Profile and Expression of Pathogenesis-Related Genes

Evangelia Stavridou [1,2], Ioannis Giannakis [3], Ioanna Karamichali [1], Nathalie N. Kamou [4], George Lagiotis [1], Panagiotis Madesis [1,5], Christina Emmanouil [6], Athanasios Kungolos [3], Irini Nianiou-Obeidat [2,*] and Anastasia L. Lagopodi [4,*]

[1] Institute of Applied Biosciences, Centre for Research and Technology Hellas, 57001 Thessaloniki, Greece; estavrid@certh.gr (E.S.); ikaramichali@certh.gr (I.K.); glagiotis@certh.gr (G.L.); pmadesis@uth.gr (P.M.)
[2] Laboratory of Genetics and Plant Breeding, School of Agriculture, Forestry and Natural Environment, Aristotle University of Thessaloniki, 54124 Thessaloniki, Greece
[3] School of Civil Engineering, Aristotle University of Thessaloniki, 54124 Thessaloniki, Greece; iogianna@civil.auth.gr (I.G.); kungolos@civil.auth.gr (A.K.)
[4] Laboratory of Plant Pathology, School of Agriculture, Forestry and Natural Environment, Aristotle University of Thessaloniki, 54124 Thessaloniki, Greece; ngkamou@agro.auth.gr
[5] Laboratory of Molecular Biology of Plants, School of Agricultural Sciences, University of Thessaly, 38221 Volos, Greece
[6] School of Spatial Planning and Development, Aristotle University of Thessaloniki, 54124 Thessaloniki, Greece; chemmanouil@plandevel.auth.gr
*   Correspondence: nianiou@agro.auth.gr (I.N.-O.); lagopodi@agro.auth.gr (A.L.L.)

Citation: Stavridou, E.; Giannakis, I.; Karamichali, I.; Kamou, N.N.; Lagiotis, G.; Madesis, P.; Emmanouil, C.; Kungolos, A.; Nianiou-Obeidat, I.; Lagopodi, A.L. Biosolid-Amended Soil Enhances Defense Responses in Tomato Based on Metagenomic Profile and Expression of Pathogenesis-Related Genes. *Plants* **2021**, *10*, 2789. https://doi.org/10.3390/plants10122789

Academic Editors: Carlos Agustí-Brisach and Eugenio Llorens

Received: 29 November 2021
Accepted: 13 December 2021
Published: 16 December 2021

Publisher's Note: MDPI stays neutral with regard to jurisdictional claims in published maps and institutional affiliations.

Abstract: Biosolid application is an effective strategy, alternative to synthetic chemicals, for enhancing plant growth and performance and improving soil properties. In previous research, biosolid application has shown promising results with respect to tomato resistance against *Fusarium oxysporum* f. sp. *radicis-lycopersici* (Forl). Herein, we aimed at elucidating the effect of biosolid application on the plant–microbiome response mechanisms for tomato resistance against Forl at a molecular level. More specifically, plant–microbiome interactions in the presence of biosolid application and the biocontrol mechanism against Forl in tomato were investigated. We examined whether biosolids application in vitro could act as an inhibitor of growth and sporulation of Forl. The effect of biosolid application on the biocontrol of Forl was investigated based on the enhanced plant resistance, measured as expression of pathogen-response genes, and pathogen suppression in the context of soil microbiome diversity, abundance, and predicted functions. The expression of the pathogen-response genes was variably induced in tomato plants in different time points between 12 and 72 h post inoculation in the biosolid-enriched treatments, in the presence or absence of pathogens, indicating activation of defense responses in the plant. This further suggests that biosolid application resulted in a successful priming of tomato plants inducing resistance mechanisms against Forl. Our results have also demonstrated that biosolid application alters microbial diversity and the predicted soil functioning, along with the relative abundance of specific phyla and classes, as a proxy for disease suppression. Overall, the use of biosolid as a sustainable soil amendment had positive effects not only on plant health and protection, but also on growth of non-pathogenic antagonistic microorganisms against Forl in the tomato rhizosphere and thus, on plant–soil microbiome interactions, toward biocontrol of Forl.

Keywords: biosolid leachates; sludge; *S. lycopersicum* L.; *Fusarium oxysporum*; PR-related genes; defense-related proteins; soil bacteria communities; 16S sequencing; biocontrol

1. Introduction

The fungus *Fusarium oxysporum* f.sp. *radicis-lycopersici* (Forl) is a destructive pathogen limiting crop productivity and causing significant losses in commercial tomato production worldwide [1,2]. Forl is a saprotrophic soilborne pathogen causing tomato foot and root rot disease (TFFR) [2] by intense colonization of the root hair zone and especially the crown of the plant [1]. Due to the resistant nature of all formae speciales of *F. oxysporum*, this fungus is extremely difficult to control with synthetic fungicides [2]. Furthermore, synthetic fungicides may also affect beneficial soil microbiota and may accumulate in the food chain. Therefore, developing alternative, more efficient methods to control Forl is necessary [3,4]. To date, several methods have been used to limit the spread of Forl in field, and greenhouse conditions, such as the use of resistant tomato hybrids and rootstocks, as well as soil disinfection [5], which is a rather expensive practice as it must be repeatedly applied, leading in turn to the increase in hazardous inputs in agriculture.

Alternative methods have been proven to trigger mechanisms of disease control in plants, such as the addition of beneficial antagonistic microorganisms applied in tomato rhizosphere [6], or of a suitable soil conditioner containing beneficial microorganisms, such as green waste compost mixtures or sewage sludge [7–9]. Additionally, soil properties and improvement of plant vitality and growth conditions have also been considered as an important factor for altering plant rhizosphere microorganism composition and thus strengthening plant defense [10]. Studies have shown that fertilization with calcium-rich soil amendments, may enhance soil fertility, strengthen the plant defense systems and thus, suppress pathogen infections [11]. This is facilitated mainly through changes in the soil properties, such as the increase in the soil pH, which favors soil microorganisms (actinomycetes and bacteria) that thrive at such pH values and are competitive with Forl.

Recently, Giannakis et al. [12] reported that sludge-based biosolids enhance tomato growth and reduce TFFR severity. Biosolids may benefit the plant's health by improving the soil properties and enhancing the diversity of the rhizosphere bacterial community [13,14]. Soil microbiota play a key role in soil suppressiveness [15], and studies have shown that they are involved in disease suppression [16,17]. Therefore, microbial community abundance, richness, evenness, and diversity have been identified as key factors involved in community functioning, soil health, and plant productivity [8]. Studies have shown that changes in the rhizosphere microbial community may affect the plant's resistance to different formae speciales of *F. oxysporum* [18–20]. Additionally, there is growing evidence that greater disease suppression is induced by a consortium of plant beneficial bacteria rather than individual strains [21]. However, identifying specific bacterial species and the underlying response mechanisms triggered by the application of biosolids against TFFR have yet to be determined. Hence, further insight into the interactions between the pathogen and the potentially suppressive soil microbiota will provide further insight into the mechanisms underlying the plant–microbe interactions against Forl and how biosolid application may impact this relationship.

Considering that *F. oxysporum* challenges not only tomato, but also several Solanaceae species of great agronomic and economic importance, the interaction between tomato and *Fusarium* has been extensively studied as a model pathosystem for disease resistance response [22]. It is therefore important to further investigate the effects of soil amendment methods on plant–pathogen systems and especially their effect on plant resistance mechanisms through monitoring gene regulation and microbiome diversity in the rhizosphere. The defense system of plants against pathogens involves regulation of gene expression, activation of signaling pathways, hormone balancing, and synthesis of defensive metabolites [23]. An investigation of the transcriptomic profile of tomato plants infected with *F. oxysporum* f. sp. *lycopersici* has revealed up-regulated gene expression related to plant response mechanisms and plant–pathogen interactions, mainly associated with maintenance of cellular structures and homeostasis [24]. Recently, Kamou et al. [4] reported induction of defense gene expression in tomato challenged with Forl. More specifically, salicylic acid (SA)-related genes such as *PR-1a* and *GLUA*, together with jasmonic acid (JA)-related ones such as the *CHI3*, were overexpressed in the presence of the pathogen but also after inoculation with beneficial *P. chlororaphis* ToZa7, revealing induction of variable defense mechanisms. Interestingly, studies have demonstrated the important role of the defense-related phytohormones SA and JA as modulators of the rhizosphere microbiome assembly of plants, such as *Phaseolus vulgaris* and *Arabidopsis thaliana* [25,26]. Moreover, plant's ability to recruit a community of beneficial microbiota and exploit protective rhizosphere processes to their advantage is genotype dependent [27,28].

As a follow-up to the report of Giannakis et al. [12] on the beneficial effects of sludge-based biosolids on tomato growth and TFFR severity, in the present work, we have investigated the plant–pathogen interactions at a molecular level and how they are affected by the addition of biosolids in the soil. More specifically, it was examined whether biosolids: (i) could act in vitro as an inhibitor of growth and sporulation of Forl, (ii) could induce gene expression related to plant response against pathogens in tomato, and (iii) would provide a beneficial substrate for the growth of non-pathogenic antagonistic microorganisms against Forl in the tomato rhizosphere. To achieve this, the relative expression analysis of genes related to defense mechanisms in tomato was analyzed; 16S sequencing analysis of the soil substrates was also performed to determine the genetic diversity and functions of microbial (bacteria and archaea) communities present in the soil substrates, which may have a beneficial effect to the plant and/or suppress the pathogen. This work may elucidate the mechanisms through which biosolid addition enhances plant resistance against pathogens. Furthermore, it is expected to contribute to deciphering the effect of biosolids on soil microbial community to sustainably suppress TFFR disease in tomato crops.

2. Results

2.1. Growth and Sporulation of Fusarium oxysporum f. sp. radicis-lycopersici

Colony diameter (cm) of Forl increased in the different mixtures of biosolid leachates with PDA compared to the control (Figure 1A). This increase was greater with elevating leachate concentration. As such, the concentrations of 5 and 10% leachate showed the largest colony diameter (an increase of 40% was noted in relation to control) followed by the 2% leachate, where an increase of 28% in colony diameter was noted in relation to control. Sporulation was also significantly affected by the biosolid leachate; the number of fungal spores increased significantly at 10% concentration of leachate (Figure 1B). Hence, the number of conidia produced per cm^2 of colony was increased with the increase in the leachate concentration (Table 1).

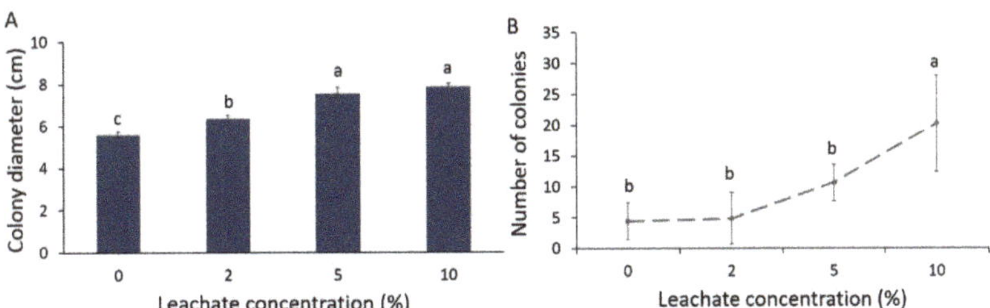

Figure 1. Growth and sporulation of *Fusarium oxysporum* f. sp. *radicis-lycopersici* (Forl) at 0, 2, 5, and 10% biosolid leachate concentrations in PDA. (**A**) Mean colony diameter; (**B**) Mean sporulation. Different letters indicate statistically significant differences between the treatments according to Tukey's post hoc test ($p < 0.05$).

Table 1. Mean number of *Fusarium oxysporum* f. sp. *radicis-lycopersici* conidia produced per cm^2 of colony at different biosolid leachate–PDA concentrations in PDA. Different letters indicate statistically significant differences between the treatments according to Tukey's post hoc test ($p < 0.05$).

Leachate Concentration (%)	Mean Initial Number of Conidia Per cm^2	Tukey's Post Hoc Test
0	$(176 \pm 119) \times 10^3$	b
2	$(184 \pm 166) \times 10^3$	b
5	$(404 \pm 119) \times 10^3$	b
10	$(808 \pm 314) \times 10^3$	a

2.2. Gene Expression Analysis

Expression patterns of defense-related genes (*GLUA, CHI3, PR1-a, LOX,* and *AOC*) following application of biosolid and inoculation with Forl were analyzed using RT-qPCR (Figure 2). The induction patterns were evaluated in tomato plants at 12, 24, 48, and 72 h after inoculation. As internal control the reference gene *β-actin* was used. A significant upregulation of *LOX* was observed at 12 h in response to Forl inoculation and the addition of biosolid (FB) as compared to the treatments C, F, and B (Figure 2A). Nevertheless, in the later time points, *LOX* was not significantly induced (Figure 2B–D). Interestingly, at 12 h an increase in the *AOC* expression levels was also observed after application of biosolid and Forl inoculation (FB), yet this was not significant in comparison to the other treatments (Figure 2A). However, at 12 h, no induction in *PR1-a, GLUA,* and *CHI3* was observed for the different treatments in relation to the control (C) (Figure 2A). At 24 h after inoculation, only *GLUA* was significantly induced in the FB treatment, whilst the *PR1-a* gene was significantly overexpressed only under biosolid application (B), yet a non-significant increase in the expression levels was also observed under FB treatment (Figure 3B). *AOC* was significantly induced at 48 and 72 h after biosolid application (Figure 2C,D). At 72 h (Figure 2D) overexpression of 2.45-, 5.5-, 3.76-, and 2.45-fold was observed for *CHI3, AOC, PR1-a,* and *GLUA,* respectively, under biosolid application (B), but not after inoculation with Forl (FB) when compared with the untreated control.

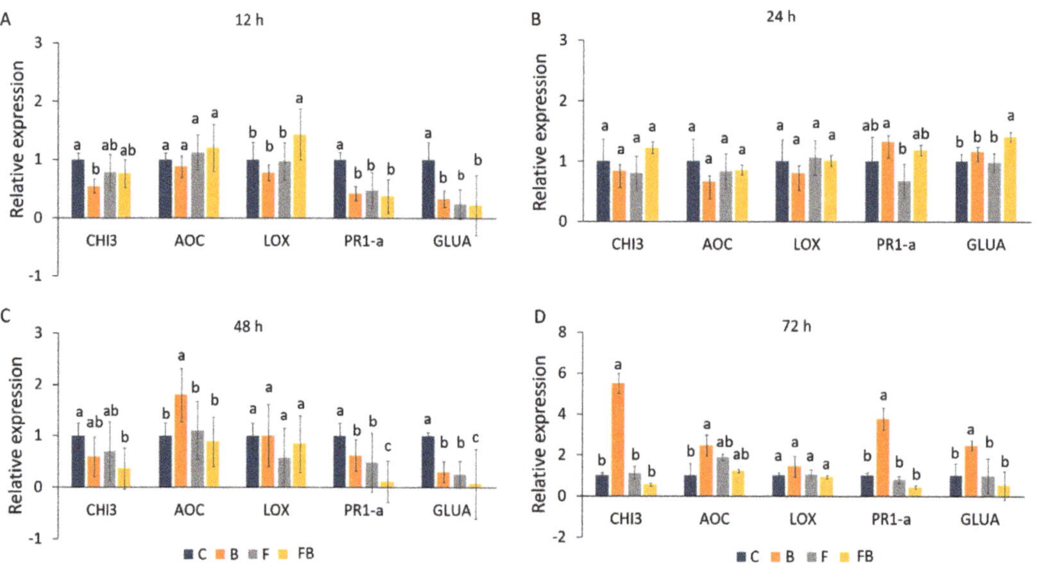

Figure 2. (**A–D**) Relative gene expression analyses of defense-related genes *GLUA, CHI3, PR1-a, LOX,* and *AOC,* in tomato

plant leaves in (**A**) 12 h, (**B**) 24 h, (**C**) 48 h, and (**D**) 72 h post inoculation in control (C), biosolid−enriched treatment (B), Forl inoculation (F), and Forl inoculation with biosolid application (FB). Different letters indicate statistically significant differences between the treatments according to Tukey's post hoc test ($p < 0.05$).

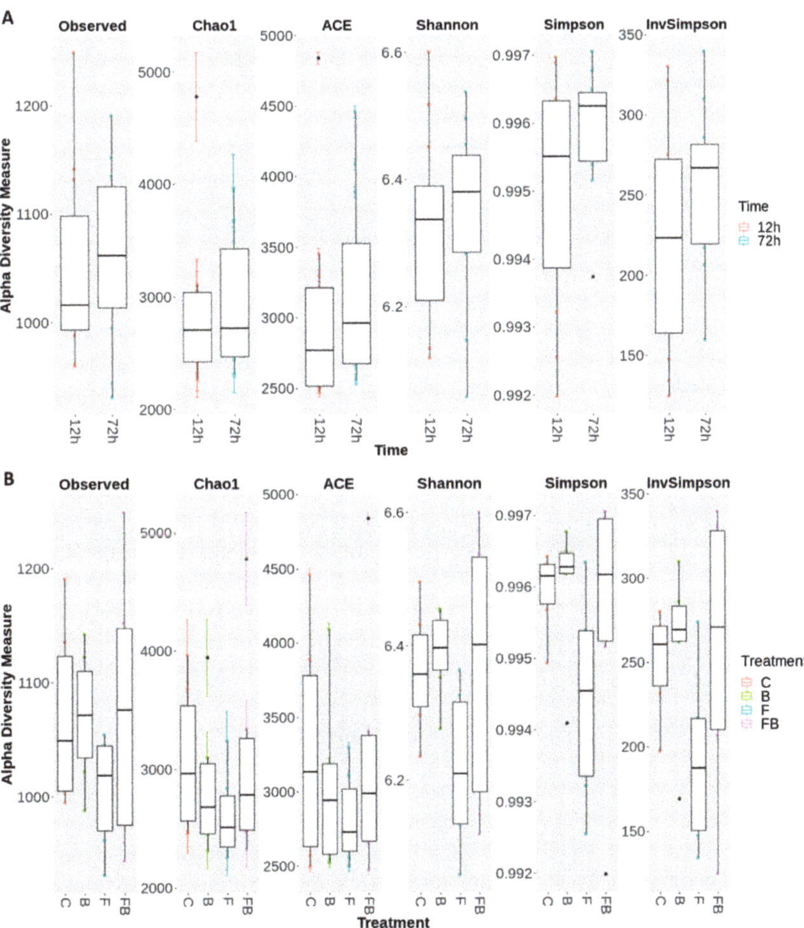

Figure 3. Boxplot demonstrating the range and the distribution of the of OTUs, evaluated using the α-diversity indices observed richness, Chao1, ACE, Shannon, Simpson, and InvSimpson, (**A**) 12 and 72 h after inoculation with *Fusarium oxysporum* f. sp. *radicis-lycopersici*; (**B**) for the treatments for the treatments control (C), biosolid−enriched treatment (B), Forl inoculation (F), and Forl inoculation with biosolid application (FB). Values represent the pooled mean of three replicates.

2.3. Characterization of Microbial Communities in the Different Soil Substrates

Based on the rank abundance curves for the top 100 OTUs (Figure S1) an even abundance was observed between the 12 and 72 h (Figure S1A). However, the treatments without the addition of biosolid, i.e., F and C, showed unevenness of abundance in soil bacterial communities due to the large difference in detected organisms, whereas in the treatments where biosolid was added (B and FB), species evenness was more homogeneous (Figure S1B).

Microbial diversity was evaluated using the following indicators: observed richness, Chao1, ACE, Shannon, Simpson, and InvSimpson (Figures 3A and 4B). The alpha diversity indices were notably increased (or did not significantly change) in soil substrate samples at 72 h, indicating that the shift pattern of the microbiome occurred at a later timepoint (Figure 3A). Respectively, treatments enriched with biosolid, regardless of the presence or absence of Forl inoculum (FB and B, respectively) showed greater microbiome diversity in α-diversity indices (Observed richness, Shannon, and Simpson), whilst the treatment inoculated with Forl (F) showed a lower species evenness (Figure 3B), indicating that the relative abundances of species within the community vary in distribution.

To further relate bacterial community variance to the different variables, such as the effect of time post Forl inoculation (F and FB) and biosolid application (B and FB) and the treatments, Canonical Correspondence Analysis was performed (Figure 4). The variation explained by the treatments and timepoints (constrained ordination) was 22.8% and the remaining 77.2% of the variation was explained by the unconstrained ordination (Table 2). From the eigenvalues of the constrained axes, it was observed that 45.65% of the variation is explained by the CCA1 and the 19.1% by CCA2 (Table 3). The factors responsible for most of the explained variation in the bacterial community were the treatments enriched with biosolid (B and FB), which were differentiated from those inoculated with Forl without the addition of biosolid (F) at 72 h (Table 4). Therefore, the addition of biosolid, with or without Forl inoculum, affected differently the bacterial community abundance compared to the control and Forl treatments.

Table 2. Proportion of inertia explained by constrained and unconstrained ordination.

	Inertia	Proportion
Total	3.9869	1
Constrained	0.9103	0.2283
Unconstrained	3.0766	0.7717

Table 3. Accumulated constrained eigenvalues.

Importance of Components:	CCA1	CCA2	CCA3	CCA4
Eigenvalue	0.4155	0.1738	0.1681	0.1529
Proportion Explained	0.4565	0.1909	0.1847	0.1679
Cumulative Proportion	0.4565	0.6474	0.8321	1

Table 4. Biplot scores for constraining variables.

Factors	CCA1	CCA2
Time 72 h	−0.1404	−0.7527
Treatment B	−0.5594	−0.2868
Treatment F	0.5453	0.3655
Treatment FB	−0.5831	0.3844

Composition of bacterial populations at the phylum and class taxonomic levels were evaluated using the β-diversity analysis for each treatment and time point tested. Overall, 38 and 94 different phyla and classes, respectively, were identified across the treatments in the two timepoints, yet relative abundance of only 13 phyla and 23 classes with OTUs representation >1% are shown in Figures 5 and 6, respectively. The most abundant amplicons were identified as *Proteobacteria, Bacteroidetes, Actinobateria, Acidobacteria, Gemmatimonadetes, Chloroflexi,* and *Planctomycetes.* The biosolid-enriched treatments (B and FB) were characterized by higher abundance of amplicons identified as *Chloroflexi* and *Bacteroidetes,* which increased with time (Figure 5, Table S1). Moreover, the phylum *Patescibacteria,* was detected only in biosolid-enriched treatments (B and FB) in both time points (Figure 5, Table S1). The phylum *Synergistetes* was initially observed in treatment B at 12 h and 72 h and in FB only at 72 h. In contrast, *Acidobacteria* and *Actinobacteria* were characterized by lower abundance in the biosolid-enriched soils (B and FB) compared to the control (C) and Forl (F) treatments (Figure 5, Table S1).

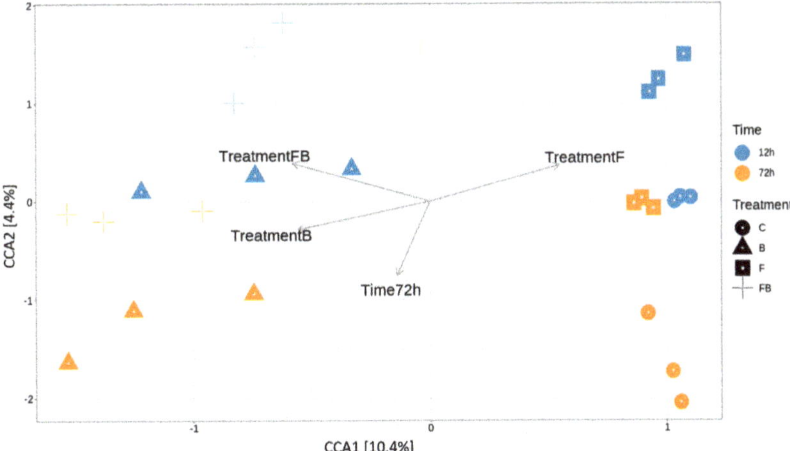

Figure 4. Canonical correspondence analysis (CCA) of the relative variance of OTUs in the soil samples subjected to the different control (C), biosolid−enriched treatment (B), Forl inoculation (F), and Forl inoculation with biosolid application (FB) treatments at 12 and 72 h. The biplots present the effect of biosolid application to the bacterial community abundance in the treatments with or without fungal inoculum at two different timepoints.

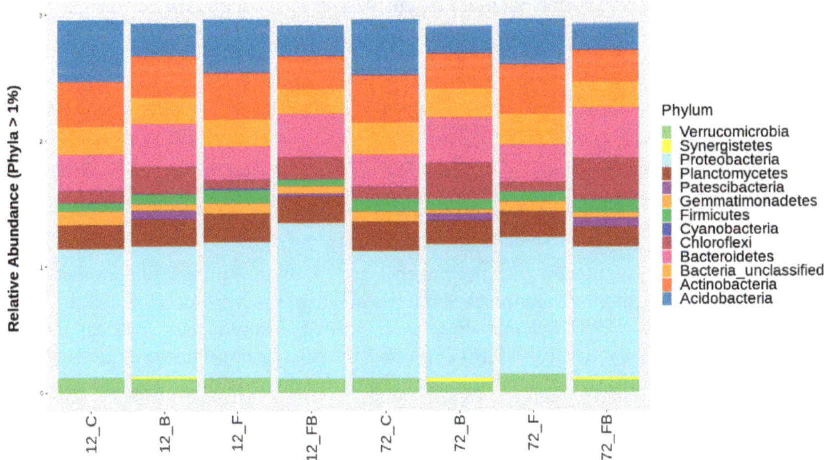

Figure 5. Classification of bacterial phyla based on relative abundance at rate greater than 1%, for the control (C), biosolid−enriched treatment (B), Forl inoculation (F), and Forl inoculation with biosolid application (FB) treatments, at 12 and 72 h after inoculation with *Fusarium oxysporum* f. sp. *radices-lycopersici*. The different phyla are depicted with different colors. Value is the pooled mean of three replicates.

At the bacterial class level, the treatments enriched with biosolid (B and FB) showed higher relative abundances of *Gammaproteobacteria*, *Bacteroidia*, *Anaerolineae*, *Deltaproteobacteria*, and *Acidimicrobiia* in both time points (Figure 6, Table S2). Interestingly, *Clostridia* were detected solely in biosolid-enriched soils and increased with time. A similar trend was observed for *Saccharimonadia* and *Synergistia*, yet these were not present in FB soils at 12 h. In contrast, the classes *Alphaproteobacteria*, *Actinobacteria*, *Planctomycetacia*, *Verrucomicrobiae*, *Acidobacteriia*, *Phycisphaerae*, *Thermoleophilia*, and *Gemmatimonadetes* were found in higher abundance in control soils (C) and soils with Forl inoculum (F) compared to biosolid-enriched treatments (Figure 6, Table S2). Unclassified actinobacteria were solely detected in C and F treatments. However, in the C treatment the call abundance was reduced over time and in F treatment increased over time. The class *Blastocatellia* (Subgroup 4) showed higher abundance under C conditions and was reduced with either biosolid application and/or Forl inoculation. Soils in C and F treatments showed higher abundance of *Bacilli*, compared to the soils treated with biosolid (B) (Figure 6, Table S2). However, at 72 h, the abundance of this class increased in C treatment, yet it was reduced in B, F, and FB treatments (Figure 6, Table S2). The class *Oxyphotobacteria* was only present in F treatment and *S0134_terrestrial_group* in C treatment at 12 h (Figure 6, Table S2).

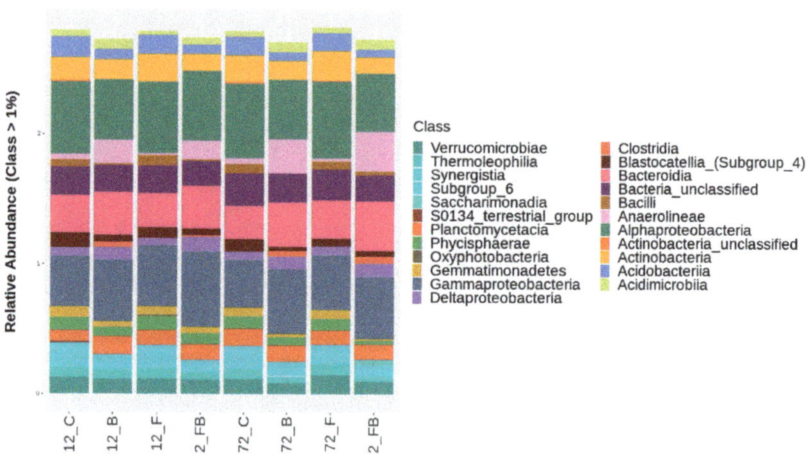

Figure 6. Classification of bacterial classes based on relative abundance at rate greater than 1%, for the treatments control (C), biosolid—enriched treatment (B), Forl inoculation (F), and Forl inoculation with biosolid application (FB) at 12 and 72 h after inoculation with *Fusarium oxysporum* f. sp. *radicis-lycopersici*. The different classes are depicted with different colors. Value is the pooled mean of three replicates.

2.4. Predicted Functional Diversity of the Microbiome Present in the Different Soil Substrates

The functional potential encoded by the soil microbial communities showed significant differences in the bacterial metabolic pathways among the treatments. The functional profiles from the 16S sequences revealed different metabolic capacity between the microbiome in biosolid-enriched and the non-biosolid-enriched soils (Figure S1), with the biosolid application explaining the 81.5% (PC1) of the functional variation and time only 12.2% (Figure S2). Out of 445 pathways, only 98 pathways passed the filter of $p < 0.05$ with an effect size >0.85, and 53 pathways with an effect size >0.9 (Figure 7; Table S3). Biosolid-enriched treatments B and BF showed higher abundance in sequences assigned to metabolic pathways such as carbohydrate biosynthesis (gluconeogenesis), vitamin biosynthesis (folate and vitamin B6) (Figure S3), electron carrier biosynthesis (quinol and quinone; menaquinol) (Figure S3), fermentation (pyruvate), and nucleic acid processing (Table S3). Lower abundances in sequences of B and BF treatments were attributed to pathways such as aromatic compound degradation, autotrophic CO_2 fixation, CMP-sugar biosynthesis, fatty acid and lipid degradation, and nitrogen compound metabolism (Table S3). High abundance of sequences in biosolid-enriched soils with Forl inoculation (BF treatment) were also assigned to metabolic pathways of vitamin biosynthesis (folate transformations III) and pathways with lower abundance, such as aromatic compound biosynthesis (chorismate), fatty acid and lipid biosynthesis, and polysaccharide biosynthesis (Table S3).

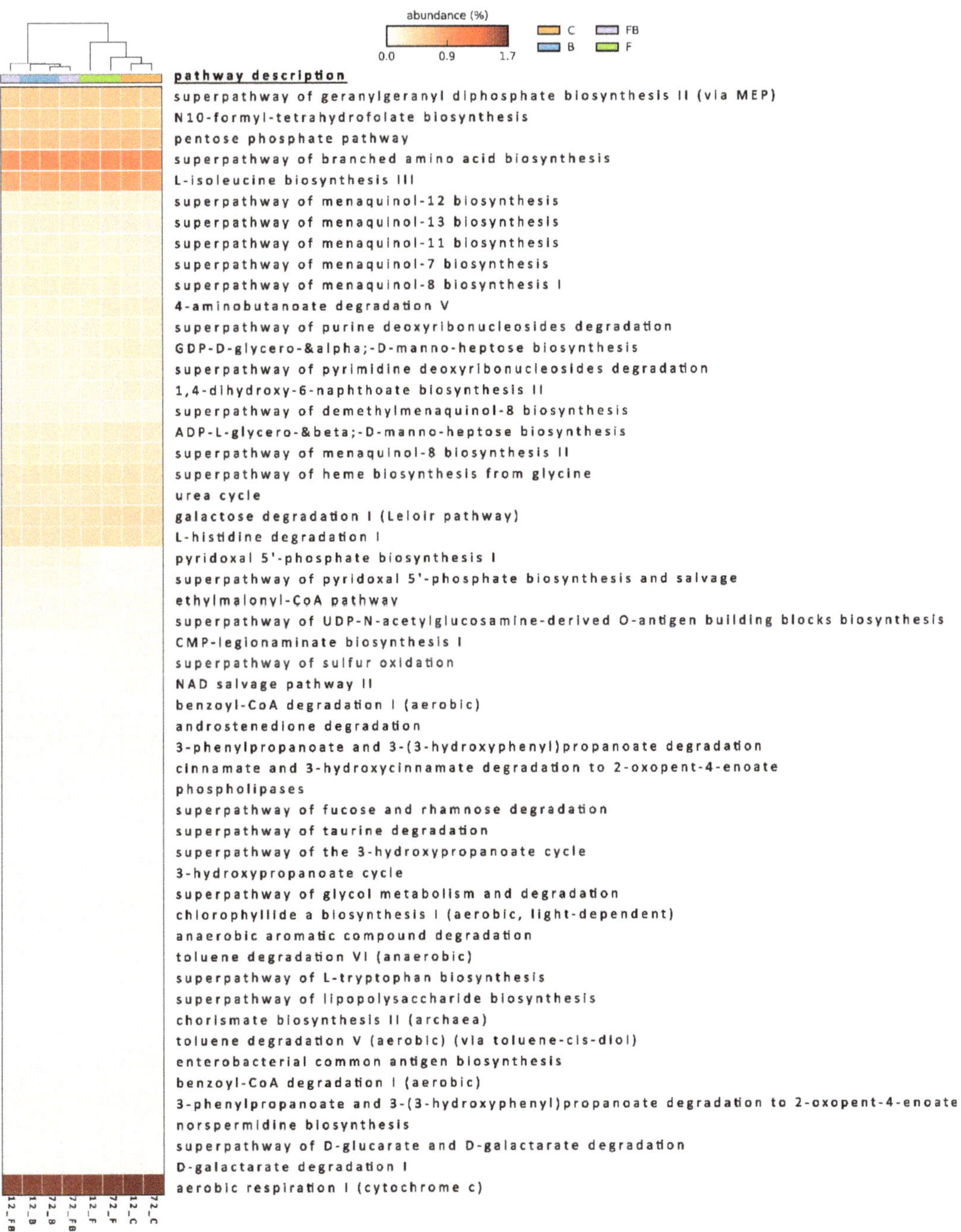

Figure 7. Heatmap of the predicted functional profile for the control (C), biosolid (B), Forl inoculation (F), and combination of Forl inoculation and biosolid application (FB) analyzed using STAMP software. The key shows the % relative abundances for *p*-value ≤ 0.05 and effect size >0.85 (n = 3).

Several common pathways were observed between control (C) and Forl inoculated (F) treatments, such as aerobic respiration, S-adenosyl-L-methionine biosynthesis, tetrapyrrole biosynthesis, nucleoside, and nucleotide degradation (purine), proteinogenic amino acid biosynthesis, sugar derivative degradation, sugar nucleotide biosynthesis and sugar degradation (galactose), terpenoid biosynthesis, nitrogen compound metabolism, menaquinol biosynthesis, heme b biosynthesis, and TCA cycle (Table S3). However, Forl inoculation treatment also had higher abundance of sequences attributed to carboxylate (sugar acid) and secondary metabolite (sugar derivative; sulfoquinovose) degradation (Table S3).

3. Discussion

Biosolids have been characterized as organic soil amendments, supplying the soil with nutrients and organic compounds and contributing to soil moisture and aeration [29–31]. However, there is limited research related to the molecular mechanisms associated with biosolid-elicited suppression of soilborne diseases, such as TFFR and the enhanced plant performance. In a previous work, biosolid application in the soil has been shown to enhance tomato growth and reduce the effects of Forl infection [12]. In the present work, a molecular approach was employed to provide further insight into the mechanistic effect of biosolid application in alleviating the negative impact of Forl on tomato plants. To achieve this, gene expression profiles related to plant response against pathogens coupled with 16S metagenome profile analysis were used to determine the genetic diversity and functions of bacterial communities present in the soil substrates and their potential impact on plant–pathogen relationships.

Biosolid application can enhance tomato tolerance against the Forl [12], which was mainly attributed to the indirect beneficial effects of biosolid application on biotic and abiotic factors [31]. Although growth and sporulation of Forl in the presence of biosolids were not studied in planta, it would have been expected that biosolids could suppress the growth and sporulation of Forl, thus reducing its aggressiveness against tomato. In contrast, based on the growth analysis and sporulation of Forl under in vitro conditions, herein, an increase in colony diameter, the number of fungal spores of Forl, and the number of conidia produced per cm^2 of colony was observed in response to the increasing concentration of biosolid-PDA leachates, compared to the control. Such positive effect of biosolid leachates on the colony growth and spore production in vitro could be explained by the fact that various biosolid leachates contained significant amounts of calcium (Ca) and magnesium (Mg) [32]. PDA used for the growth of Forl is a universally used medium providing the necessary nutrients for the growth of fungi for laboratory purposes. However, nutritional requirements in macro- and microelements vary among different fungal species [33]. Magnesium is considered a macro element necessary for enzyme activation and ATP metabolism. On the other hand, Ca is generally accepted as a micronutrient required for enzyme activity and membrane structure in fungi [33] and plays a key role in hyphal tip growth [34]. It is assumed that nutrients provided in the leachates exert positive effects on hyphal growth and production of conidia in Forl. However, this calls for further investigation, as the nutritional requirements for the improvement of growth and sporulation has not been studied in this fungus.

Nevertheless, biosolid application in soils induced an upregulation of defense-related genes in tomato plants post Forl inoculation. Increase in transcript levels of defense-related genes indicate activation of the tomato response mechanisms against the pathogen [35–37]. The observed increase in mean transcript levels of *LOX*, *AOC*, and *GLUA* in the biosolid-enriched soils (B and FB) as compared to the non-amended soils (C, F), indicated that biosolid application may play a key role in the early (12 and 24 h) activation of the tomato response mechanisms against Forl. Activation of the ethylene (ETH) and jasmonic acid (JA)/ETH signaling pathways even at 72 h in FB treatment may indicate an attempt to limit pathogen progression under the effect of biosolids. JA elicitor is a signaling molecule involved in various plant developmental processes and defense mechanisms [38]. The role of JA pathway in protection against Forl was confirmed in tomato after biochar application [39]. More specifically, biochar application induced upregulation of the pathways and genes associated with plant defense and growth such as JA, yet biosynthesis and signaling of the salicylic acid (SA) pathway was downregulated (Jaiswal et al., 2020). In contrast, herein, genes involved in the SA biosynthesis and signaling pathways, such as *GLUA* and *PR1-a*, were upregulated in the biosolid-enriched treatments (B and FB). However, the *PR1-a* gene was significantly upregulated only under biosolid application (B) at early time points and only at 72 h in both B and FB treatments. PR proteins are elicited in many plant species by the attack of different pathogens. Plants inoculated with Forl (F) did not show any significant upregulation of the analyzed defense-related genes, which is consistent with other studies showing a delayed induction of defense-related gene expression only 2–3 weeks post inoculation [40–42]. This suggests that either Forl needs to be in close association with tomato roots for upregulation of *PR1-a* to take place, that the accumulation of PR proteins requires longer period post inoculation [42], or even that the selected genotype was rather tolerant to Forl [43]. It has been shown that Forl first interacts with tomato roots 48 h after inoculation [1]. In addition, it has been shown that the developmental stage seems to play an important role in the induction of resistance genes [35].

Nevertheless, genes associated with the JA and SA biosynthesis and signaling pathways such as *CHI3*, *AOC*, *PR1-a*, and *GLUA* were upregulated 2.45-, 5.5-, 3.76-, and 2.45-fold, respectively, at 72 h in biosolid without Forl (B), but not after inoculation with Forl (FB), indicating that the plant's innate defense mechanisms were induced even without the presence of the pathogen. Therefore, such effects imply indirect interactions possibly via the induction of systemic acquired resistance (SAR) [44,45], given that SAR is mediated by pathways that are dependent on, but not only to, SA, JA, and ethylene [46,47]. Overall, the upregulation of genes involved in plant defense and plant growth may indicate defense priming that could explain the significant improvement in plant performance and Forl suppression observed in the presence of biosolid.

Elicitors of biosolid-mediated plant defenses include chemical compounds that are beneficial to the plant along with biosolid-induced microorganisms with potentially direct antagonistic effects towards Forl [4,35,36]. The potential induction of systemic acquired resistance in plants by compost mixtures, used as soil conditioners in plant growth substrates, has been demonstrated in several studies [48,49]. Therefore, it was hypothesized that these indirect plant defense mechanisms could also be induced by other biodegradable and non-composted materials, such as the anaerobically digested biosolid used in the present study.

Enhanced tomato performance against Forl could also be attributed to different types of interactions between the phytopathogenic fungus and the beneficial to the plant soil microflora, which is induced by biosolid application. Such interactions include competition for nutrients, production of antifungal metabolites, parasitism, and enzymatic hydrolysis of fungal cell walls [50–52]. Although the interaction mechanisms between microbial communities and plants are very complex, intense microbial diversity usually has a beneficial effect on both plant diversity and their growth and productivity [48,53,54]. Most of the microorganisms in the soil and the rhizosphere develop beneficial and mutual symbiotic relationships with the plant and among other microorganisms [55].

The bacterial biodiversity observed in the biosolid-enriched treatments, seems to have had a beneficial effect on both growth of the tomato plants and protection against Forl. The relation of the relative abundance of bacterial groups and pathogen inhibition along with development of disease suppression in soils has been recently demonstrated [16,17]. Herein, four phyla of bacteria were identified in greater abundance in biosolid-enriched soils, namely *Chloroflexi* (class *Anaerolineae*), *Bacteroidetes* (class *Bacteroidia*), *Patescibacteria* (class *Saccharimonadia*), and *Synergistetes* (class *Synergistia*). Interestingly, *Clostridia* (phylum *Firmicutes*), were detected solely in biosolid-enriched treatments and increased with time. Several genera within the phylum *Firmicutes*, have been shown to positively affect disease suppressiveness of soil amendments, such as compost [56]. *Clostridia* occur mainly in the rhizosphere and perform beneficial functions for the plants, such as atmospheric nitrogen fixation, phosphate solubilization and the reduction of Fe^{3+} to the more readily available iron form Fe^{2+} [57]. The higher abundance of *Bacilli* class was increased at 72 h under control conditions, yet it was reduced in B, FB, and F treatments. Bacillus species have exhibited ability for plant growth promotion [58] and have also demonstrated the ability to excrete exopolysaccharides, biosurfactants, and chelating agents, which are important for the remediation of heavy metals from soils [59]. *Bacillus* species have also demonstrated broad functions, especially in various enzymatic activities [60], indicating a possible role in pathogen control [61,62]. Nevertheless, the reduced abundance over time in the biosolid-enriched treatments indicate that possibly the presence or absence of mineral fertilization affects the structure of the bacterial community in the soil [63]. A possible explanation could be the acidification of soil by higher NPK content [64] present in the sewage sludge [65] and therefore the biosolid applied herein [12].

Other phyla of anaerobic bacteria, such as *Patescibacteria* and *Synergistetes*, that were observed mainly in the biosolid enriched treatment (B) have been shown to contribute to the degradation of organic matter [66,67]. Bacterial phyla such as *Chloroflexi* (class *Anaerolineae*) thrive under anaerobic conditions and are considered to play an important role in the vital process of photosynthesis [68]. In addition, these bacteria can degrade a large number of organic compounds and producing acetic acid [69,70], which has been shown to stimulate plant growth and inhibit the growth of pathogenic fungi [71,72].

The phylum *Proteobacteria* has a key role in anaerobic digestion by metabolizing volatile fatty acids [73]. In addition, bacteria in this phylum are known to remove a broad range of synthetic, as well as natural organic pollutants [74,75]. Within this phylum, the class *Alphaproteobacteria* were in higher abundance in C and F treatments compared to the biosolid-enriched ones, whereas *Gammaproteobacteria* were in higher abundance in B (increased over time) and FB (decreased over time) treatments. This indicates that different classes of the same phylum are affected differently by the addition of biosolid in the soils and in response to Forl, and further supports the evidence that consortia of beneficial microorganisms, rather than specific taxa, may drive disease suppression and lead to plant protection [76].

Similarly, we observed changes in abundance of *Actinobacteria*, *Firmicutes*, and *Acidobacteria* with time among the different treatments (Table S1). The higher abundance of the classes *Actinobacteria* and *Acidobacteria* have been associated with antagonistic activity toward several phytopathogens, such as *Fusarium* [77], and with suppressiveness in compost [78]. Additionally, high abundance in members of bacterial phyla, such as *Actinobacteria*, *Firmicutes*, and *Acidobacteria*, have been shown to directly antagonize pathogens through various mechanisms [17]. The example of *Firmicutes* abundance being increased over time under control (C) and Forl inoculation with biosolid-enriched soil (FB) treatment, remaining unchanged in the B treatment, and being reduced in the Forl inoculation (F) treatment indicates not only the presence, but also the change of abundance over time is significant. Additionally, *Acidobacteria* decreased in the B, F, and FB treatments and *Actinobacteria* increased in F, decreased in B, and remained unchanged in FB treatments. On the other hand, the abundance of *Actinobacteria*, *Firmicutes*, and *Acidobacteria* in the control treatment, where plants and microorganisms in the rhizosphere were undisturbed, was increased. These observations suggest that: (i) possibly different consortia of beneficial microorganisms, rather than specific species, may provide plant protection against Forl by suppressing the pathogen, which is in accordance with other studies [21,79], and (ii) the relative change (in the concept of increase or decrease) in abundance of these consortia over time may also play a regulatory role in the biocontrol of Forl.

Functional analysis was further performed to acquire information about the potential community functions. Different pathways associated with various aspects of metabolism were activated in biosolid-enriched (B and FB) compared to the non-enriched (C and F) treatments, indicating the impact of biosolid application on various microbial functions. For instance, in the secondary metabolism, vitamin biosynthesis pathway and especially folate biosynthesis pathway were activated in biosolid enriched soils. Potentially, the abundance of nutrients in the biosolid-enriched soil may have enhanced the biosynthesis of plant secondary metabolites, such as folates (reviewed by Kołton et al. [80]). Interestingly, studies have highlighted the importance of folates in inducing plant tolerance to several biotic and abiotic stresses [81,82]. Folates in plants are involved in redox homeostasis, physiological processes, epigenetic regulation, cell proliferation, and mitochondrial respiration, as well as photosynthesis [80,83,84], and have been shown to have antifungal functions [85].

The menaquinone (Vitamin K2) biosynthesis pathway was also activated in B and BF treatments. Menaquinones are involved in bacterial electron transport and in sensing environmental changes such as alterations in redox state; they have also been implicated in sporulation and proper regulation of cytochrome formation in all Gram-positive bacteria and anaerobically respiring Gram-negative bacteria [86–88]. In plants, vitamin K functions as a priming agent against biotic and abiotic stresses given its redox properties. Menadione (pro-vitamin K) was found to induce resistance by priming in Arabidopsis against the virulent strain *Pseudomonas syringae* pv. tomato DC3000, with more than two-fold PR1 expression in MSB-pretreated plants as compared to non-treated plants [89]. Similarly, in our study, an increase in PR1-a expression by 3.76-fold was observed 72 h in the B treatment, but not in the BF treatment, indicating the potential priming effect of biosolid in tomato plants.

The autotrophic CO_2 fixation pathway found in the biosolid-enriched treatments indicated the presence of autotrophic microorganisms, which contribute significantly to CO_2 fixation in the soil carbon sink of agricultural soils [90]. Other pathways activated in the biosolid-enriched treatments include the carbohydrate biosynthesis (gluconeogenesis) indicative of sugar synthesis [91], which is the primary source of energy for all eukaryotic organisms [92]. Specifically, in plants, they are involved in most metabolic and signaling pathways controlling growth, development, and fitness [92]. Additionally, in the biosolid-enriched treatments, sequences were attributed to the nitrogen metabolism pathway which, according to Jacoby et al. [93], when active in the rhizosphere, is an indicator that the microbiome plays an important role in mediating plant nutrition. Recent studies have shown that amino acids play a key role in plant root growth and microbial colonization, symbiotic interactions, and pathogenesis in the rhizosphere [94]. Amino acids are considered a key intermediary in the soil nitrogen cycle, and function as carbon and nitrogen sources for both microorganisms and plants, in synthesis and regulation of auxin activity and biofilm formation and disassembly [95]. Therefore, the nucleic acid processing for protein synthesis pathway, which was found active in biosolid-enriched soils, could be an indicator of intrinsic amino acid biosynthesis.

Control and Forl treatments shared common pathways that were active under such conditions including: aerobic respiration and the related TCA cycle pathways, important processes in the global carbon cycle and of crucial importance in the partitioning of energy in soil [96]; the tetrapyrrole biosynthesis (TBS) and heme b biosynthesis pathways, important for oxidative and energy metabolism in a variety of biological functions, such as gas transport, respiration, and nitrite and sulphite reduction [97]; and also in chlorophyll synthesis in plants and algae [98]. In S-adenosyl-L-methionine (SAM) biosynthesis, SAM functions as a methyl donor and plays a key role in antibiotic production [99], and inhibits sporulation and cellular differentiation in *Streptomyces* spp., *Bacillus subtilis*, and *Saccharomyces cerevisiae* [100]. Other functions detected include the sugar nucleotide biosynthesis and sugar degradation (galactose). It has been previously shown that galactose metabolism plays a central role in biofilm formation by *B. subtilis* and other bacteria [101]. Nevertheless, Forl inoculation treatment also showed higher abundance of sequences attributed to Carboxylate (sugar acid) and secondary metabolism (sugar derivative; sulfoquinovose) degradation. Sulfoquinovose biosynthesis is largely conserved within plants, algae, and photosynthetic bacteria and plays a major role in the global biogeochemical sulfur cycle by serving as a sulfur reservoir that can be mobilized in the early stages of sulfur starvation [102]. Nevertheless, a sulfoglycolytic pathway is being employed by a diverse collection of bacterial species, such as γ-*Proteobacteria*, as well as α- and β-*Proteobacteria* [103].

Overall, based on the results of this research, the biosolid application seems to result in a successful priming of tomato plants inducing resistance mechanisms against Forl. This effect was also associated with the microbiome diversity in the biosolid-enriched treatments and the changes in abundance with time in response to Forl. Organic amendments, such as green manures, stable manures, and composts, have long been recognized to facilitate biological control within the context of bacterial communities [104]. Microbe-microbe associations and microbe–plant interactions are important in the context of pathogen inhibition via direct antagonism and mediating processes involved in nutrient dynamics [105,106]. Therefore, the use of biosolid as a soil amendment had a positive effect not only on plant health, but also on the bacterial diversity, relative abundance and predicted soil functioning, toward enhancing tomato resistance against Forl.

4. Materials and Methods

4.1. Culture of Fusarium oxysporum f. sp. radicis-lycopersici and Inoculum Preparation

A virulent strain of Forl provided by the Institute Biology, Leiden University, Leiden, the Netherlands and deposited in the Centraalbureau voor Schimmelcultures, the Netherlands (CBS 101587), was used for artificial inoculations of tomato plants. The fungus was routinely kept on potato dextrose agar (PDA; Lab M, Lancashire, UK), at 4 °C and was often inoculated on a surface-sterilized tomato fruit and re-isolated on PDA to maintain its virulence. For inoculum preparation, Forl was grown in Czapek Dox Broth (Ducheta Biochemie, Haarlem, The Netherlands) as described in Kamou et al. [3]. Conidia were separated from mycelium, washed with sterile distilled water, and adjusted to 10^6 spores mL^{-1}, as previously described by Giannakis et al. [12].

4.2. In Vitro Growth and Sporulation of Fusarium oxysporum f. sp. radicis-lycopersici in the Presence of Biosolid Leachates

Anaerobically digested and dehydrated sludge [12] was subjected to the one-step static leaching method EN 12457-2 as described in Giannakis et al. [107] and was sterilized by filtration through 0.2 μm pore antimicrobial filters (Whatman Puradisk™, Buckinghamshire, UK) before incorporation into PDA at a temperature of 40 °C, just before pouring plates. Growth substrates of biosolid leachates and PDA mixtures were prepared at concentrations of 0% (control), 2%, 5%, and 10% v/v in Petri dishes. Forl discs, 4 mm in diameter, taken from the periphery of a fresh colony on PDA, were used to inoculate the different mixtures of PDA-biosolid leachates and cultures were incubated for 6 days. After incubation, fungal growth was expressed as colony diameter in centimeters, as described by Bardas et al. [108]. More specifically, two vertical lines were drawn at the bottom of each plate with an intersection point at the center of the inoculation point. The diameter of the colony was measured on each line and an average value of 10 measurements derived from 5 replications (Petri dishes) was calculated.

Sporulation of the fungus in the PDA-biosolid leachates media of different concentrations was assessed as follows: 10 pieces of 0.5 cm^2, from the periphery of each colony, were added to 20 mL sterile distilled water in screw-capped sterile tubes. Conidia were detached from mycelium by vortexing each tube for 1 min and were harvested by filtering through sterile Miracloth (Calbiochem, San Diego, CA, USA) followed by a 10^3-fold dilution with sterile distilled water under sterile conditions. Then, 100 μL of the diluted mixtures were placed on PDA plates and incubated for 2 days at 20–25 °C. The number of single spore colonies obtained was counted and the initial number of conidia per cm^2 of colony was calculated for the different media. The fungal growth and sporulation assays were repeated 5 times.

4.3. In Planta Experiment and Relative Gene Expression Analysis in Tomato Leaves

4.3.1. Substrate Preparation, Tomato Plant Growth, and Inoculation Procedure

For the *in planta* experiment, four different substrates were prepared from peat and clay soil with two levels of biosolid (absence or presence), to which two levels of inoculum were applied; without (−) or with (+) Forl, as described by Giannakis et al. [12]. Specifically, growth substrates were: (i) peat and clay soil 1:1 (*w/w*) (treatment C), (ii) peat and clay soil 1:1 (*w/w*), supplemented with biosolid 2% (*w/w*) (treatment B), (iii) peat and clay soil 1:1 (*w/w*) inoculated with Forl (treatment F), and (iv) peat and clay soil 1:1 (*w/w*) supplemented with biosolid 2% (*w/w*) and inoculated with Forl (treatment FB). Clay soil was characterized as 40% silt, 30% clay, 30% sand, 2.5% organic matter, and pH 7.9. Inoculations and plant growth conditions were as described by Giannakis et al. [12]. Specifically, Forl inoculum, prepared as described above, was mixed thoroughly into the substrates at a 1/10 volume ratio, to a final concentration of 10^5 spores g^{-1} substrate. Non-inoculated substrates were mixed with water. Tomato seedlings of the cv. ACE 55 at the two-leaf stage were transplanted in the different substrates in 0.1 L pots. All plants were placed in the same growth chamber and were grown under controlled conditions at 20–25 °C with 14/10 h photoperiod light/dark, respectively, and 60% RH.

4.3.2. Gene Expression Analysis

Gene expression analysis was monitored at 12, 24, 48, and 72 h in leaves of tomato after transplantation and exposure to biosolids and Forl. The relative gene expression analysis of *GLUA, CHI3, PR1-a, LOX,* and *AOC* genes (Table 5), associated with enhanced resistance against pathogen infection in tomato plants, was performed using the reverse-transcription quantitative polymerase chain reaction (RT-qPCR). Total RNA from tomato leaves was extracted using the Monarch Total RNA miniprep kit (New England Biolabs Inc., Ipswich, MA, USA) in three biological replications at each time point and treatment. RNA quality and quantity were assessed using the UV-Vis Spectrophotometer Q5000 (Quawell Technology Inc., San Jose, CA, USA) and optically with gel electrophoresis in 1.5% agarose gel. Pooling of biological replicates was performed equimolarly with 1 μg per replicate at a working concentration of 100 ng $μL^{-1}$.

Table 5. Primers used in RT-qPCR of the 5 genes (*GLUA, CHI3, PR1-a, LOX,* and *AOC*) associated with response mechanisms to pathogens and the 3 housekeeping genes (*β-actin, CyOXID,* and *Gapdh*).

Gene		Gene Sequence	Encoding Protein	Defense Pathway
GLUA	F	GTCTCAACCGCGACATATT	PR-2 (β-1,3 glucanase, basic type)	SA signaling pathway
	R	CACAAGGGCATCGAAAAGAT		
CHI3	F	TGCAGGAACATTCACTGGAG	PR-3 (Chitinase)	JA/ETH signaling pathway
	R	TAACGTTGTGGCATGATGGT		
PR1-a	F	TCTTGTGAGGCCCAAAATTC	PR-1 (acidic type)	SA signaling pathway
	R	TAGTCTGGCCTCTCGGACA		
LOX	F	CCTGAAATCTATGGCCCTCA	Lipoxygenase	ETH signaling pathway
	R	ATGGGCTTAAGTGTGCCAAC		
AOC	F	CTCGGAGATCTTGTCCCCTTT	Allene oxide cyclase	JA/ETH signaling pathway
	R	CTCCTTTCTTCTCTTCTTCGTGCT		
β-actin	F	GAAATAGCATAAGATGGAGACG	Actin	Reference gene
	R	ATACCCACCATCACACCAGTAT		
CyOXID	F	TGGTAATTGGTCTGTTCCGATT	Cytochrome oxidase subunit I	Reference gene
	R	TGGAGGCAACAACCAGAATG		
Gapdh	F	GAAATGCATCTTGCACTACCAACTGTCTTGC	Glyceraldehyde-3-phosphate-dehydrogenase	Reference gene
	R	CTGTGAGTAACCCCATTCATTATCATACCAAGC		

The cDNA was prepared with 1 μg pooled RNA using the PrimeScript™ Reverse Transcriptase kit (TAKARA BIO Inc., Kusatsu, Shiga, Japan) and random hexamers according to the manufacturers' instructions. Relative gene expression was assessed by real-time quantitative reverse transcriptase PCR (RT-qPCR) performed on a Rotor-Gene 6000 real-time 5-Plex HRM PCR Thermocycler (Corbett Research, Sydney, Australia) using the Rotor-Gene Q software version 2.0.2 (Corbett Life Science, Cambridge, UK) and melt curve analysis. The reaction mixtures were prepared in a total volume of 20 μL per reaction consisting of 50 ng cDNA, 1× PCR buffer, 0.5 µM forward and reverse primers, 0.2 mM dNTPs, 1.5 mM SYTO™ 9 Green Fluorescent Nucleic Acid Stain (Invitrogen, Eugene, OR, USA), and 1 U Kapa Taq DNA polymerase (Kapa Biosystems, Wilmington, MA, USA). The amplification was performed according to the following thermal cycling conditions: initial denaturation at 95 °C for 2 min, followed by 35 cycles of 95 °C for 10 s, 54 °C for 25 s, and 72 °C for 30 s. Fluorescence was acquired at the end of each PCR cycle. Melting curve analysis was performed at temperature range between 65–95 °C and in increments of 0.3 °C every 2 s; fluorescence was measured at the end of each increment step.

The tomato gene-specific primers (Table 5) *GLUA* [35], *CHI3* [35,37], *PR1-a, LOX* [35], and *AOC* [7,36,109] were used for the relative gene expression analysis. The housekeeping genes encoding for actin (*β-actin*) [35], mitochondrion cytochrome oxidase subunit I (*Cy-OXID*) [110], and Glyceraldehyde-3-phosphate-dehydrogenase (*Gapdh*) [51] were used as reference genes for normalization in tomato. Data analysis was carried out with relative quantification in three technical replicates for each pool, using the $2^{-\Delta\Delta CT}$ method [111], and data normalization was achieved using the expression levels of the reference genes.

4.4. Characterization of the Soil Substrate Microbiome Using 16S Sequencing

Soil substrate samples near the roots were collected from the same pots that the leaf tissue was collected, at 12 and 72 h after treatment application, in 3 biological replications. The total microbial genomic DNA was extracted with the NucleoSpin Soil, Mini kit (Macherey-Nagel GmbH & Co.KG., Düren, Germany). The DNA quantity and quality were assessed using the UV-Vis Spectrophotometer Q5000 (Quawell Technology Inc., San Jose, CA, USA). The 16S rRNA libraries were constructed after amplification of the 16S subunit of the prokaryotic ribosomal 16S RNA gene (16S rRNA), using primers that amplify between the V3 and V4 regions of the gene for each sample. The 2 × 300 bp paired-end reads were generated using the Illumina MiSeq platform (Illumina Inc., San Diego, CA, USA).

For the identification and quantification of the detected microorganisms, as well as the comparison between the samples, bioinformatics analysis was performed using Mothur programs (1.44.1) and R packages (4.0.3) [112], Phyloseq (1.34.0), ggplot2 (3.3.2), DESeq2 (1.28.1), and vegan (2.5.6). The raw sequencing data were processed to remove low-quality sequences as well as DNA adapters. Pure sequences were grouped into Operational Taxonomic Unit (OTUs) with a sequence similarity rate of 97%. The classification of OTUs was characterized by SILVA multiple sequencing (SILVA alignment Release 132, 8517 bacteria, 147 archaea, and 2516 eukarya sequences). The Mothur analysis was performed following the Standard Operating Procedure (SOP) for the Mothur Metagenomics analysis as described in Schloss et al. [113].

Variation of microbial communities within a single soil substrate or between the soil substrates was found using alpha and beta diversity, respectively. More specifically, the α-diversity was evaluated by rarefaction curves measuring the Shannon, Chao1, Abundance-based Coverage Estimator (ACE), and Simpson diversity indices. The β-diversity was assessed by non-metric multidimensional scaling (NMDS) to define the structure of the microbiome. Canonical correspondence analysis (CCA) was used to relate species abundance to the treatment and time variables. A graphic interpretation of the main principal axes by tri-plot on the two dimensions was obtained with the Phyloseq software. The functional potential of the detected microbial communities was predicted based on the 16s rRNA marker using the PICRUSt software (2.3.0_b) [114]. The functional prediction was based on the unique OTU sequences and the biom file produced by Mothur (1.44.11). The analysis returned the relative abundance of the predicted EC codes and their related pathways description. The statistical analysis and final visualization of the results were performed using the Statistical Analysis of Metagenomic Profiles (STAMP) software (2.1.3) [115].

4.5. Statistical Analysis

Data from fungal growth analysis and sporulation assay (n = 4) and relative gene expression analysis (n = 3) were subjected to analysis of variance (ANOVA) based on the completely randomized design (CRD). Differences among the treatments were assessed using Tukey's post hoc test at a significance level predetermined at $p \leq 0.05$. All statistical analyses were performed using SPSS Statistics for Windows, v.25 (IBM Corp., Armonk, NY, USA). For the functional analysis, the Statistical Analysis of Metagenomic Profiles (STAMP) [115] software was used to provide a statistical view of differences in abundant features. The data were subjected to analysis of variance (ANOVA) and differences among the treatments were assessed using Tukey–Kramer post hoc test at a significance level at $p \leq 0.05$.

5. Conclusions

This is the first attempt to better understand the plant–microbiome interactions in the presence of biosolid application and the biocontrol mechanism against Forl in tomato plants. More specifically, the effect of biosolid application on the biocontrol of Forl was investigated based on the enhanced plant resistance measured as expression of pathogen-response genes and the pathogen suppression in the context of soil microbiome diversity, abundance, and predicted functions. Plants and rhizosphere microbiome share complex interactions required for optimal root and soil functioning. When this balance is disturbed, changes occur in microbial communities, soil functioning, and soil abiotic properties interactively [106], shaping the resistance potential of plants and the biocontrol of the pathogen. Our results suggest that biosolid application alters microbial diversity and the predicted soil functioning, along with the relative abundance of specific phyla and classes, as a proxy for disease suppression. Further research is required to identify the biochemical and molecular mechanisms of the priming effect induced by the biosolid and specific functional genes associated with bacterial consortia as biological indicators for the identification of the biocontrol potential of biosolid application.

Supplementary Materials: The following are available online at https://www.mdpi.com/article/10.3390/plants10122789/s1, Figure S1: Rank abundance curves. Abundances of the top 100 OTUs for bacterial communities in: A. Between two time points at 12 and 72 h after inoculation with *Fusarium oxysporum* f. sp. *radicis-lycopersici* and biosolid application. B. amongst the four different treatments control (C), biosolid (B), Forl (F), and Forl and biosolid (FB), Figure S2: Principal Component Analysis (PCA) of the functional diversity for the four different treatments control (C), biosolid application (B), Forl inoculation (F), and Forl inoculation + biosolid application (FB) at 12 and 72 h post inoculation and biosolid application. The different treatments are depicted with different colors and shapes, Figure S3: Post hoc plots for the predicted pathways demonstrating greater abundance of sequences (%) in biosolid-enriched treatments compared to the Control (C) and Forl inoculation + biosolid application (FB) treatments, indicating: (i) the mean proportion of sequences within each treatment, (ii) the difference in mean proportions for each pair of treatments, and (iii) a p-value indicating whether the mean proportion is equal for a given pair of treatments. The analysis was performed in the STAMP software using ANOVA with p-value ≤ 0.05 and effect size >0.9, Table S1: Relative abundance (%) of bacterial phyla at rate greater than 1%, for the treatments C, B, F, and FB at 12 and 72 h after inoculation with *Fusarium oxysporum* f. sp. *radicis-lycopersici*. Value is the pooled mean of three replicates, Table S2: Relative abundance (%) of bacterial classes at rate greater than 1%, for the treatments C, B, F, and FB at 12 and 72 h after inoculation with *Fusarium oxysporum* f. sp. *radicis-lycopersici*. Value is the pooled mean of three replicates, Table S3: Multiple group statistics table for the predicted functional diversity in the different treatments: Control (C), Biosolid application (B), Forl inoculation (F), and Forl inoculation + biosolid application (FB) based on the abundances of sequences associated with specific metabolic pathways. The analysis was performed in the STAMP software using ANOVA with p-value ≤ 0.05 and effect size >0.9. The order of pathways is according to the abundance (%) as indicated in Figure 7.

Author Contributions: Conceptualization, A.L.L. and I.N.-O.; methodology, E.S., N.N.K., A.L.L., and I.N.-O.; software, I.K. and E.S.; validation, E.S. and I.K.; formal analysis, E.S. and I.K.; investigation, E.S. (principal investigator), I.G. (principal investigator) and G.L.; resources P.M., I.N.-O., and A.L.L.; data curation, E.S., I.G., and I.K.; writing—original draft preparation, E.S. and I.G.; writing—review and editing, E.S., A.L.L., I.N.-O., C.E. and N.N.K.; visualization, E.S. and I.K.; supervision, P.M.,

A.L.L., and I.N.-O.; project administration, C.E.; and funding acquisition, C.E., A.K., and I.N.-O. All authors have read and agreed to the published version of the manuscript.

Funding: Part of the project was funded by I. Giannakis' PhD Scholarship: European Social Fund— ESF through the Operational Program "Human Resources Development, Education and Lifelong Learning in the context of the project "Strengthening Human Resources Research Potential via Doctorate Research" (MIS-5000432), implemented by the State Scholarships Foundation (IKΥ)".

Data Availability Statement: Data is contained within the article or Supplementary Materials.

Acknowledgments: The authors are grateful to Ben Lugtenberg, Institute Biology, Leiden University, The Netherlands for providing the isolate of Forl, and EYATH SA for providing biosolids of municipal sludge. All individuals in the Acknowledgments have consented to the acknowledgement.

Conflicts of Interest: The authors declare no conflict of interest.

References

1. Lagopodi, A.L.; Ram, A.F.J.; Lamers, G.E.M.; Punt, P.J.; Van den Hondel, C.A.M.J.J.; Lugtenberg, B.J.J.; Bloemberg, G.V. Novel aspects of tomato root colonization and infection by *Fusarium oxysporum* f. sp. *radicis-lycopersici*. Revealed by confocal laser scanning microscopic analysis using the green fluorescent protein as a marker. *Mol. Plant-Microbe Interact.* **2002**, *15*, 172–179. [CrossRef]

2. Szczechura, W.; Staniaszek, M.; Habdas, H. *Fusarium oxysporum* f. sp. *radicis-lycopersici*—The cause of fusarium crown and root rot in tomato cultivation. *J. Plant Prot. Res.* **2013**, *53*, 172–176. [CrossRef]

3. Kamou, N.N.; Karasali, H.; Menexes, G.; Kasiotis, K.M.; Bon, M.C.; Papadakis, E.N.; Tzelepis, G.D.; Lotos, L.; Lagopodi, A.L. Isolation screening and characterisation of local beneficial rhizobacteria based upon their ability to suppress the growth of *Fusarium oxysporum* f. sp. *radicis-lycopersici* and tomato foot and root rot. *Biocontrol Sci. Technol.* **2015**, *25*, 928–949. [CrossRef]

4. Kamou, N.N.; Cazorla, F.; Kandylas, G.; Lagopodi, A.L. Induction of defense-related genes in tomato plants after treatments with the biocontrol agents *Pseudomonas chlororaphis* ToZa7 and Clonostachys rosea IK726. *Arch. Microbiol.* **2020**, *202*, 257–267. [CrossRef]

5. Koike, S.T.; Gladders, P.; Paulus, A.O. Fusarium oxysporum f.sp. *radicis-lycopersici*: Fusarium crown and root rot. In *Vegetable Diseases: A Color Handbook*; Manson Publishing Ltd.: London, UK, 2007; p. 448. ISBN 0123736757.

6. Minchev, Z.; Kostenko, O.; Soler, R.; Pozo, M.J. Microbial consortia for effective biocontrol of root and foliar diseases in tomato. *Front. Plant Sci.* **2021**, *12*, 756368. [CrossRef] [PubMed]

7. Song, Y.Y.; Zeng, R.S.; Xu, J.F.; Li, J.; Shen, X.; Yihdego, W.G. Interplant communication of tomato plants through underground common mycorrhizal networks. *PLoS ONE* **2010**, *5*, e13324. [CrossRef]

8. De Corato, U. Disease-suppressive compost enhances natural soil suppressiveness against soil-borne plant pathogens: A critical review. *Rhizosphere* **2020**, *13*, 100192. [CrossRef]

9. De Corato, U.; Patruno, L.; Avella, N.; Lacolla, G.; Cucci, G. Composts from green sources show an increased suppressiveness to soilborne plant pathogenic fungi: Relationships between physicochemical properties, disease suppression, and the microbiome. *Crop Prot.* **2019**, *124*, 104870. [CrossRef]

10. Glick, B.R.; Gamalero, E. Recent developments in the study of plant microbiomes. *Microorganisms* **2021**, *9*, 1553. [CrossRef] [PubMed]

11. Bonanomi, G.; Zotti, M.; Idbella, M.; Di Silverio, N.; Carrino, L.; Cesarano, G.; Assaeed, A.M.; Abd-ElGawad, A.M. Decomposition and organic amendments chemistry explain contrasting effects on plant growth promotion and suppression of *Rhizoctonia solani* damping off. *PLoS ONE* **2020**, *15*, e0230925. [CrossRef]

12. Giannakis, I.; Manitsas, C.; Eleftherohorinos, I.; Menexes, G.; Emmanouil, C.; Kungolos, A.; Lagopodi, A.L. Use of biosolids to enhance tomato growth and tolerance to *Fusarium oxysporum* f. sp. *radicis-lycopersici*. *Environ. Process.* **2021**, *8*, 1415–1431. [CrossRef]

13. Zhao, J.; Liu, J.; Liang, H.; Huang, J.; Chen, Z.; Nie, Y.; Wang, C.; Wang, Y. Manipulation of the rhizosphere microbial community through application of a new bioorganic fertilizer improves watermelon quality and health. *PLoS ONE* **2018**, *13*, e0192967.

14. Curci, M.; Lavecchia, A.; Cucci, G.; Lacolla, G.; De Corato, U.; Crecchio, C. Short-term effects of sewage sludge compost amendment on semiarid soil. *Soil Syst.* **2020**, *4*, 48. [CrossRef]

15. Garbeva, P.; Van Veen, J.A.; Van Elsas, J.D. Microbial diversity in soil: Selection of microbial populations by plant and soil type and implications for disease suppressiveness. *Annu. Rev. Phytopathol.* **2004**, *42*, 243–270. [CrossRef] [PubMed]

16. Mendes, R.; Kruijt, M.; De Bruijn, I.; Dekkers, E.; Van Der Voort, M.; Schneider, J.H.M.; Piceno, Y.M.; DeSantis, T.Z.; Andersen, G.L.; Bakker, P.A.H.M.; et al. Deciphering the rhizosphere microbiome for disease-suppressive bacteria. *Science* **2011**, *332*, 1097–1100. [CrossRef] [PubMed]

17. Cha, J.Y.; Han, S.; Hong, H.J.; Cho, H.; Kim, D.; Kwon, Y.; Kwon, S.K.; Crusemann, M.; Bok Lee, Y.; Kim, J.F.; et al. Microbial and biochemical basis of a Fusarium wilt-suppressive soil. *ISME J.* **2016**, *10*, 119–129. [CrossRef]

18. García-Ruiz, R.; Ochoa, V.; Hinojosa, M.B.; Carreira, J.A. Suitability of enzyme activities for the monitoring of soil quality improvement in organic agricultural systems. *Soil Biol. Biochem.* **2008**, *40*, 2137–2145. [CrossRef]

19. Ou, Y.; Penton, C.R.; Geisen, S.; Shen, Z.; Sun, Y.; Lv, N.; Wang, B.; Ruan, Y.; Xiong, W.; Li, R.; et al. Deciphering underlying drivers of disease suppressiveness against pathogenic *Fusarium oxysporum*. *Front. Microbiol.* **2019**, *10*, 2535. [CrossRef]

20. Zhang, H.; Wang, R.; Chen, S.; Qi, G.; He, Z.; Zhao, X. Microbial taxa and functional genes shift in degraded soil with bacterial wilt. *Sci. Rep.* **2017**, *7*, 39911. [CrossRef]

21. Berendsen, R.L.; Vismans, G.; Yu, K.; Song, Y.; De Jonge, R.; Burgman, W.P.; Burmølle, M.; Herschend, J.; Bakker, P.A.H.M.; Pieterse, C.M.J. Disease-induced assemblage of a plant-beneficial bacterial consortium. *ISME J.* **2018**, *12*, 1496–1507. [CrossRef] [PubMed]

22. Takken, F.; Rep, M. The arms race between tomato and *Fusarium oxysporum*. *Mol. Plant Pathol.* **2010**, *11*, 309–314. [CrossRef]

23. Andersen, E.J.; Ali, S.; Byamukama, E.; Yen, Y.; Nepal, M.P. Disease resistance mechanisms in plants. *Genes* **2018**, *9*, 339. [CrossRef]

24. Andolfo, G.; Ferriello, F.; Tardella, L.; Ferrarini, A.; Sigillo, L.; Frusciante, L.; Ercolano, M.R. Tomato genome-wide transcriptional responses to Fusarium wilt and tomato mosaic virus. *PLoS ONE* **2014**, *9*, e94963. [CrossRef]

25. Lebeis, S.L.; Paredes, S.H.; Lundberg, D.S.; Breakfield, N.; Gehring, J.; McDonald, M.; Malfatti, S.; Del Rio, T.G.; Jones, C.D.; Tringe, S.G.; et al. Salicylic acid modulates colonization of the root microbiome by specific bacterial taxa. *Science* **2015**, *349*, 860–864. [CrossRef] [PubMed]

26. Carvalhais, L.C.; Dennis, P.G.; Badri, D.V.; Kidd, B.N.; Vivanco, J.M.; Schenk, P.M. Linking jasmonic acid signaling, root exudates, and rhizosphere microbiomes. *Mol. Plant-Microbe Interact.* **2015**, *28*, 1049–1058. [CrossRef]

27. Pérez-Jaramillo, J.E.; Carrión, V.J.; Bosse, M.; Ferrão, L.F.V.; De Hollander, M.; Garcia, A.A.F.; Ramírez, C.A.; Mendes, R.; Raaijmakers, J.M. Linking rhizosphere microbiome composition of wild and domesticated *Phaseolus vulgaris* to genotypic and root phenotypic traits. *ISME J.* **2017**, *11*, 2244–2257. [CrossRef] [PubMed]

28. Haney, C.H.; Samuel, B.S.; Bush, J.; Ausubel, F.M. Associations with rhizosphere bacteria can confer an adaptive advantage to plants. *Nat. Plants* **2015**, *1*, 15051. [CrossRef]

29. Vergara Cid, C.; Ferreyroa, G.V.; Pignata, M.L.; Rodriguez, J.H. Biosolid compost amendment increases soil fertility and soybean growth. *J. Plant Nutr.* **2020**, *44*, 1131–1140. [CrossRef]

30. Antonelli, P.M.; Fraser, L.H.; Gardner, W.C.; Broersma, K.; Karakatsoulis, J.; Phillips, M.E. Long term carbon sequestration potential of biosolids-amended copper and molybdenum mine tailings following mine site reclamation. *Ecol. Eng.* **2018**, *117*, 38–49. [CrossRef]

31. Brown, S.; Ippolito, J.A.; Hundal, L.S.; Basta, N.T. Municipal biosolids—A resource for sustainable communities. *Curr. Opin. Environ. Sci. Health* **2020**, *14*, 56–62. [CrossRef]

32. Giannakis, I.; Emmanouil, C.; Mitrakas, M.; Manakou, V.; Kungolos, A. Chemical and ecotoxicological assessment of sludge-based biosolids used for corn field fertilization. *Environ. Sci. Pollut. Res.* **2021**, *28*, 3797–3809. [CrossRef]

33. Griffin, D.H. Chapter 5. Chemical requirements for growth. In *Fungal Physiology*; John Wiley & Sons Inc.: New York, NY, USA, 1996; pp. 130–157. ISBN 0471166154.

34. Jackson, S.L.; Heath, I.B. Roles of calcium ions in hyphal tip growth. *Microbiol. Rev.* **1993**, *57*, 367–382. [CrossRef]

35. Aimé, S.; Alabouvette, C.; Steinberg, C.; Olivain, C. The endophytic strain *Fusarium oxysporum* Fo47: A good candidate for priming the defense responses in tomato roots. *Mol. Plant-Microbe Interact.* **2013**, *26*, 918–926. [CrossRef] [PubMed]

36. Song, Y.; Chen, D.; Lu, K.; Sun, Z.; Zeng, R. Enhanced tomato disease resistance primed by arbuscular mycorrhizal fungus. *Front. Plant Sci.* **2015**, *6*, 786. [CrossRef]

37. Vilasinee, S.; Toanuna, C.; McGovern, R.J.; Nalumpang, S. Expression of pathogenesis-related (PR) genes in tomato against Fusarium wilt by challenge inoculation with Streptomyces NSP3. *Int. J. Agric. Technol.* **2019**, *15*, 157–170.

38. Santino, A.; Taurino, M.; De Domenico, S.; Bonsegna, S.; Poltronieri, P.; Pastor, V.; Flors, V. Jasmonate signaling in plant development and defense response to multiple (a)biotic stresses. *Plant Cell Rep.* **2013**, *32*, 1085–1098. [CrossRef]

39. Jaiswal, A.K.; Alkan, N.; Elad, Y.; Sela, N.; Philosoph, A.M.; Graber, E.R.; Frenkel, O. Molecular insights into biochar-mediated plant growth promotion and systemic resistance in tomato against Fusarium crown and root rot disease. *Sci. Rep.* **2020**, *10*, 13934. [CrossRef] [PubMed]

40. Houterman, P.M.; Speijer, D.; Dekker, H.L.; De Koster, C.G.; Cornelissen, B.J.C.; Rep, M. The mixed xylem sap proteome of *Fusarium oxysporum*-infected tomato plants: Short communication. *Mol. Plant Pathol.* **2007**, *8*, 215–221. [CrossRef]

41. Rep, M.; Dekker, H.L.; Vossen, J.H.; De Boer, A.D.; Houterman, P.M.; Speijer, D.; Back, J.W.; De Koster, C.G.; Cornelissen, B.J.C. Mass spectrometric identification of isoforms of PR proteins in xylem sap of fungus-infected tomato. *Plant Physiol.* **2002**, *130*, 904–917. [CrossRef]

42. Aimé, S.; Cordier, C.; Alabouvette, C.; Olivain, C. Comparative analysis of PR gene expression in tomato inoculated with virulent *Fusarium oxysporum* f. sp. *lycopersici* and the biocontrol strain *F. oxysporum* Fo47. *Physiol. Mol. Plant Pathol.* **2008**, *73*, 9–15. [CrossRef]

43. Vitale, A.; Rocco, M.; Arena, S.; Giuffrida, F.; Cassaniti, C.; Scaloni, A.; Lomaglio, T.; Guarnaccia, V.; Polizzi, G.; Marra, M.; et al. Tomato susceptibility to Fusarium crown and root rot: Effect of grafting combination and proteomic analysis of tolerance expression in the rootstock. *Plant Physiol. Biochem.* **2014**, *83*, 207–216. [CrossRef]

44. Graber, E.R.; Frenkel, O.; Jaiswal, A.K.; Elad, Y. How may biochar influence severity of diseases caused by soilborne pathogens? *Carbon Manag.* **2014**, *5*, 169–183. [CrossRef]

45. Jaiswal, A.K.; Elad, Y.; Paudel, I.; Graber, E.R.; Cytryn, E.; Frenkel, O. Linking the belowground microbial composition, diversity and activity to soilborne disease suppression and growth promotion of tomato amended with biochar. *Sci. Rep.* **2017**, *7*, 44382. [CrossRef] [PubMed]

46. Balmer, A.; Pastor, V.; Gamir, J.; Flors, V.; Mauch-Mani, B. The "prime-ome": Towards a holistic approach to priming. *Trends Plant Sci.* **2015**, *20*, 443–452. [CrossRef]

47. Mauch-Mani, B.; Baccelli, I.; Luna, E.; Flors, V. Defense Priming: An adaptive part of induced resistance. *Annu. Rev. Plant Biol.* **2017**, *68*, 485–512. [CrossRef]

48. He, M.; Jahan, M.S.; Wang, Y.; Sun, J.; Shu, S.; Guo, S. Compost amendments based on vinegar residue promote tomato growth and suppress bacterial wilt caused by *Ralstonia Solanacearum*. *Pathogens* **2020**, *9*, 227. [CrossRef]

49. Tubeileh, A.M.; Stephenson, G.T. Soil amendment by composted plant wastes reduces the *Verticillium dahliae* abundance and changes soil chemical properties in a bell pepper cropping system. *Curr. Plant Biol.* **2020**, *22*, 100148. [CrossRef]

50. Bonilla, N.; Gutiérrez-Barranquero, J.A.; De Vicente, A.; Cazorla, F.M. Enhancing soil quality and plant health through suppressive organic amendments. *Diversity* **2012**, *4*, 475–491. [CrossRef]

51. Kavroulakis, N.; Papadopoulou, K.K.; Ntougias, S.; Zervakis, G.I.; Ehaliotis, C. Cytological and other aspects of pathogenesis-related gene expression in tomato plants grown on a suppressive compost. *Ann. Bot.* **2006**, *98*, 555–564. [CrossRef]

52. Lugtenberg, B.J.J.; Chin-a-woeng, T.F.C.; Bloemberg, G.V. Microbe-plant interactions: Principles and mechanisms. *Antonie Van Leeuwenhoek* **2002**, *81*, 373–383. [CrossRef] [PubMed]

53. Chen, Q.L.; Ding, J.; Zhu, Y.G.; He, J.Z.; Hu, H.W. Soil bacterial taxonomic diversity is critical to maintaining the plant productivity. *Environ. Int.* **2020**, *140*, 105766. [CrossRef] [PubMed]

54. Liu, H.; Brettell, L.E.; Qiu, Z.; Singh, B.K. Microbiome-mediated stress resistance in Plants. *Trends Plant Sci.* **2020**, *25*, 733–743. [CrossRef] [PubMed]

55. Binyamin, R.; Nadeem, S.M.; Akhtar, S.; Khan, M.Y.; Anjum, R. Beneficial and pathogenic plant-microbe interactions: A review. *Soil Environ.* **2019**, *38*, 127–150. [CrossRef]

56. Antoniou, A.; Tsolakidou, M.D.; Stringlis, I.A.; Pantelides, I.S. Rhizosphere microbiome recruited from a suppressive compost improves plant fitness and increases protection against vascular wilt pathogens of tomato. *Front. Plant Sci.* **2017**, *8*, 2022. [CrossRef]

57. Figueiredo, G.G.O.; Lopes, V.R.; Romano, T.; Camara, M.C. Clostridium. In *Beneficial Microbes in Agro-Ecology*; Academic Press: Amsterdam, The Netherlands, 2020; pp. 477–491. ISBN 9780128234143.

58. Sarawaneeyaruk, S.; Lorliam, W.; Krajangsang, S.; Pringsulaka, O. Enhancing plant growth under municipal wastewater irrigation by plant growth promoting rhizospheric *Bacillus* spp. *J. King Saud Univ. Sci.* **2019**, *31*, 384–389. [CrossRef]

59. Li, Y.; Ali, A.; Jeyasundar, P.G.S.A.; Azeem, M.; Tabassum, A.; Guo, D.; Li, R.; Mian, I.A.; Zhang, Z. *Bacillus subtilis* and saponin shifted the availability of heavy metals, health indicators of smelter contaminated soil, and the physiological indicators of *Symphytum officinale*. *Chemosphere* **2021**, *285*, 131454. [CrossRef]

60. Hariprasad, P.; Divakara, S.T.; Niranjana, S.R. Isolation and characterization of chitinolytic rhizobacteria for the management of Fusarium wilt in tomato. *Crop Prot.* **2011**, *30*, 1606–1612. [CrossRef]

61. Mowlick, S.; Inoue, T.; Takehara, T.; Kaku, N.; Ueki, K.; Ueki, A. Changes and recovery of soil bacterial communities influenced by biological soil disinfestation as compared with chloropicrin-treatment. *AMB Express* **2013**, *3*, 46. [CrossRef]

62. Ueki, A.; Kaku, N.; Ueki, K. Role of anaerobic bacteria in biological soil disinfestation for elimination of soil-borne plant pathogens in agriculture. *Appl. Microbiol. Biotechnol.* **2018**, *102*, 6309–6318. [CrossRef] [PubMed]

63. Schmid, C.A.O.; Schröder, P.; Armbruster, M.; Schloter, M. Organic amendments in a long-term field trial—Consequences for the bulk soil bacterial community as revealed by network analysis. *Microb. Ecol.* **2018**, *76*, 226–239. [CrossRef] [PubMed]

64. Li, F.; Chen, L.; Zhang, J.; Yin, J.; Huang, S. Bacterial community structure after long-term organic and inorganic fertilization reveals important associations between soil nutrients and specific taxa involved in nutrient transformations. *Front. Microbiol.* **2017**, *8*, 187. [CrossRef]

65. Dhanker, R.; Chaudhary, S.; Goyal, S.; Kumar, R. Soil microbial properties and functional diversity in response to sewage sludge amendments. *Arch. Agron. Soil Sci.* **2020**, *66*, 1–14. [CrossRef]

66. Hosokawa, S.; Kuroda, K.; Narihiro, T.; Aoi, Y.; Ozaki, N.; Ohashi, A.; Kindaichi, T. Cometabolism of the superphylum patescibacteria with anammox bacteria in a long-term freshwater anammox column reactor. *Water* **2021**, *13*, 208. [CrossRef]

67. Jumas-Bilak, E.; Marchandin, H. The phylum synergistetes BT—The prokaryotes: Other major lineages of bacteria and the archaea. In *The Prokaryotes*; Rosenberg, E., DeLong, E.F., Lory, S., Stackebrandt, E., Thompson, F., Eds.; Springer: Berlin/Heidelberg, Germany, 2014; pp. 931–954. ISBN 978-3-642-38954-2.

68. Jagannathan, B.; Golbeck, J.H. Understanding of the binding interface between PsaC and the PsaA/PsaB heterodimer in photosystem I. *Biochemistry* **2009**, *48*, 5405–5416. [CrossRef] [PubMed]

69. Rivière, D.; Desvignes, V.; Pelletier, E.; Chaussonnerie, S.; Guermazi, S.; Weissenbach, J.; Li, T.; Camacho, P.; Sghir, A. Towards the definition of a core of microorganisms involved in anaerobic digestion of sludge. *ISME J.* **2009**, *3*, 700–714. [CrossRef] [PubMed]

70. Xu, S.; Yao, J.; Ainiwaer, M.; Hong, Y.; Zhang, Y. Analysis of bacterial community structure of activated sludge from wastewater treatment plants in winter. *Biomed Res. Int.* **2018**, *2018*, 8278970. [CrossRef] [PubMed]

71. Abd-El-Kareem, F. Effect of acetic acid fumigation on soil-borne fungi and cucumber root rot disease under greenhouse conditions. *Arch. Phytopathol. Plant Prot.* **2009**, *42*, 213–220. [CrossRef]

72. Allen, M.M.; Allen, D.J. Biostimulant potential of acetic acid under drought stress is confounded by pH-dependent root growth inhibition. *Front. Plant Sci.* **2020**, *11*, 647. [CrossRef] [PubMed]

73. Guo, J.; Peng, Y.; Ni, B.J.; Han, X.; Fan, L.; Yuan, Z. Dissecting microbial community structure and methane-producing pathways of a full-scale anaerobic reactor digesting activated sludge from wastewater treatment by metagenomic sequencing. *Microb. Cell Fact.* **2015**, *14*, 33. [CrossRef]

74. Shah, N.; Tang, H.; Doak, T.G.; Ye, Y. Comparing bacterial communities inferred from 16S rRNA gene sequencing and shotgun metagenomics. *Pac. Symp. Biocomput.* **2011**, *2011*, 165–176.

75. Khan, A.H.; Topp, E.; Scott, A.; Sumarah, M.; Macfie, S.M.; Ray, M.B. Biodegradation of benzalkonium chlorides singly and in mixtures by a *Pseudomonas* sp. isolated from returned activated sludge. *J. Hazard. Mater.* **2015**, *299*, 595–602. [CrossRef]

76. Lutz, S.; Thuerig, B.; Oberhaensli, T.; Mayerhofer, J.; Fuchs, J.G.; Widmer, F.; Freimoser, F.M.; Ahrens, C.H. Harnessing the microbiomes of suppressive composts for plant protection: From metagenomes to beneficial microorganisms and reliable diagnostics. *Front. Microbiol.* **2020**, *11*, 1810. [CrossRef]

77. Cuesta, G.; García-de-la-Fuente, R.; Abad, M.; Fornes, F. Isolation and identification of actinomycetes from a compost-amended soil with potential as biocontrol agents. *J. Environ. Manag.* **2012**, *95*, 280–284. [CrossRef]

78. Yu, D.; Sinkkonen, A.; Hui, N.; Kurola, J.M.; Kukkonen, S.; Parikka, P.; Vestberg, M.; Romantschuk, M. Molecular profile of microbiota of Finnish commercial compost suppressive against Pythium disease on cucumber plants. *Appl. Soil Ecol.* **2015**, *92*, 47–53. [CrossRef]

79. Tsolakidou, M.D.; Stringlis, I.A.; Fanega-Sleziak, N.; Papageorgiou, S.; Tsalakou, A.; Pantelides, I.S. Rhizosphere-enriched microbes as a pool to design synthetic communities for reproducible beneficial outputs. *FEMS Microbiol. Ecol.* **2019**, *95*, fiz138. [CrossRef] [PubMed]

80. Kołton, A.; Długosz-Grochowska, O.; Wojciechowska, R.; Czaja, M. Biosynthesis regulation of folates and phenols in plants. *Sci. Hortic.* **2022**, *291*, 110561. [CrossRef]

81. Ibrahim, M.F.M.; Ibrahim, H.A.; Abd El-Gawad, H.G. Folic acid as a protective agent in snap bean plants under water deficit conditions. *J. Hortic. Sci. Biotechnol.* **2021**, *96*, 94–109. [CrossRef]

82. Wittek, F.; Kanawati, B.; Wenig, M.; Hoffmann, T.; Franz-Oberdorf, K.; Schwab, W.; Schmitt-Kopplin, P.; Vlot, A.C. Folic acid induces salicylic acid-dependent immunity in Arabidopsis and enhances susceptibility to *Alternaria brassicicola*. *Mol. Plant Pathol.* **2015**, *16*, 616–622. [CrossRef]

83. Gorelova, V.; Ambach, L.; Rébeillé, F.; Stove, C.; Van Der Straeten, D. Folates in plants: Research advances and progress in crop biofortification. *Front. Chem.* **2017**, *5*, 21. [CrossRef]

84. Zheng, Y.; Cantley, L.C. Toward a better understanding of folate metabolism in health and disease. *J. Exp. Med.* **2019**, *216*, 253–266. [CrossRef] [PubMed]

85. Meir, Z.; Osherov, N. Vitamin biosynthesis as an antifungal target. *J. Fungi* **2018**, *4*, 72. [CrossRef]

86. Boersch, M.; Rudrawar, S.; Grant, G.; Zunk, M. Menaquinone biosynthesis inhibition: A review of advancements toward a new antibiotic mechanism. *RSC Adv.* **2018**, *8*, 5099–5105. [CrossRef]

87. Cenci, U.; Qiu, H.; Pillonel, T.; Cardol, P.; Remacle, C.; Colleoni, C.; Kadouche, D.; Chabi, M.; Greub, G.; Bhattacharya, D.; et al. Host-pathogen biotic interactions shaped vitamin K metabolism in Archaeplastida. *Sci. Rep.* **2018**, *8*, 15243. [CrossRef] [PubMed]

88. Johnston, J.M.; Bulloch, E.M. Advances in menaquinone biosynthesis: Sublocalisation and allosteric regulation. *Curr. Opin. Struct. Biol.* **2020**, *65*, 33–41. [CrossRef]

89. Borges, A.A.; Jiménez-Arias, D.; Expósito-Rodríguez, M.; Sandalio, L.M.; Pérez, J.A. Priming crops against biotic and abiotic stresses: MSB as a tool for studying mechanisms. *Front. Plant Sci.* **2014**, *5*, 642. [CrossRef] [PubMed]

90. Wu, X.; Ge, T.; Wang, W.; Yuan, H.; Wegner, C.E.; Zhu, Z.; Whiteley, A.S.; Wu, J. Cropping systems modulate the rate and magnitude of soil microbial autotrophic CO_2 fixation in soil. *Front. Microbiol.* **2015**, *6*, 379. [CrossRef]

91. Ramachandran, V.K.; East, A.K.; Karunakaran, R.; Downie, J.A.; Poole, P.S. Adaptation of *Rhizobium leguminosarum* to pea, alfalfa and sugar beet rhizospheres investigated by comparative transcriptomics. *Genome Biol.* **2011**, *12*, R106. [CrossRef] [PubMed]

92. Hennion, N.; Durand, M.; Vriet, C.; Doidy, J.; Maurousset, L.; Lemoine, R.; Pourtau, N. Sugars en route to the roots. Transport, metabolism and storage within plant roots and towards microorganisms of the rhizosphere. *Physiol. Plant.* **2019**, *165*, 44–57. [CrossRef]

93. Jacoby, R.P.; Succurro, A.; Kopriva, S. Nitrogen substrate utilization in three rhizosphere bacterial strains investigated using proteomics. *Front. Microbiol.* **2020**, *11*, 784. [CrossRef]

94. Moe, L.A. Amino acids in the rhizosphere: From plants to microbes. *Am. J. Bot.* **2013**, *100*, 1692–1705. [CrossRef] [PubMed]

95. Grzyb, A.; Wolna-Maruwka, A.; Niewiadomska, A. The significance of microbial transformation of nitrogen compounds in the light of integrated crop management. *Agronomy* **2021**, *11*, 1415. [CrossRef]

96. Dilly, O. Regulation of the respiratory quotient of soil microbiota by availability of nutrients. *FEMS Microbiol. Ecol.* **2003**, *43*, 375–381. [CrossRef]

97. Layer, G. Heme biosynthesis in prokaryotes. *Biochim. Biophys. Acta Mol. Cell Res.* **2021**, *1868*, 118861. [CrossRef]

98. Brzezowski, P.; Richter, A.S.; Grimm, B. Regulation and function of tetrapyrrole biosynthesis in plants and algae. *Biochim. Biophys. Acta Bioenerg.* **2015**, *1847*, 968–985. [CrossRef]

99. Okamoto, S.; Lezhava, A.; Hosaka, T.; Okamoto-Hosoya, Y.; Ochi, K. Enhanced expression of S-adenosylmethionine synthetase causes overproduction of actinorhodin in streptomyces coelicolor A3(2). *J. Bacteriol.* **2003**, *185*, 601–609. [CrossRef] [PubMed]

100. Kim, D.J.; Huh, J.H.; Yang, Y.Y.; Kang, C.M.; Lee, I.H.; Hyun, C.G.; Hong, S.K.; Suh, J.W. Accumulation of S-adenosyl-L-methionine enhances production of actinorhodin but inhibits sporulation in *Streptomyces lividans* TK23. *J. Bacteriol.* **2003**, *185*, 592–600. [CrossRef] [PubMed]

101. Chai, Y.; Beauregard, P.B.; Vlamakis, H.; Losick, R.; Kolter, R. Galactose metabolism plays a crucial role in biofilm formation by *Bacillus subtilis*. *mBio* **2012**, *3*, e00184-12. [CrossRef]

102. Goddard-Borger, E.D.; Williams, S.J. Sulfoquinovose in the biosphere: Occurrence, metabolism and functions. *Biochem. J.* **2017**, *474*, 827–849. [CrossRef] [PubMed]

103. Felux, A.-K.; Spiteller, D.; Klebensberger, J.; Schleheck, D. Entner–Doudoroff pathway for sulfoquinovose degradation in *Pseudomonas putida* SQ1. *Proc. Natl. Acad. Sci. USA* **2015**, *112*, E4298–E4305. [CrossRef] [PubMed]

104. Hoitink, H.; Boehm, M. Biocontrol within the context of soil microbial communitites: A substrate-dependent phenomenon. *Annu. Rev. Phytopathol.* **1999**, *37*, 427–446. [CrossRef] [PubMed]

105. Lugtenberg, B.; Kamilova, F. Plant-growth-promoting rhizobacteria. *Annu. Rev. Microbiol.* **2009**, *63*, 541–556. [CrossRef] [PubMed]

106. Trivedi, P.; Delgado-Baquerizo, M.; Trivedi, C.; Hamonts, K.; Anderson, I.C.; Singh, B.K. Keystone microbial taxa regulate the invasion of a fungal pathogen in agro-ecosystems. *Soil Biol. Biochem.* **2017**, *111*, 10–14. [CrossRef]

107. Giannakis, I.; Emmanouil, C.; Kungolos, A. Evaluation of effects of municipal sludge leachates on water quality. *Water* **2020**, *12*, 2046. [CrossRef]

108. Bardas, G.A.; Lagopodi, A.L.; Kadoglidou, K.; Tzavella-Klonari, K. Biological control of three *Colletotrichum lindemuthianum* races using *Pseudomonas chlororaphis* PCL1391 and *Pseudomonas fluorescens* WCS365. *Biol. Control* **2009**, *49*, 139–145. [CrossRef]

109. Hu, T.; Wang, Y.; Wang, Q.; Dang, N.; Wang, L.; Liu, C.; Zhu, J.; Zhan, X. The tomato 2-oxoglutarate-dependent dioxygenase gene SlF3HL is critical for chilling stress tolerance. *Hortic. Res.* **2019**, *6*, 45. [CrossRef]

110. Papayiannis, L.C.; Harkou, I.S.; Markou, Y.M.; Demetriou, C.N.; Katis, N.I. Rapid discrimination of tomato chlorosis virus, Tomato infectious chlorosis virus and co-amplification of plant internal control using real-time RT-PCR. *J. Virol. Methods* **2011**, *176*, 53–59. [CrossRef]

111. Livak, K.J.; Schmittgen, T.D. Analysis of relative gene expression data using real-time quantitative PCR and the $2^{-\Delta\Delta CT}$ method. *Methods* **2001**, *25*, 402–408. [CrossRef] [PubMed]

112. R Core Team. R: A language and environment for statistical computing. In *R Foundation for Statistical Computing*; R Core Team: Vienna, Austria, 2020; Available online: https://www.R-project.org/ (accessed on 25 November 2021).

113. Schloss, P.D.; Westcott, S.L.; Ryabin, T.; Hall, J.R.; Hartmann, M.; Hollister, E.B.; Lesniewski, R.A.; Oakley, B.B.; Parks, D.H.; Robinson, C.J.; et al. Introducing mothur: Open-source, platform-independent, community-supported software for describing and comparing microbial communities. *Appl. Environ. Microbiol.* **2009**, *75*, 7537–7541. [CrossRef]

114. Douglas, G.M.; Maffei, V.J.; Zaneveld, J.R.; Yurgel, S.N.; Brown, J.R.; Taylor, C.M.; Huttenhower, C.; Langille, M.G.I. PICRUSt2 for prediction of metagenome functions. *Nat. Biotechnol.* **2020**, *38*, 685–688. [CrossRef]

115. Parks, D.H.; Tyson, G.W.; Hugenholtz, P.; Beiko, R.G. STAMP: Statistical analysis of taxonomic and functional profiles. *Bioinformatics* **2014**, *30*, 3123–3124. [CrossRef]

Article

Bacillus velezensis Strains for Protecting Cucumber Plants from Root-Knot Nematode *Meloidogyne incognita* in a Greenhouse

Anzhela M. Asaturova [1], Ludmila N. Bugaeva [2], Anna I. Homyak [1,*], Galina A. Slobodyanyuk [2], Evgeninya V. Kashutina [2], Larisa V. Yasyuk [2], Nikita M. Sidorov [1], Vladimir D. Nadykta [1] and Alexey V. Garkovenko [3,4]

[1] Federal Research Center of Biological Plant Protection, p/o 39, 350039 Krasnodar, Russia; biocontrol-vniibzr@yandex.ru (A.M.A.); elisitor@mail.ru (N.M.S.); vnadykta@mail.ru (V.D.N.)

[2] Lazarevskaya Experimental Plant Protection Station, the Affiliated Branch of the Federal Research Centre of Biological Plant Protection, l. 200, Sochi Highway-77, 354200 Sochi, Russia; bugaevaln@mail.ru (L.N.B.); bugaeval@mail.ru (G.A.S.); kashutinaev@mail.ru (E.V.K.); gnu_oszr@mail.ru (L.V.Y.)

[3] Shemyakin-Ovchinnikov Institute of Bioorganic Chemistry of the Russian Academy of Sciences, Miklukho-Maklaya Str. 16/10, 117997 Moscow, Russia; garkovenko@gmail.com

[4] Laboratory of Molecular Genetic Research in the Agroindustrial Complex, Department of Biotechnology, Biochemistry and Biophysics, Trubilin Kuban State Agrarian University, Kalinina Str. 13, 350044 Krasnodar, Russia

* Correspondence: HomyakAI87@mail.ru; Tel.: +7-(967)-311-58-10

Citation: Asaturova, A.M.; Bugaeva, L.N.; Homyak, A.I.; Slobodyanyuk, G.A.; Kashutina, E.V.; Yasyuk, L.V.; Sidorov, N.M.; Nadykta, V.D.; Garkovenko, A.V. *Bacillus velezensis* Strains for Protecting Cucumber Plants from Root-Knot Nematode *Meloidogyne incognita* in a Greenhouse. *Plants* **2022**, *11*, 275. https://doi.org/10.3390/plants 11030275

Academic Editors: Carlos Agustí-Brisach and Eugenio Llorens

Received: 29 December 2021
Accepted: 17 January 2022
Published: 20 January 2022

Publisher's Note: MDPI stays neutral with regard to jurisdictional claims in published maps and institutional affiliations.

Abstract: *Meloidogyne incognita* Kofoid et White is one of the most dangerous root-knot nematodes in greenhouses. In this study, we evaluated two *Bacillus* strains (*Bacillus velezensis* BZR 86 and *Bacillus velezensis* BZR 277) as promising microbiological agents for protecting cucumber plants from the root-knot nematode *M. incognita* Kof. The morphological and cultural characteristics and enzymatic activity of the strains have been studied and the optimal conditions for its cultivation have been developed. We have shown the nematicidal activity of these strains against *M. incognita*. Experiments with the cucumber variety Courage were conducted under greenhouse conditions in 2016–2018. We determined the effect of plant damage with *M. incognita* to plants on the biometric parameters of underground and aboveground parts of cucumber plants, as well as on the gall formation index and yield. It was found that the treatment of plants with *Bacillus* strains contributed to an increase in the height of cucumber plants by 7.4–43.1%, an increase in leaf area by 2.7–17.8%, and an increase in root mass by 3.2–16.1% compared with the control plants without treatment. The application of these strains was proved to contribute to an increase in yield by 4.6–45.8% compared to control. Our experiments suggest that the treatment of cucumber plants with two *Bacillus* strains improved plant health and crop productivity in the greenhouse. *B. velezensis* BZR 86 and *B. velezensis* BZR 277 may form the basis for bionematicides to protect cucumber plants from the root-knot nematode *M. incognita*.

Keywords: *B. velezensis*; cultivation conditions; cucumber; root-knot nematode; *Meloidogyne incognita*; greenhouse

1. Introduction

In greenhouses, the diversity of harmful species is less than in the field, however, the specific microclimate in greenhouses, as well as the absence of natural enemies as regulatory factors, lead to pest accumulation and increase its harmfulness to cultivated plants. Parasitic phytonematodes are among the most serious pests that negatively affect the quality of vegetables in a greenhouse [1,2]. They are obligate parasites that feed mostly on plant roots with common aboveground symptoms of stunting, yellowing, wilting, and yield losses and belowground root malformation due to direct feeding damage. These factors lead to yield losses [3]. Parasitic phytonematodes feed on many crops worldwide and they can cause enormous yield losses with an estimation of 100 billion dollars a year [4].

Root-knot nematodes are one of the most harmful groups of phytophagous organisms in greenhouses. About 5% of world crop production is annually destroyed by *Meloidogyne* spp. Gall nematodes induce overgrowth of root cells, which leads to the formation of galls on plant roots. The nematode damages the vascular tissue of the roots, thereby interfering with the normal movement of water and nutrients. Infected plants show signs of nutrient deficiency: slow growth, yellowing of leaves, wilting, and dying off of the plant. In such a condition, the yield of plants can drop by 50–80%, and in some cases, there is an entire yield loss [5]. Control of nematodes in greenhouses is expensive and time-consuming, and none of the existing methods completely relieves the plant from its presence. To reduce the number of nematodes in greenhouses, agricultural producers use resistant varieties and steam the soil. In addition, soil fumigation with highly toxic substances prohibited in greenhouses is often used. [6].

In recent years, there have been numerous studies worldwide concerning the possibility of reducing the harmfulness of phytopathogenic nematodes using antagonistic microorganisms [7–9]. Active strains of antagonistic bacteria with high nematicidal activity in combination with high biological and economic efficiency have been identified. Special attention of the experts is paid to *Bacillus*, *Pseudomonas*, and *Pasteuria* bacterial strains [10–13]. In addition, many researchers have found the nematicidal activity of the metabolites of *Trichoderma*, *Paecilomyces*, *Arthrobotrys*, *Beauveria*, *Pochonia*, *Fusarium*, and *Myrothecium* fungi [14–16].

Currently, successful research has been carried out and technologies have been developed for the development of biological products, such as Bioact WG (*Purpureocillium lilacinus*), KlamiC (*Pochonia chlamydosporia*), Econem (*Pasteuria penetrans*), Deny, Blue Circle (*Burkholderia cepacia*), DiTera (*Myrothecium Verruotia*), and Nortica VOTiVO (*Bacillus firmus*). Giant corporations, such as BASF and Bayer CropScience, which are key suppliers of plant protection products in the world market, became interested in the production of these preparations. It should be noted that some products are produced only for use in the country of manufacture [17]. However, none of these biological products are registered for use in Russia.

Currently, only one formulation, Phytoverm, based on *Streptomyces avermectilis* metabolites, is registered against phytonematodes in Russia. However, this group of products belongs to the category of agrochemicals and is prohibited for use in organic farming technologies. So far, there is no registered biological product based on microorganisms against phytonematodes in Russia [18]. The reasons for such a poor assortment of biological products are the insufficient knowledge of the biological characteristics of bacterial and fungal strains as the basis of nematicidal biological products, as well as the study of their trophic needs and resistance to various factors.

Thus, one of the promising solutions to the problem of protecting plants from *M. incognita* can be biological control of the pest population based on the use of effective microbiological agents that are currently not available in Russia.

This study focused on the isolation, identification, and characterization of *B. velezensis* BZR 86 and *B. velezensis* BZR 277 isolated from winter wheat roots. They showed high fungicidal activity against *F. graminearum* [19] and high nematicidal activity in vitro against the root-knot nematode *M. incognita* Kofoid et White [20]. The strains were selected from the BRC «State collection of beneficial insects, mites and microorganisms» of the Laboratory for the creation of microbiological plant protection products and the collection of microorganisms (FSBSI FRCBPP, Russia), since they have enzymatic, growth-stimulating, and nematicidal activity against the root-knot nematode *M. incognita*. In addition, the conditions for cultivating *Bacillus* strains were optimized.

2. Results

2.1. Characteristics and Identification of Bacillus strains

The cells of the *B. velezensis* BZR 86 strain are rod-shaped with rounded ends; single or paired; cells are motile; and have sizes of 2.8–4.1 × 6.6–9.4 microns. The cells usually contain spores. Coloring indicates positive results in the Gram stain. On the MPA, the shape of the colonies is round with a scalloped edge. The colonies are shiny and colorless. The profile of the colonies is flat, the edge is wavy, the structure is fine-grained, the texture is soft, and the colonies do not stick to the loop. The diameter of the colonies is 2–4 mm. On the PGA, rhizoid colonies with a smooth edge are formed. The colonies are matte and cream-colored. The profile of the colonies is convex, the structure is streaky in the center, fine-grained along the edge, the texture is soft, mucous, and the colonies adhere to the loop. The diameter of the colonies is 1–3 mm.

B. velezensis BZR 277 cells are rod-shaped with rounded ends; single or paired; cells are motile; and have sizes of 1.3–1.8 × 4.5–5.4 microns. The cells usually contain spores. Coloring indicates positive results in the Gram stain. On the MPA, the shape of the colonies is round with a scalloped edge. Colonies are opaque and colorless. The profile of the colonies is flat, the structure is fine-grained, the texture is soft, and the mucous colonies adhere to the loop. The diameter of the colonies is 0.5–2 mm. On the PGA, rhizoid colonies with a smooth edge are formed. Colonies are matte and cream-colored or yellowish-brown. The profile of the colonies is curved, the structure is streaky, the consistency is soft, mucous, and the colonies adhere to the loop. The diameter of the colonies is 1–3 mm.

MALDI-TOF MS analysis of bacterial colonies of the studied strains showed score values of ≥0.65 (BactoSCREEN ID) and ≥1.952 (Bruker Autoflex). These data confirm taxonomic identification only up to the genus Bacillus (Figure 1).

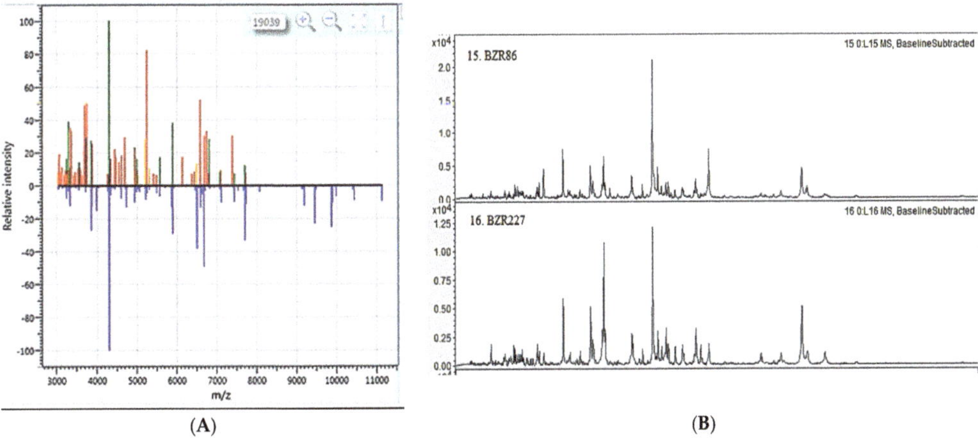

(A)   (B)

Figure 1. Identification of *Bacillus* strains BZR 86 and BZR 277: (**A**) Unique spectrum of ribosomal proteins of strain BZR 86 (blue peaks belong to the reference strain *Bacillus subtilis*, the rest of the peaks in the upper part of the profile belong to the studied strain BZR 86 (BactoSCREEN)). Green and yellow peaks coincide with the data of the reference strain *B. subtilis*, red ones do not match. Similar data were obtained by this method for the BZR 277 strain (data not shown); (**B**) Unique ribosomal protein spectra of strains BZR 86 and BZR 277 showing taxonomic identity of strains (Bruker).

A phylogenetic tree based on genome-wide sequencing of 120 conserved marker genes shows that strains BZR 86 and BZR 277 clearly cluster with *B. velezensis* NRRL B-41580 (*B. velezensis* GCF 001461 825.1 at the phylogenetic level); the average nucleotide identity (ANI) is 97.59% (Figure 2). Strains BZR 86 and BZR 277 are closer to *B. siamensis* and *B. amyloliquefaciens* than to *B. subtilis* (as determined using 16S rRNA).

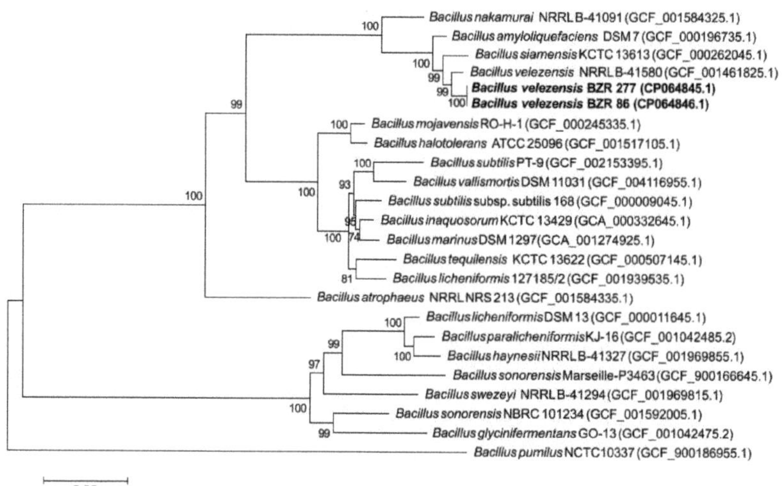

Figure 2. Maximum likelihood phylogenetic tree constructed using amino acid sequences of 120 conserved marker genes. The tree was constructed using PhyML v.3.3.

2.2. Enzymatic Activity of B. velezensis BZR 86 and B. velezensis BZR 277 Strains

It is known that effective lysis of the cell walls of pests is associated with the complex action of various hydrolytic enzymes. Therefore, the ability of *Bacillus* strains to produce various hydrolytic enzymes was studied. We revealed a different level of synthesis of lytic enzymes in the bacterial strains (Table 1).

Table 1. Enzymatic activity of *B. velezensis* BZR 86 and *B. velezensis* BZR 277 strains.

Strain	Enzymatic Activity			
	Lipase	Chitinase	Protease	Gelatinase
B. velezensis BZR 86	-	+	-	+
B. velezensis BZR 277	+++	-	+++	+

- no activity; + low activity; ++ mean activity; +++ high activity.

The *B. velezensis* BZR 277 strain showed the ability to synthesize protease and lipolytic enzymes, while *B. velezensis* BZR 86 strain showed the ability to synthesize chitinases only. Both strains produce gelatinase. When inoculating with a prick on a gelatinous medium in a test tube, it was noted that the *B. velezensis* BZR 86 strain forms a crater-like liquefaction of the gelatinase medium, and the *B. velezensis* BZR 277 strain forms a turnip-like one.

2.3. Cultivation Conditions for B. velezensis BZR 86 and B. velezensis BZR 277 Strains

In our studies, five parameters were identified that affect the growth of strains in the process of periodic cultivation: cultivation temperature, the acidity of the medium, sources of carbon and nitrogen nutrition, and cultivation time. During the study, we found that some parameters, such as temperature and acidity of the environment, have a significant effect on cell growth. When the incubation temperature changed from 20 to 35 °C, and the acidity of the medium from 3 to 10, there was significant variation in the titer (Table 2).

Table 2. The number of colony-forming units in liquid cultures is based on strains *B. velezensis* BZR 86 and *B. velezensis* BZR 277, depending on the cultivation conditions.

Parameter	Titer, CFU/ml	
	B. velezensis **BZR 86**	*B. velezensis* **BZR 277**
Temperature, °C		
20.0	$(9.6 \pm 0.14)\,[1] \times 10^6$ b [2]	$(3.4 \pm 0.3) \times 10^5$ b
25.0	$(8.6 \pm 0.42) \times 10^6$ a	$(6.2 \pm 0.14) \times 10^5$ a
30.0	$(8.3 \pm 0.67) \times 10^6$ a	$(1.4 \pm 0.04) \times 10^6$ c
35.0	$(1 \pm 0.05) \times 10^7$ c	$(6.6 \pm 0.17) \times 10^5$ a
pH		
3.0	$(3.2 \pm 0.06) \times 10^7$ c	$(4.3 \pm 0.2) \times 10^6$ d
6.0	$(7.6 \pm 0.3) \times 10^6$ a	$(1.7 \pm 0.3) \times 10^6$ c
8.0	$(1.1 \pm 0.14) \times 10^7$ b	$(1.2 \pm 0.02) \times 10^6$ b
10.0	$(1.1 \pm 0.4) \times 10^7$ b	$(1.1 \pm 0.05) \times 10^6$ a
Carbon sources		
sucrose	$(2.3 \pm 0.36) \times 10^6$ a	$(6.2 \pm 0.6) \times 10^5$ a
glucose	$(3.1 \pm 0.22) \times 10^6$ a	$(6.6 \pm 0.75) \times 10^5$ a
glycerol	$(2.3 \pm 0.25) \times 10^6$ a	$(1 \pm 0.02) \times 10^6$ a
molasses	$(1.6 \pm 0.03) \times 10^9$ b	$(5.8 \pm 0.39) \times 10^8$ b
Nitrogen sources		
NaNO3	$(3.7 \pm 0.4) \times 10^6$ a	$(7 \pm 0.66) \times 10^7$ a
peptone	$(1.8 \pm 0.07) \times 10^8$ e	$(4.7 \pm 0.4) \times 10^8$ c
yeast extracts	$(4 \pm 0.2) \times 10^7$ b	$(5.2 \pm 0.5) \times 10^7$ a
corn extracts	$(9.4 \pm 0.3) \times 10^7$ c	$(1.1 \pm 0.05) \times 10^8$ b
Cultivation time, h		
8	$(2.5 \pm 0.15) \times 10^6$ a	$(1.2 \pm 0.05) \times 10^7$ a
16	$(7.4 \pm 0.37) \times 10^8$ b	$(4.7 \pm 0.35) \times 10^8$ b
24	$(1.3 \pm 0.11) \times 10^9$ d	$(1.7 \pm 0.02) \times 10^9$ d
36	$(1.2 \pm 0.06) \times 10^9$ c	$(1.6 \pm 0.04) \times 10^9$ c
48	$(2.3 \pm 0.35) \times 10^7$ a	$(4 \pm 0.02) \times 10^7$ a
72	$(1.1 \pm 0.2) \times 10^7$ a	$(9.4 \pm 0.3) \times 10^7$ a

[1] The error corresponds to the standard deviation of three independent analyses. [2] Between the options marked with the same letters, when comparing within the columns there are no statistically significant differences according to the Duncan criterion at a 95% probability level-% increase to control. Each optimal parameter is determined while keeping the other parameters unchanged.

A high spore or cell titer for the *B. velezensis* BZR 86 strain was noted at a temperature of 35 °C and for the *B. velezensis* BZR 277 strain at a temperature of 30 °C. The optimal pH for both strains was three. Different nutrient sources had a more significant effect on cell growth. Thus, molasses had appeared to be the optimal carbon source for both strains. The highest titer was recorded on a nutrient medium, where peptone was used as a nitrogen source. The maximum cell titer for both strains was noted in the range of 24–36 h of cultivation. When studying the influence of the cultivation time on the dynamics of the growth of the strains, it was noted that the phase of initial growth (the period when the volume of cells increased, but not their number) began after the introduction of the parent culture into the medium and lasted up to 8 h. The phase of the most active growth began 16 h after the introduction of the parent culture and reached its maximum in 24–36 h for both strains. This was followed by a phase of withering away.

2.4. Study of B. velezensis BZR 86 and B. velezensis BZR 277 Strains under Greenhouse Conditions

As a result of comprehensive monitoring, it was found that the cause of the formation of the galls on plant roots is the root-knot nematode M. incognita. In 2016–2018, *B. velezensis* BZR 277 and *B. velezensis* BZR 86 strains were tested against the M. incognita on cucumber plants in the greenhouse. In 2016, a single soil treatment under the root during planting with the studied strains led to plant growth retardation, shredding, and necrosis of the leaves, and then plants wilting. The double treatment under the root with Bacillus strains during the entire growing season caused intensive plant development and no visual signs of plant damage by the nematode were observed. Growth and development indicators of plants after treatment with Bacillus strains significantly exceeded the control. The vegetative period of plants lasted for three months. To the end of the period, complete drying of the control plants was noted. Phytopathological analysis of cucumber plant samples showed that no signs of damage by peronosporosis and powdery mildew were discovered on the plants (Figure 3).

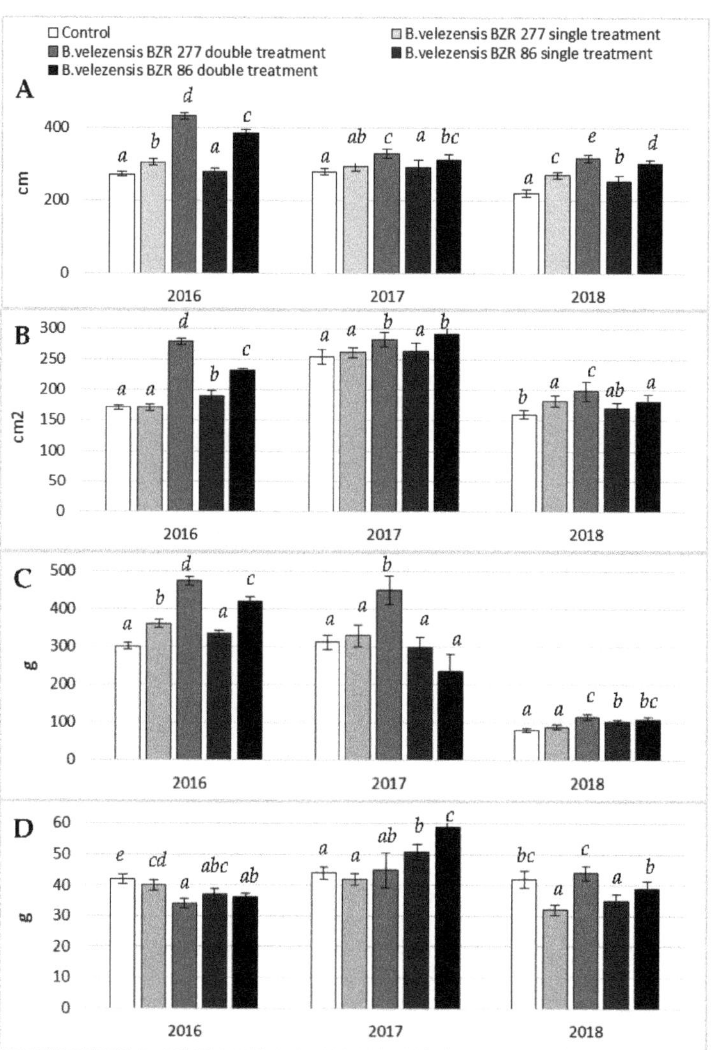

Figure 3. Effect of *B. velezensis* BZR 86 and *B. velezensis* BZR 277 treatment on the growth and development of cucumber plants in the greenhouse, 2016–2018: (**A**) plant height, cm; (**B**) leaf area, cm^2; (**C**) shoot mass, g; (**D**) root mass, g. The results are presented as the mean ± standard deviation. Different letters in each column indicate significant difference (p ≤ 0.05).

In 2016, the gall index for the *B. velezensis* BZR 86 strain was five times less, and for the *B. velezensis* BZR 277 strain was 14 times less, than in the control. However, yield is provided not only by a decrease in the number of galls, but also by a direct effect on the growth and development of plants. Therefore, the yield of cucumber plants treated with the studied strains was 5.2% higher for *B. velezensis* BZR 277 (a double treatment), and 21.4% higher for *B. velezensis* BZR 86 (a double treatment) compared to the control (Figure 4).

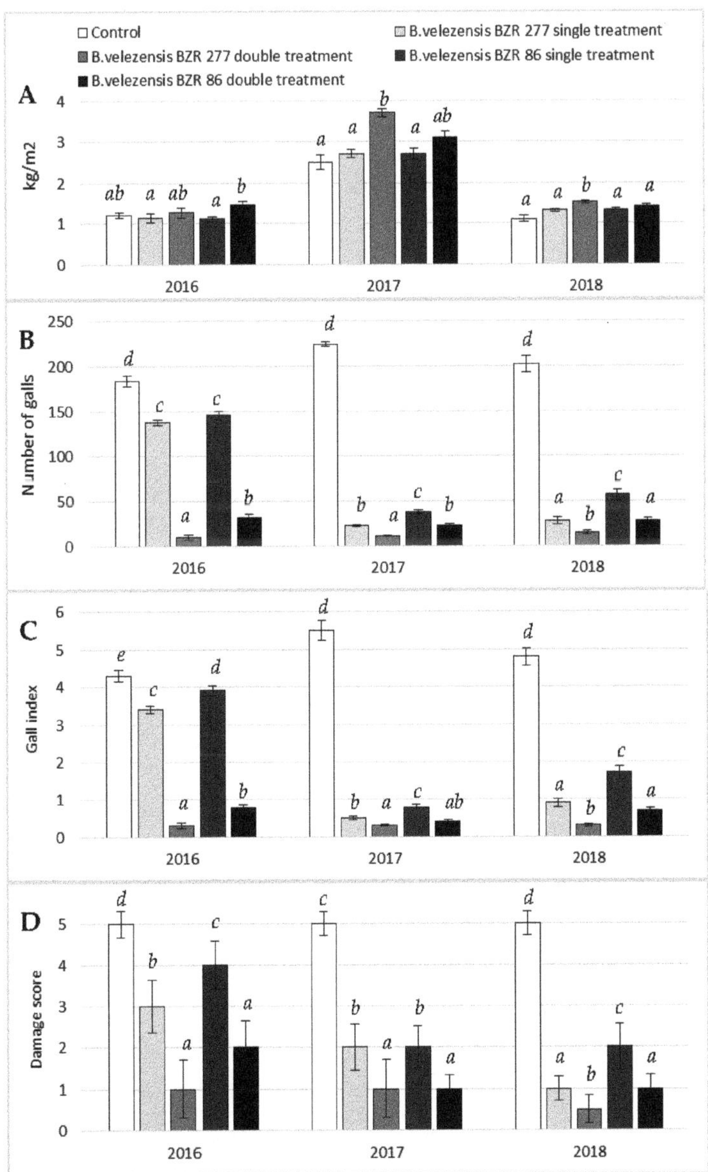

Figure 4. Effect of *B. velezensis* BZR 86 and *B. velezensis* BZR 277 treatment on cucumber yield and gall index in a greenhouse, 2016–2018: (**A**) yield, kg/m^2; (**B**) number of galls; (**C**) gall index; (**D**) damage score. The results are presented as the mean ± standard deviation. Different letters in each column indicate significant difference ($p \leq 0.05$).

In 2017, a double treatment of cucumber plants with *B. velezensis* BZR 86 and *B. velezensis* BZR 277 strains was statistically more effective than a single treatment. It was noted that the *B. velezensis* BZR 277 strain had a greater effect on the development of the aerial part of cucumber plants, while the *B. velezensis* BZR 86 strain stimulated the development of the root system (Figures 3 and 5). Due to the increase in the underground part, a higher percentage of nematode damage was noted for *B. velezensis* BZR 86 strain: two times compared to *B. velezensis* BZR 277. Both strains were noted to influence an increase in yield by 17.8–45.8% compared to the control (Figure 4).

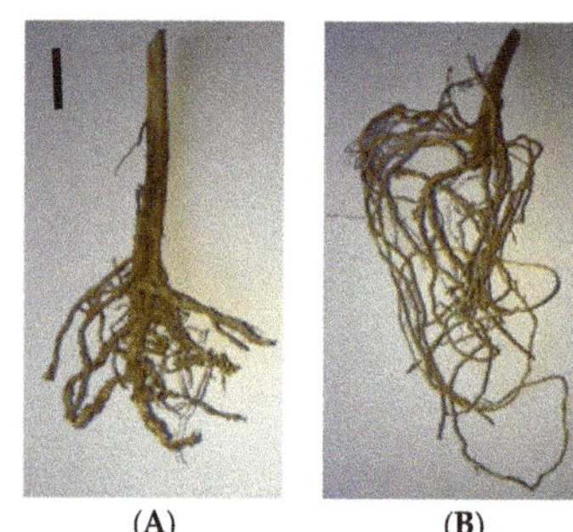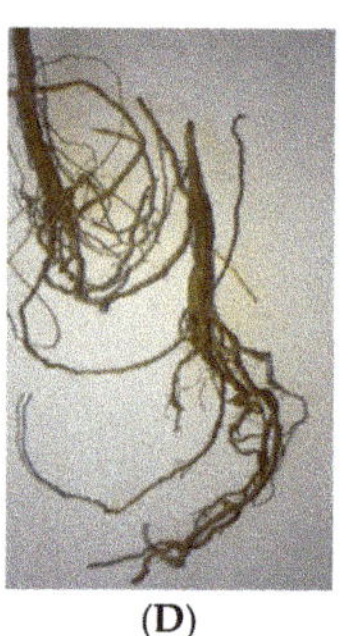

(A)　　　　**(B)**　　　　**(C)**　　　　**(D)**　　　　**(E)**

Figure 5. Effect of *B. velezensis* BZR 86 and *B. velezensis* BZR 277 strains on the degree of gall formation on the roots of cucumber plants: (**A**) *B. velezensis* BZR86, single treatment; (**B**) *B. velezensis* BZR86, double treatment; (**C**) *B. velezensis* BZR277, single treatment; (**D**) *B. velezensis* BZR277, double treatment; (**E**) control. Black scale bars represent 1 cm.

In 2018, a double treatment of cucumber plants with the *B. velezensis* BZR 86 and *B. velezensis* BZR 277 strains also turned out to be more effective than a single treatment. This may indirectly indicate the ability of the studied strains to enhance the induced systemic resistance of plants. The maximum biometric parameters were recorded for plants treated with the *B. velezensis* BZR 277 strain. It also showed greater activity against M. incognita when it was applied to the root system; two times fewer galls were noted compared to the strain B velezensis BZR 86. As a result, the maximum cucumber yield was recorded with double application of the *B. velezensis* BZR 277 strain—32.7% more than in the control, and 12.7% more than in the option with a double application of the *B. velezensis* BZR 86 strain (Table 1).

3. Discussion

In this study, we assessed the disease control efficacy of *B. velezensis* BZR 86 and *B. velezensis* BZR 277 strains against the root-knot nematode *M. incognita* of cucumber plants in the greenhouse.

Many researchers note the ability of *Bacillus* strains bacteria to reduce the number and harmfulness of phytoparasitic nematodes that attack agricultural crops [3,21]. One of the mechanisms of the nematicidal action of Bacillus strains is the synthesis of extracellular enzymes [22–25]. It is likely that extracellular enzymes, such as proteases, lipases, and chitinases, are capable of destroying the physical and physiological integrity of the nematode cuticle and eggs [26]. It was shown that enzymes can damage the egg membrane of nematodes, which consists of a protein matrix and a chitinous layer [27]. In addition, chitin is the main component of the nematode pharynx. Therefore, any disturbance in the synthesis or hydrolysis of chitin can lead to the death of nematode embryos, the laying of defective eggs, or a violation of molting [28]. The enzymatic activity (chitinase and protease) of the *B. megaterium* culture supernatant provided a nematicidal effect in the range of 21–30% against the larvae of *Meloidogyne* sp. and 30–37% against its eggs [29].

Some studies revealed the ability of Bacillus firmus to synthesize toxins that reduced the number of eggs of *Meloidogyne* spp. [30]. It is known that effective lysis of the cell walls of pests is associated with the complex action of various hydrolytic enzymes. Geng et al. [31] found the ability of *B. firmus* to produce peptidase group enzymes capable of destroying intestinal tissue of nematodes. That research served as the basis for the development of biological products Nortica and VOTiVO, which were successfully used in the United States. Other studies revealed 100% mortality of the second age larvae of the root-knot nematodes *M. incognita* and *M. javanica* within 72 h caused by liquid cultures based on *B. thuringiensis* and *P. fluorescens* bacterial strains. Furthermore, these bacteria stimulated plant growth [32].

Our results demonstrated the activity of *Bacillus* strains against *M. incognita* on cucumber plants (up to 76.4%). According to the study, we assumed the ability of Bacillus strains to synthesize proteolytic enzymes that caused the death of nematodes [10]. In Iran, the treatment of tomato plants with the filtrate of the bacterial culture *Bacillus* sp. in vitro caused mortality of *M. incognita* juveniles by 72% [33]. The use of chitinase and protease synthesized by the *B. pumilus* L1 strain in vitro reduced the hatching of *M. arenaria* eggs by 78%, and the mortality of juveniles was 98.6% compared to the control. In this study, structural damage of nematode bodies and eggs was noted [34]. Under in vitro conditions, *B. subtilis* culture filtrates inhibited the hatching of M. incognita eggs by 94.6% and caused mortality of juveniles by 91.3% [35]. Laboratory studies of *B. subtilis* strains exhibiting nematicidal activity against *M. incognita* showed that the use of a cell suspension provided the mortality of juveniles at the level of 39.3%, while the mortality rate when using the supernatant was 82.3–90.7%, which may be caused by the ability of *B. subtilis* to produce lytic enzymes [36]. These data are confirmed by the studies carried out in 2020, during which liquid culture filtrate caused the death of 90% of juveniles of the second stage and reduced egg hatchability by 97% in vitro [37].

Our studies confirmed the ability of *B. velezensis* BZR 86 and *B. velezensis* BZR 277 to produce a number of enzymes. The *B. velezensis* BZR 86 strain discovered chitinolytic activity. As to the *B. velezensis* BZR 277 strain, a high level of synthesis of lipases and proteases was noted, which may be partially involved in the process of suppression of the nematode *M. incognita*. Various factors have a significant effect on the growth of the number of cells and the synthesis of enzymes: temperature, acidity, the composition of the nutrient medium, and the incubation period [38]. Therefore, it is important to optimize the cultivation conditions for improving the nematicidal activity of bacterial strains. For example, Cheba et al. [39] showed that incubation of the *Bacillus* sp. R2 at 30 °C led to the highest level of chitinase synthesis. *B. cereus* JK-XZ3 filtrate cultivated at 30 °C showed the highest nematicidal activity, resulting in 100% mortality of Bursaphelenchus xylophilus [40]. However, Dukariya and Kumar [41] reported maximum chitinase productivity at 37 °C.

In our studies, we improved the cultivation conditions for strains exhibiting nematicidal activity according to a criterion, such as the number of colony-forming units. It is known that *Bacillus* strains are capable of growing in a temperature range of 5.5 °C to 55.7 °C [42–44]. According to the data obtained for the *B. velezensis* BZR 86 strain, the optimum temperature is 35 °C, and for the *B. velezensis* BZR 277 strain—30 °C. Our study is fully consistent with the conclusions of Park et al. [45], according to whom, when cultivating the B. velezensis GH1-13 strain, a titer of 7.5×10^9 CFU/mL was obtained at 37 °C. A similar pattern was noted in the study by Usanov et al. [46], according to which an increase in the cultivation temperature to 40 °C has a positive effect on the growth dynamics of the *B. subtilis*, and the number of colony-forming units exceeds the control by 88.9%.

In our study, the maximum titer of both strains was observed at pH 3. However, it should be noted that when the strains were cultivated on media with different levels of acidity, no significant jumps in cell density were observed. These data are confirmed by Mohapatra et al. [47] 2015, according to whom, insignificant fluctuations in CFU were observed during cultivation of the *Bacillus* sp. P1on media with pH 6–9 (5.0–8.4×10^5 CFU/mL). *B. velezensis* XT1 strain was able to grow in a wide pH range of 5–10 [48]. A relatively high cell density at pH 3 was noted during the cultivation of the *B. velezensis* CE 100 strain, the initial pH of the medium of 7.5 decreased to 4.7 over 24 h of cultivation, which may indicate the ability of some *B. velezensis* strains to grow in an acidic medium [49].

Molasses has been shown to optimize the parameters of growing strains-producers of biological products [50,51]. Being a source of not only sugars, but also vitamins, macro- and microelements, this ingredient ensures the active growth of microorganism cultures. For strains *B. velezensis* BZR 86 and *B. velezensis* BZR 277, the cell titer in liquid cultures using molasses as media component was 2–3 times higher than on media with sucrose, glucose, and glycerol and amounted up to $(1.6 \pm 0.03) \times 10^9$ CFU/mL for the *B. velezensis* BZR 86 strain and $(5.8 \pm 0.39) \times 10^8$ CFU/mL for the *B. velezensis* BZR 277 strain. Peptone was the most preferred source of nitrogen nutrition for both strains. It should be noted that the B. velezensis BZR 86 strain turned out to be more sensitive to the source of nitrogen nutrition: when $NaNO_3$ was added to the nutrient medium, the titer was $(3.7 \pm 0.4) \times 10^6$ CFU/mL, and when peptone was added—$(1.8 \pm 0.07) \times 10^8$ CFU/mL. These results are consistent with the studies on the optimization of the nutrient medium for the cultivation of the *B. velezensis* NRC-1 strain, according to which the maximum cell growth and mannanase synthesis were achieved on the medium with peptone [52].

The cultivation time is an important parameter for the growth of biocontrol agents. If bacterial cultures are incubated for too long, some metabolites may be converted to other compounds. In contrast, if the incubation period is not long enough, it is possible that important metabolites have not yet been synthesized (for example, enzymes that are formed during the stationary growth phase). This demonstrates the importance of making bacterial growth curves [53]. During our research, it was noted that the phase of initial growth began after the introduction of a parent culture with a titer of $(1.4 \pm 0.05) \times 10^8$ CFU/mL in the *B. velezensis* BZR 277 strain and $(8.1 \pm 0.01) \times 10^7$ CFU/mL in the *B. velezensis* 86 strain and lasted up to 8 h. The phase of the most active growth began 16 h after the introduction of the mother culture and began to decay after 36 h for the *B. velezensis* 277 strain and from 8 to 36 h for the *B. velezensis* 86 strain. The maximum cell titer was observed during this period and amounted to $(1.7 \pm 0.02) \times 10^9$ CFU/mL in the *B. velezensis* BZR 277 strain and $(1.3 \pm 0.11) \times 10^9$ CFU/mL in the B. velezensis BZR 86 strain. The phase of bacterial cell death occurred 36 h after the start of cultivation for both strains, which contradicts the data obtained in the study of the cultivation of the *B. velezensis* IP22 strain, according to which the stationary phase lasted up to 60 h [54].

Three-year vegetation tests on the Courage variety of cucumber plants under the greenhouse conditions showed that plants treated with the B. velezensis BZR 86 and *B. velezensis* BZR 277 strains developed more intensively; the growth and development indicators significantly exceeded the control ones. This tendency was observed in research by Sahebani and Omranzade [55], according to which the introduction of a liquid culture based on *B. megaterium* wr101 into the soil infected with *M. javanica* contributed to an increase in the mass of cucumber shoots twice as compared to the control. Other authors showed that the decrease in the number of M. incognita contributed to the improvement of plant development, which was manifested by an increase in biometric indicators and in cucumbers compared with the control. It was noted that the use of two Bacillus strains contributed to the formation of an additional cucumber yield from 4.6% to 45.8%, which confirms the results of the 2019 studies, in which the treatment of cucumber plants with a liquid culture based on the B. subtilis Bs-1 strain provided an additional yield of up to 53.1% in combination with a high nematicidal effect against *M. incognita* [56].

In our research, we showed that in three consecutive years, application of two Bacillus strains in the greenhouse demonstrated that the number of galls per gram of cucumber root mass decreased by 13 times for the *B. velezensis* BZR 277 strain, and by 7.2 times for the *B. velezensis* BZR 86 strain. In general, the *B. velezensis* BZR 277 strain exhibited a more pronounced nematicidal and plant growth-promoting effect. Probably, such difference is due to the selectivity of the nematicidal action of lipases and proteases synthesized by the *B. velezensis* BZR 277 strain against predominantly adult nematodes. It should be noted that the cuticle of nematodes contains different types of structural proteins and their proportions change throughout their life cycle as other authors have shown [54,57].

4. Materials and Methods

4.1. Microorganisms

B. velezensis BZR 86 and *B. velezensis* BZR 277 were isolated from the field soil of the Krasnodar region (45°1′ N, 38°59′ E). Winter wheat plants were selected under aseptic conditions, placed in sterile bags, and stored in a refrigerator at +4 °C. The strains were isolated from the root zone of winter wheat plants by the method of successive washing of the roots from adjacent soil particles [58]. The roots with soil were placed in a flask with 100 mL of sterile water and shaken for 40 min at 180 rpm. The roots were removed from the flask with sterile tweezers and transferred to the next flask containing 100 mL of sterile tap water. The procedure was repeated, successively washing the roots in three flasks (40 min each). The bacteria were inoculated by streaking from the second and third flasks. The *B. velezensis* BZR 277 strain was inoculated on potato-glucose agar and the *B. velezensis* BZR 86 strain on a medium with chitinase. Petri dishes with *B. velezensis* BZR 86 strain were incubated at +28 °C for three days. Petri dishes with *B. velezensis* BZR 277 strain were incubated at +4 °C for 21 days. Isolated colonies were streaked onto potato-glucose agar. This procedure was repeated until a pure culture of the strain was obtained in the Petri dish.

Strains deposited in the Bioresource collection "State collection of beneficial insects, mites and microorganisms" of the Federal State Budgetary Scientific Institution "Federal Research Center of Biological Plant Protection" (FSBSI "FRCBPP") (registry number 585858).

4.2. MALDI-TOF MS Analysis

MALDI-TOF MS analysis was performed to identify bacterial cultures. The taxonomic status of bacterial cultures was determined by the BactoScreen analyzer. The spectra were analyzed using the BactoScreen-ID software version 2.4. In addition, the BRUKER autoflex III smartbeam and flexcontrol 3.0 software was used. The analysis was carried out using a database containing the spectra of 3995 microorganisms. The values obtained were expressed as a logarithmic score. In particular, a score of 2.0—identification at the species level is allowed, a score in the range of 1.7 ± 2.0—identification only at the genus level, a score below 1.7—the absence of significant similarity of the obtained spectrum with any record of the database (not reliable identification). *E. coli* DH5a proteins were used as a bacterial standard.

4.3. Molecular Genetic Identification of Strains

Isolation of genomic DNA preparations of *B. velezensis* BZR 86 and *B. velezensis* BZR 277 strains was performed using the DNeasy PowerSoil Kit, QIAGEN (Hilden, Germany) according to standard protocols. The amount of isolated DNA was determined by the fluorometric method using Qubit dsDNA HS Assay Kit, ThermoFisher Scientific (Waltham, MA, USA) according to the manufacturer's protocols.

To determine the complete genomes of *B. velezensis* BZR 86 and *B. velezensis* BZR 277 strains, a combined strategy was used, including the use of two high-productive sequencing platforms—Illumina (MiSeq) and monomolecular sequencing on MinIon (Oxford Nanopore). At the first stage, a genomic library of "random fragments" was prepared, suitable for sequencing on the MiSeq device (Illumina) using the NEBNext ultra II DNA Library kit (NEB) and then read on the MiSeq genomic analyzer. At the second stage, the genome was additionally sequenced using monomolecular nanopore sequencing technology (MinION instrument from Oxford Nanopore). To prepare genomic libraries suitable for sequencing on the MinION device, a Ligation Sequencing kit 1D (Oxford Nanopore) was used according to the manufacturer's recommendations. Sequencing on the MinION was performed using the Ligation Sequencing kit 1D protocol using FLO-MIN110 wells. The sequencing results were saved in a FAST5 file. Using the flash program [59], paired intersecting reads obtained on MiSeq (Illumina) were combined and the poor-quality ends of the reads were cut using the Sickle program. Structural (protein-coding) genes and ribosomal RNA genes were identified and their functions were theoretically predicted using the RAST server [60].

Multiple alignments of the concatenated amino acid sequences of 120 bacterial single-copy marker genes were performed using the Genome Taxonomy Data Base (GTDB-Tk v. 1.3.0 toolkit software) from RefSeq and Genbank genomes (USA) [61]. This multiple alignment was used to construct a maximum similar phylogenetic tree using PhyML v.3.3 with default parameters [62]. Internal branching support was assessed using a Bayesian test in PhyML.

The obtained *B. velezensis* BZR 86 sequences were deposited in the NCBI database under accession numbers PRJNA677970 (BioProject), SRX9502286 (SRA), and SAMN16784691 (BioSample). The obtained *B. velezensis* BZR 277 sequences were deposited into the NCBI database under accession numbers PRJNA588983 (BioProject), SRX9502288 (SRA), and SAMN16784690 (BioSample).

4.4. Cultural and Morphological Characteristics of Bacillus strains

Cultural and morphological characteristics of the strains were studied on meat-peptone agar (MPA) and potato-glucose agar (PGA) using an Axio Scope A1 microscope (Jena, Germany). The shape and size of the cells, the ability to form spores, the location of spores in the cells, the ability to move, and coloring were determined according to Gram. We studied the growth characteristics, shape, size, surface, profile, color, shine, and transparency of the colonies, as well as their edge, structure, and consistency [63].

4.5. Enzymatic Activity of Bacillus strains

The enzymatic activity of antagonistic bacterial strains was carried out using generally accepted methods [64]. Chitinase, lipase, and protease activity was determined. To determine the chitinolytic activity, a synthetic medium of the following composition was used, g/l: sucrose—20.0; $NaNO_3$—3.0; KH_2PO_4—1.0; MgSO4—0.3; chalk—10.0; and agar—20.0. The medium was sterilized by autoclaving, poured into Petri dishes, and cooled. Inoculating of bacterial strains was performed by streaking. Petri dishes were incubated for 7–10 days at a temperature of +28.0 °C. Chitinase activity was judged by the formation of clearing zones around the colonies.

Lipolytic activity was determined on yolk agar of the following composition, g/l: peptone—40.0; glucose—2.0; Na_2HPO_4—5.0; NaCl—2.0; $MgSO_4$ 0.5% solution—2.0 mL; and agar—25.0. 4. The medium was sterilized by autoclaving and cooled to +60.0 °C. The egg shell was disinfected with alcohol and allowed to dry. The egg was broken and the yolk was separated from the egg white. The yolk, in compliance with the rules of asepsis, was transferred into molten agar and stirred until a homogeneous suspension was obtained, which was poured into Petri dishes and left to solidify. Inoculating of bacterial strains was performed by streaking. Petri dishes were incubated for 14 days at a temperature of +28.0 °C. Then the lid of the Petri dish was removed and the surface was carefully examined under oblique illumination. Lipolytic activity was judged by the formation of an oily, glistening, or nacreous layer above and around the bacterial colony on the agar surface.

To determine protease activity, sterile (autoclaved) skim milk was mixed at +50.0 °C with an equal volume of 4% molten aqueous agar. Inoculating of bacterial strains was performed by streaking. Petri dishes were incubated for 14 days. Protease activity was judged by the formation of clearing zones around the colonies.

Gelatinase activity was tested on meat-peptone gelatin, g/l: meat-peptone broth—39.0; and gelatin—150.0. The medium was poured into test tubes, sterilized by autoclaving, and cooled at room temperature. Inoculating of bacterial strains was carried out by injection. The tubes were incubated for 7–10 days at room temperature. The liquefaction of the gelatin was observed visually. The intensity and form of liquefaction were indicated.

4.6. Optimal Conditions for the Cultivation of Bacillus strains

To determine the optimal cultivation temperature, the strains were incubated at 20.0, 25.0, 30.0, and 35.0 °C. Czapek medium for bacteria was used as a basis [63]. Carbon sources sucrose, glucose, molasses, and glycerol were added in the test media. In the study of carbon sources nitrogen sodium nitrate served as the unchanged nutrition component. In determining the optimal sources of nitrogen nutrition peptone, $NaNO_3$, yeast and corn extracts were tested with glucose as the only (constant) carbon source. To choose the optimal acidity of the medium, the strains were grown on a liquid medium, g/L KCl—0.5, $MgSO_4 \times 7H_2O$—0.5, $K_2HPO_4 \times 3H_2O$—1.0, $CaCO_3$—3.0, $FeSO_4 \times 7H2O$—0.01, corn extract—2.0, molasses—20.0 at optimum temperatures. By adding lactic acid or alkali (4N NaOH solution), the medium pH was adjusted to 3.0, 6.0, 8.0, and 10.0 using a Sartorius PB-11 pH-meter (Goettingen, Germany). To determine the optimal cultivation time, samples for analysis were taken after 8, 16, 24, 36, 48, and 72 h from the start of cultivation. All experiments were replicated three times. For all experiments, a liquid culture was obtained by the method of periodic cultivation. Incubation was carried out in thermostat cell cultivation systems (180 rpm) "New Brunswick Scientific Excella E25" (Enfield, CT, USA) for 48 h. Periodic cultivation was carried out in conical flasks (350 mL) with a nutrient medium volume of 100 mL and preliminary introduction of stock culture (2% of the nutrient medium volume). The stock culture was obtained by introducing agar blocks with the studied strains into conical flasks and subsequent cultivation.

At the end of cultivation, the number of bacterial cells was determined by the Koch method in all experiments on MPA [63]. The grown colonies were counted with the Color Qcount, Spiral Biotech, 530 (Canton, MA, USA) system for the automatic counting colonies.

In our research we used Unique Scientific Facility "New generation technological line for developing microbiological plant protection products" of Federal Research Center of Biological Plant Protection, Krasnodar, Russia (http://ckp-rf.ru/%E2%84%96671367, accessed on 21 August 2019).

4.7. Greenhouse Evaluation of Bacillus strains

A liquid culture of bacterial strains was obtained in New Brunswick Scientific Excella E25 cell culture systems (180 rpm) on a potato-glucose medium, g/L: potato broth—500.0 and glucose—20.0. Cultivation was carried out for 48 h.

The average temperature in the greenhouse was 27.5 °C in May, 35.4 °C in June, 35.6 °C in July, and 37.5 °C in August.

The tests were carried out in conditions of protected ground in a greenhouse with a total area of 100 m^2. Five variants of the experiment were randomized in three times repetition and the area of each plot was 5 m^2. There were eight sample plants in each plot. The cucumber plants of Courage variety with natural infection by root-knot nematode were used. The number of galls was determined by the Guskova method [65]. The starting titer of the liquid culture preparation was 1×10^9 CFU/mL. The application rate of the preparation was 360 L/ha (50–60 mL per plant). The liquid culture was diluted at the rate of 60 mL per liter of water. The consumption rate of the working fluid is up to 6000 L/ha. In total, 200 mL of the preparation was added under the root of each cucumber plant.

Assessment of nematicidal activity of the strains was carried to the scheme:

1. Single treatment by suspension of *B. velezensis* BZR 86 before planting.
2. Double treatment by suspension of *B. velezensis* BZR 86 during planting followed by treatment under the plant root 3 weeks after planting.
3. Single treatment by suspension of *B. velezensis* BZR 277 before planting.
4. Double treatment by suspension of *B. velezensis* BZR 277 during planting followed by treatment under the plant root 3 weeks after planting.
5. Plants without treatment were used as control.

The effectiveness of the bacterial strains was determined by plant height, leaf area, shoot mass and root mass, the ratio of the number of galls to the mass of roots (damage score), and yield.

The root gall severity was based on gall indices (GI) measured on 0 to 5 scales: 0-GI = 0%; 1-GI = 10 to 20%; 2-GI = 21 to 50%; 3-GI = 51 to 80%; 4-GI = 81 to 100%., 5-GI = > 100% on roots [66].

4.8. Statistical Analysis

Statistical data processing was performed by standard methods using MS Excel and ANOVA program for Windows. All data were expressed as mean from triplicate samples ± standard deviation. Duncan test was used, and differences were considered statistically significant at $p < 0.05$ level.

5. Conclusions

BZR 277 and BZR 86 strains are endophytic bacteria associated with the roots of winter wheat plants from Krasnodar Krai (Russia). They are representatives of *B. velezensis*. They possess the properties of PGPR, exhibiting enzymatic activity and promoting the growth of aboveground and underground parts of the cucumber variety Courage and increasing its yield. In addition, BZR 277 and BZR 86 strains have nematicidal activity against the root-knot nematode *M. incognita*. However, further research is needed on the mechanisms associated with the growth-stimulating and biocontrol activity of the strains.

Author Contributions: Project administration, A.M.A. and V.D.N.; writing—original draft preparation, A.I.H.; resources, L.N.B.; investigation, G.A.S., L.V.Y., and N.M.S.; data curation, E.V.K.; software, A.V.G. All authors have read and agreed to the published version of the manuscript.

Funding: The study of bacterial strains, cultivation conditions, and greenhouse evaluation was carried out within the framework of the State Assignment of the Ministry of Science and Higher Education of the Russian Federation, No. FGRN-2022-0005. DNA sequencing of the strain BZR 277, BZR 86, and the construction of the phylogenetic tree based on the whole genome sequencing was supported by a project of the Kuban Science Foundation (grant MFI-20.1/68).

Institutional Review Board Statement: Not applicable.

Informed Consent Statement: Not applicable.

Data Availability Statement: Data is contained within the article.

Conflicts of Interest: The authors declare no conflict of interest.

References

1. Jones, J.T.; Haegeman, A.; Danchin, E.G.; Gaur, H.S.; Helder, J.; Jones, M.G.; Kikuchi, T.; Manzanilla-López, R.; Palomares-Rius, J.E.; Wesemael, W.M.L.; et al. Top 10 plant-parasitic nematodes in molecular plant pathology. *Mol. Plant Pathol.* **2013**, *14*, 946–961. [CrossRef]
2. Aydinli, G.; Mennan, S. Identification of root-knot nematodes (*Meloidogyne* spp.) from greenhouses in the Middle Black Sea Region of Turkey. *Turk. J. Zool.* **2016**, *40*, 675–685. [CrossRef]
3. Briar, S.S.; Wichman, D.; Reddy, G.V. Plant-parasitic nematode problems in organic agriculture. In *Organic Farming for Sustainable Agriculture. Sustainable Development and Biodiversity;* Nandwani, D., Ed.; Springer: Cham, Switzerland, 2016; Volume 9, pp. 107–122. [CrossRef]
4. Topalović, O.; Bredenbruch, S.; Schleker, A.S.S.; Heuer, H. Microbes attaching to endoparasitic phytonematodes in soil trigger plant defense upon root penetration by the nematode. *Front. Plant Sci.* **2020**, *11*, 138. [CrossRef] [PubMed]
5. Tagiev, M.M. Gall nematodes (*Meloidogyne*) on the Apsheron Peninsula and their control. *Succes. Mod. Sci. Educ.* **2015**, *5*, 22–24.
6. Singh, M.C.; Yousuf, A.; Singh, J.P. Greenhouse microclimate modeling under cropped conditions: A review. *Res. Environ. Life Sci.* **2016**, *9*, 1552–1557.
7. Abd-Elgawad, M.M.M. Optimizing safe approaches to manage plant-parasitic nematodes. *Plants* **2021**, *10*, 1911. [CrossRef]
8. Peng, D.; Lin, J.; Huang, Q.; Zheng, W.; Liu, G.; Zheng, J.; Zhu, L.; Sun, M. A novel metalloproteinase virulence factor is involved in *Bacillus thuringiensis* pathogenesis in nematodes and insects. *Environ. Microbiol.* **2016**, *18*, 846–862. [CrossRef] [PubMed]
9. Zheng, Z.; Zheng, J.; Zhang, Z.; Peng, D.; Sun, M. Nematicidal spore-forming *Bacilli* share similar virulence factors and mechanisms. *Sci. Rep.* **2016**, *6*, 31341. [CrossRef]
10. Aballay, E.; Ordenes, P.; Mårtensson, A.M.; Persson, P. Effects of rhizobacteria on parasitism by *Meloidogyne ethiopica* on grapevines. *Eur. J. Plant Pathol.* **2012**, *135*, 137–145. [CrossRef]
11. Ann, Y.C. Screening for Nematicidal activities of *Bacillus* species against root knot nematode (*Meloidogyne incognita*). *Am. J. Exp. Agric.* **2013**, *3*, 794–805. [CrossRef]

12. Mahesha, H.S.; Ravichandra, N.G.; Rao, M.S.; Narasegowda, N.C.; Shreeshail, S.; Shivalingappa, H. Bio-efficacy of different strains of *Bacillus* spp. against *Meloidogyne incognita* under in vitro. *Int. J. Curr. Microbiol. Appl. Sci.* **2017**, *6*, 2511–2517. [CrossRef]
13. Abdel-Salam, M.S.; Ameen, H.H.; Soliman, G.M.; Elkelany, U.S.; Asar, A.M. Improving the nematicidal potential of *Bacillus velezensis* and *Lysinibacillus sphaericus* against the root-knot nematode *Meloidogyne incognita* using protoplast fusion technique. *Egypt. J. Biol. Pest Control* **2018**, *28*, 31. [CrossRef]
14. Jang, J.Y.; Choi, Y.H.; Shin, T.S.; Kim, T.H.; Shin, K.-S.; Park, H.W.; Kim, Y.H.; Kim, H.; Choi, G.J.; Jang, K.S.; et al. Biological control of *Meloidogyne incognita* by *Aspergillus niger* F22 producing oxalic acid. *PLoS ONE* **2016**, *11*, e0156230. [CrossRef]
15. Nguyen, V.T.; Yu, N.H.; Lee, Y.; Hwang, I.M.; Bui, H.X.; Kim, J.-C. Nematicidal activity of cyclopiazonic acid derived from *Penicillium* commune against root-knot nematodes and optimization of the culture fermentation process. *Front. Microbiol.* **2021**, *12*, 726504. [CrossRef] [PubMed]
16. Kim, Y.J.; Duraisamy, K.; Jeong, M.-H.; Park, S.-Y.; Kim, S.; Lee, Y.; Nguyen, V.T.; Yu, N.H.; Park, A.R.; Kim, J.-C. Nematicidal activity of chemical intermediates of grammicin biosynthesis pathway in *Xylaria grammica* EL000614 against *Meloidogyne incognita*. *Molecules* **2021**, *26*, 4675. [CrossRef]
17. Stoytcheva, M. *Pesticides in the Modern World—Pesticides Use and Management*; InTech: London, UK, 2011; 532p.
18. The State Catalog of pesticides and agrochemicals approved for use on the territory of the Russian Federation. Ministry of Agriculture: Moscow, Russia, 2021; pp. 143–144.
19. Asaturova, A.M.; Dubyaga, V.M.; Tomashevich, N.S.; Zharnikova, M.D. Selection of promising biological control agents to protect winter wheat against *Fusarium* pathogens. *Electron. Polythem. Sci. J. KubSAU* **2012**, *75*. Available online: http://ej.kubagro.ru/2012/01/pdf/37.pdf (accessed on 18 November 2021).
20. Lychagina, S.V.; Migunova, V.D.; Asaturova, A.M. The influence of Bacillus bacteria on the development of cucumber plants affected by the gall nematode. *Theor. Pract. Control. Parasit. Dis.* **2017**, *18*, 227–229.
21. Abd El-Rahman, A.F.; Shaheen, H.A.; Abd El-Aziz, R.M.; Ibrahim, D.S.S. Influence of hydrogen cyanide-producing rhizobacteria in controlling the crown gall and root-knot nematode, *Meloidogyne incognita*. *Egypt. J. Biol. Pest Control* **2019**, *29*, 41. [CrossRef]
22. Chahal, P.P.K.; Chahal, V.P.S. Effect of thuricide on the hatching of eggs root-knot nematode, *Meloidogyne incognita*. *Curr. Nematol.* **1993**, *4*, 247.
23. Dhawan, S.C.; Sarvjeet, K.; Aqbal, S. Effect of *Bacillus thuringiensis* on the mortality of root-knot nematode, *Meloidogyne incognita*. *Indian J. Nematol.* **2004**, *34*, 98–99.
24. Vijayalakshmi, S.; Ranjitha, J.; Rajeswari, V.D. Enzyme production ability by *Bacillus subtilis* and *Bacillus licheniformis*— A comparative study. *Asian J. Pharm. Clin. Res.* **2013**, *6*, 29–32.
25. Mota, M.S.; Gomes, C.B.; Souza Júnior, I.T.; Moura, A.B. Bacterial selection for biological control of plant disease: Criterion determination and validation. *Brazil. J. Microbiol.* **2017**, *48*, 62–70. [CrossRef]
26. Dihingia, S.; Das, D.; Bora, S. Effect of microbial secretion on inhibitory effect of phytonematode: A review. *Int. J. Inform. Res. Rev.* **2017**, *04*, 4275–4280.
27. Li, X.; Hu, H.J.; Li, J.Y.; Wang, C.; Chen, S.L.; Yan, S.Z. Effects of the endophytic bacteria *Bacillus cereus* BCM2 on tomato root exudates and *Meloidogyne incognita* infection. *Plant Dis.* **2019**, *103*, 1551–1558. [CrossRef]
28. Chen, Q.; Peng, D. Nematode Chitin and Application. *Adv. Exp. Med. Biol.* **2019**, *1142*, 209–219.
29. Tran, T.P.H.; Wang, S.-L.; Nguyen, V.B.; Tran, D.M.; Nguyen, D.S.; Nguyen, A.D. Study of novel endophytic bacteria for biocontrol of black pepper root-knot nematodes in the central highlands of Vietnam. *Agronomy* **2019**, *9*, 714. [CrossRef]
30. Mendoza, A.R.; Kiewnick, S.; Sikora, R.A. In vitro activity of *Bacillus firmus* against the burrowing nematode *Radopholus similis* the root-knot nematode *Meloidogyne incognita* and the stem nematode *Ditylenchus dipsaci*. *Biocontrol Sci. Technol.* **2008**, *18*, 377–389. [CrossRef]
31. Geng, C.; Nie, X.; Tang, Z.; Zhang, Y.; Lin, J.; Sun, M.; Peng, D. A novel serine protease, Sep1, from *Bacillus firmus* DS-1 has nematicidal activity and degrades multiple intestinal-associated nematode proteins. *Sci. Rep.* **2016**, *6*, 25012. [CrossRef]
32. Mekete, T.J.; Kiewnick, S.; Sikora, R. Endophytic bacteria from Ethiopian coffee plants and their potential to antagonize *Meloidogyne incognita*. *Nematology* **2009**, *11*, 117–127. [CrossRef]
33. Ramezani, M.M.; Mahdikhani, M.E.; Baghaee, R.S. The nematicidal potential of local *Bacillus* species against the root-knot nematode infecting greenhouse tomatoes. *Biocontrol Sci. Technol.* **2014**, *24*, 279–290. [CrossRef]
34. Lee, Y.S.; Kim, K.Y. Antagonistic potential of *Bacillus pumilus* L1 against root-knot nematode, *Meloidogyne arenaria*. *J. Phytopathol.* **2016**, *164*, 29–39. [CrossRef]
35. Rao, M.S.; Kamalnath, M.; Umamaheswari, R.; Rajinikanth, R.; Prabu, P.; Priti, K.; Grace, G.N.; Chaya, M.K.; Gopalakrishnan, C. *Bacillus subtilis* IIHR BS-2 enriched vermicompost controls root knot nematode and soft rot disease complex in carrot. *Sci. Hortic.* **2017**, *218*, 56–62. [CrossRef]
36. Basyony, A.G.; Abo-Zaid, G.A. Biocontrol of the root-knot nematode, *Meloidogyne incognita*, using an eco-friendly formulation from *Bacillus subtilis*, lab and greenhouse studies. *Egypt. J. Biol. Pest Control* **2018**, *28*, 87. [CrossRef]
37. Hussain, T.; Haris, M.; Shakeel, A.; Ahmad, G.; Khan, A.A.; Khan, M.A. Bio-nematicidal activities by culture filtrate of *Bacillus subtilis* HussainT-AMU: New promising biosurfactant bioagent for the management of Root Galling caused by *Meloidogyne incognita*. *Vegetos* **2020**, *33*, 229–238. [CrossRef]
38. Tyc, O.; Song, C.; Dickschat, J.S.; Vos, M.; Garbeva, P. The ecological role of volatile and soluble secondary metabolites produced by soil bacteria. *Trends Microbiol.* **2017**, *25*, 280–292. [CrossRef] [PubMed]

39. Cheba, B.A.; Zaghloul, T.I.; El-Mahdy, A.R.; El-Massry, M.H. Effect of nitrogen sources and fermentation conditions on *Bacillus* sp. R2 chitinase production. *Procedia Manuf.* **2018**, *22*, 280–287. [CrossRef]

40. Zhang, W.J.; Wu, X.Q.; Wang, Y.H. Nematicidal activity of bacteria against *Bursaphelenchus xylophilus* and its fermentation and culture characteristics. *J. Biotechnol. Bull.* **2019**, *35*, 76–82.

41. Dukariya, G.; Kumar, A. Chitinase production from locally isolated *Bacillus cereus* GS02 from chitinous waste enriched soil. *J. Adv. Biol. Biotechnol.* **2020**, *23*, 39–48. [CrossRef]

42. Khadka, S.; Adhikari, S.; Thapa, A.; Panday, R.; Adhikari, M.; Sapkota, S.; Regmi, R.S.; Adhikari, N.P.; Proshad, R.; Koirala, N. Screening and optimization of newly isolated thermotolerant *Lysinibacillus fusiformis* strain SK for protease and antifungal activity. *Curr. Microbiol.* **2020**, *77*, 1558–1568. [CrossRef] [PubMed]

43. Gauvry, E.; Mathot, A.G.; Couvert, O.; Leguérinel, I.; Coroller, L. Effects of temperature, pH and water activity on the growth and the sporulation abilities of *Bacillus subtilis* BSB1. *Int. J. Food Microbiol.* **2020**, *337*, 108915. [CrossRef] [PubMed]

44. Pant, G.; Prakash, A.; Pavani, J.V.P.; Bera, S.; Deviram, G.V.N.S.; Kumar, A.; Panchpuri, M.; Prasuna, R.G. Production, optimization and partial purification of protease from *Bacillus subtilis*. *J. Taibah Univ. Sci.* **2015**, *9*, 50–55. [CrossRef]

45. Park, J.-K.; Kim, J.; Lee, C.-W.; Song, J.; Seo, S.-I.; Bong, K.-M.; Kim, D.-H.; Kim, P.I. Mass cultivation and characterization of multifunctional *Bacillus velezensis* GH1-13. *Korean J. Org. Agric.* **2019**, *27*, 65–76. [CrossRef]

46. Usanov, V.S.; Penzin, A.A.; Shishkin, V.V.; Tatarenko, I.Y. Influence of cultivation temperature and active acidity of soybean-corn substrate on the growth dynamics of the *Bacillus subtilis* bacterium. *Far East. Agrar. Bull.* **2020**, *3*, 117–124. [CrossRef]

47. Mohapatra, S.; Samantaray, D.P.; Samantaray, S.M. Study on polyhydroxyalkanoates production using rhizospheric soil bacterial isolates of sweet potato. *Ind. J. Sci. Technol.* **2015**, *8* (Suppl. 7), 57–62. [CrossRef]

48. Torres, M.; Llamas, I.; Torres, B.; Toral, L.; Sampedro, I.; Béjar, V. Growth promotion on horticultural crops and antifungal activity of *Bacillus velezensis* XT1. *App. Soil Ecol.* **2020**, *150*, 103453. [CrossRef]

49. Lee, D.R.; Maung, C.E.H.; Choi, T.G.; Kim, K.Y. Large scale cultivation of *Bacillus velezensis* CE 100 and effect of its culture on control of *Citrus Melanose* caused by *Diaporthe citri*. *Korean J. Soil Sci. Fertil.* **2021**, *54*, 297–310.

50. Garcha, S.; Kansal, R.; Gosal, S.K. Molasses growth medium for production of *Rhizobium* sp. based biofertilizer. *Ind. J. Biochem. Biophys.* **2019**, *56*, 378–383.

51. Gojgic-Cvijovic, G.D.; Jakovljevic, D.M.; Loncarevic, B.D.; Todorovic, N.M.; Pergal, M.V.; Ciric, J.; Loosc, K.; Beskoski, V.P.; Vrvic, M.M. Production of levan by *Bacillus licheniformis* NS032 in sugar beet molasses-based medium. *Int. J. Biol. Macromol.* **2019**, *121*, 142–151. [CrossRef]

52. Mazeed, T. Optimization of nutrient composition medium and culture condition for mannanase production by *Bacillus velezensis* nrc-1 using Taguchi method. *Al-Azhar J. Pharm. Sci.* **2012**, *45*, 198–207. [CrossRef]

53. Horak, I.; Engelbrecht, G.; van Rensburg, P.J.J.; Claassens, S. Microbial metabolomics: Essential definitions and the importance of cultivation conditions for utilizing *Bacillus* species as bionematicides. *J. App. Microbiol.* **2019**, *127*, 326–343. [CrossRef]

54. Pajčin, I.; Vlajkov, V.; Frohme, M.; Grebinyk, S.; Grahovac, M.; Mojićević, M.; Grahovac, J. Pepper Bacterial Spot Control by *Bacillus velezensis*. *Bioproc. Solut. Microorg.* **2020**, *8*, 1463. [CrossRef]

55. Sahebani, N.; Omranzade, F. Assessment of plant defence induction and biocontrol potential of *Bacillus megaterium* wr101 against *Meloidogyne javanica*. *Nematology* **2020**, *22*, 1091–1099. [CrossRef]

56. Cao, H.; Jiao, Y.; Yin, N.; Li, Y.; Ling, J.; Mao, Z.; Yang, Y.; Xie, B. Analysis of the activity and biological control efficacy of the *Bacillus subtilis* strain Bs-1 against *Meloidogyne incognita*. *Crop Protect.* **2019**, *122*, 125–135. [CrossRef]

57. Castaneda-Alvarez, C.; Aballay, E. Rhizobacteria with nematicide aptitude: Enzymes and compounds associated. *World J. Microbiol. Biotechnol.* **2016**, *32*, 203. [CrossRef] [PubMed]

58. Marfenina, O.E. *Anthropogenic Ecology of Soil Fungi*; Medicine for Everyone: Moscow, Russia, 2005; 196p.

59. Magoc, T.; Salzberg, S. FLASH: Fast length adjustment of short reads to improve genome assemblies. *Bioinformatics* **2011**, *27*, 2957–2963. [CrossRef] [PubMed]

60. Aziz, R.K.; Bartels, D.; Best, A.A.; DeJongh, M.; Disz, T.; Edwards, R.A.; Formsma, K.; Gerdes, S.; Glass, E.M.; Kubal, M.; et al. The RAST Server: Rapid annotations using subsystems technology. *BMC Genom.* **2008**, *9*, 75. [CrossRef]

61. Chaumeil, P.A.; Mussig, A.J.; Hugenholtz, P.; Parks, D.H. GTDB-Tk: A toolkit to classify genomes with the genome taxonomy database. *Bioinformatics* **2020**, *36*, 1925–1927. [CrossRef]

62. Guindon, S.; Dufayard, J.F.; Lefort, V.; Anisimova, M.; Hordijk, W.; Gascuel, O.; Notes, A. New algorithms and methods to estimate maximum-likelihood phylogenies: Assessing the performance of PhyML 3.0. *Syst. Biol.* **2010**, *59*, 307–321. [CrossRef] [PubMed]

63. Netrusov, F.I. *Workshop on Microbiology*; Publishing Center Academy: Moscow, Russia, 2005; pp. 105–107.

64. Lysak, L.V.; Dobrovol'skaya, D.G.; Skvortsova, I.N. *Methods for Assessing Soil Bacterial Diversity and Identifying Soil Bacteria*; Max Press: Moscow, Russia, 2003; pp. 56–58.

65. Guskova, L.A.; Metlitskiy, O.Z.; Danilov, L.G. *Guidelines for Conducting State Tests of Nematicides*; VNIESKH: Moscow, Russia, 1983; 35p.

66. Barker, K.R. An advanced treatise on Meloidogyne. In *Methodology*; North Carolina State University: Raleigh, NC, USA, 1985; Volume II, pp. 19–35.

Article

Influence of Cultivar and Biocontrol Treatments on the Effect of Olive Stem Extracts on the Viability of *Verticillium dahliae* Conidia

Ana López-Moral, Carlos Agustí-Brisach, Francisco M. Leiva-Egea and Antonio Trapero *

Departamento de Agronomía, María de Maeztu Unit of Excellence 2020-23, Campus de Rabanales, Universidad de Córdoba, Edif. C4, 14071 Córdoba, Spain; b92lomoa@uco.es (A.L.-M.); cagusti@uco.es (C.A.-B.); g42leegf@uco.es (F.M.L.-E.)
* Correspondence: trapero@uco.es

Abstract: The effect of olive (*Olea europaea*) stem extract (OSE) on the viability of conidia of *Verticillium dahliae*, the causal agent of Verticillium wilt of olive (VWO), is not yet well understood. Thus, the aim of this study was to determine the influence of the olive genotype (cultivar resistance) and the interaction between olive cultivars and biocontrol treatments on the effect of OSE on conidial germination of *V. dahliae* by *in vitro* sensitivity tests. To this end, OSE from cultivars Frantoio, Arbequina, and Picual, respectively tolerant, moderately susceptible, and highly susceptible to *V. dahliae*, were tested alone or after treatments with biological control agents (BCAs) and commercial products efficient at reducing the progress of VWO. *Aureobasidium pullulans* strain AP08, *Phoma* sp. strain ColPat-375, and *Bacillus amyloliquefaciens* strain PAB-24 were considered as BCAs. Aluminium lignosulfonate (IDAI Brotaverd®), copper phosphite (Phoscuprico®), potassium phosphite (Naturfos®), and salicylic acid were selected as commercial products. Our results indicate that the influence of biological treatments against the pathogen depends on the genotype, since the higher the resistance of the cultivar, the lower the effect of the treatments on the ability of OSE to inhibit the germination of conidia. In 'Picual', the BCA *B. amyloliquefaciens* PAB024 and copper phosphite were the most effective treatments in inhibiting conidia germination by the OSE. This work represents a first approach to elucidate the role of cultivar and biological treatments in modifying the effect on the pathogen of the endosphere content of olive plants.

Keywords: biocontrol; endosphere; *Olea europaea*; plant–pathogen interactions; vascular pathogen; Verticillium wilt

Citation: López-Moral, A.; Agustí-Brisach, C.; Leiva-Egea, F.M.; Trapero, A. Influence of Cultivar and Biocontrol Treatments on the Effect of Olive Stem Extracts on the Viability of *Verticillium dahliae* Conidia. *Plants* **2022**, *11*, 554. https://doi.org/10.3390/plants11040554

Academic Editor: Artur Alves

Received: 2 January 2022
Accepted: 18 February 2022
Published: 20 February 2022

Publisher's Note: MDPI stays neutral with regard to jurisdictional claims in published maps and institutional affiliations.

1. Introduction

Verticillium wilt of olive (*Olea europaea* subsp. *europaea*; VWO) causes high levels of tree mortality and reduces fruit yield in most olive-growing areas worldwide and is considered the main limiting factor of olive in Mediterranean-type climate regions [1–3]. In southern Spain, the disease is one of the major concerns for olive growers. Although the global disease incidence in this region is around 0.5%, it can reach values higher than 20%, together with high levels of disease severity and tree mortality in certain areas across the Guadalquivir valley. The causal agent of VWO is the hemibiotrophic soil-borne fungus *Verticillium dahliae*, from which two populations, defoliating (D) and nondefoliating (ND) pathotypes, have been identified in olive, with D pathotype causing the most severe damage [1,4]. The pathogen develops microsclerotia (MS), which are dormant structures that not only confer the ability to survive in the soil for a long time, but also serve as the primary inoculum source in natural infections [1,2,4]. Regarding the life cycle of *V. dahliae*, MS germinate by the stimuli of the root exudates from the susceptible hosts, giving rise to infectious hyphae that penetrate the plant roots and grow until they reach the vascular system. Then, the xylem vessels of the infected plants are colonized by the pathogen through mycelia and conidia development, contributing to the occlusion of the vascular system, as well as the production of gels and tyloses in the xylem vessels. Altogether cause a reduction in water flow, leading to water stress, and, consequently, plants become wilted and eventually die [5–7].

The innate biology of the pathogen, besides of the agronomical factors related to the intensification of olive crop, favours a year-by-year increase in disease incidence and have made it difficult to control VWO, making this disease one of the largest threats to olive groves worldwide [1,8,9]. Likewise, no truly efficient method to control VWO has been reported. Thus, there is no doubt that an integrated disease management (IDM) strategy is needed to prevent *V. dahliae* infections by both pre- and post-planting control measures in olive groves [1,2,10].

Considering IDM against VWO, both genetic resistance and biological control methods must be combined to achieve synergistic effectiveness to reduce both pathogen dispersal and disease incidence in the field. It is well known that disease severity varies depending on the olive cultivars, with the selection of cultivar being essential to avoid serious infections in the field. For instance, 'Picual' is considered one of the most susceptible olive cultivars, whereas 'Arbequina' and 'Hojiblanca' have shown moderate susceptibility, and 'Changlot Real', 'Empeltre', and Frantoio have high levels of tolerance [11,12]. On the other hand, important advances in biological control against VWO have been achieved in the last two decades [2,13]. In this regard, a broad diversity of essential oils, organic amendments, and biological control agents (BCAs), including endophytic bacteria and fungi, plant biostimulants, and host resistance inducers, have been evaluated both under controlled and natural field conditions towards selection of the best candidates for the control of VWO [3,14–18]. Foliar or root applications with two beneficial microorganisms, the fungus *Aureobasidium pullulans* AP08 and the bacterium *Bacillus amyloliquefaciens* PAB-024; and two phosphonate salts, one of copper and the other of potassium, were effective in reducing disease progression in artificially inoculated olive plants [3].

Due to the different behaviour patterns previously observed on the effect of all these treatments against VWO, their mechanisms of action have been explored, elucidating both direct effects, such as dual culture assays and *in vitro* sensitivity tests [3], and indirect effects, such as the effect on root exudates inhibiting MS and conidia germination enhancing the natural plant defence response [7,19] on the viability of the infectious structures of *V. dahliae* and VWO progress. These previous studies contributed to better understanding of how these biocontrol products can act against the pathogen; however, knowledge generated here opens new paths to be explored concerning such a topic. In this way, determining whether these products can modulate the effect of endosphere contents of the treated olive plants on *V. dahliae* infection would suppose other relevant knowledge towards to better elucidating their mode of action.

The xylem vessels are an ideal niche for microbial endophytes such as the Verticillium wilt pathogen by providing an effective internal pathway for whole-plant colonization and by acquiring the scarce nutrients available in xylem sap, either by enzymatic digestion of host cell walls, by invading neighbouring cells, or by inducing leakage of nutrients from surrounding tissues [20,21]. Indeed, xylem sap contains a wide range of compounds beyond water and minerals, such as amino acids [22], organic acids [23], and vitamins [24]. The xylem sap composition of woody plant species, including olive tree, has been characterized in several previous studies [25]. However, this composition can be influenced by multiple factors, such as the water content of the soil [26]; the cultivar; the type and age of organs selected; and the incidence of microbial interactions, including infection by plant pathogens, among others [25]. Related to this aspect, recently, Anguita-Maeso et al. [27] determined that the xylem microbiome of olive plants inoculated with *V. dalhiae* increases the diversity of bacterial communities compared to non-inoculated plants. In addition, these same authors also showed a breakdown of resistance to *V. dahliae* in wild olive trees related to a modification of their xylem microbiome [27]. However, to our knowledge, there are no scientific studies that address the effect of the content of the olive endosphere, including xylem sap, on the viability of infectious structures of *V. dahliae*. Thus, the aim of this study was to determine the influence of the olive genotype (cultivar resistance) and the interaction between cultivars and biocontrol treatments on the effect of olive stem extract (OSE) on the viability of conidia of *V. dahliae in vitro*.

2. Results

2.1. Effect of Olive Stem Extract on Conidial Viability of Verticillium dahliae

2.1.1. Experiment I: Effect of Olive Cultivars

The results on the effect of OSE from plants of cvs. Frantoio (tolerant), Arbequina (moderately susceptible), and Picual (highly susceptible) at 5, 10, 15, 20, 25, and 50% concentration on the conidial viability of *V. dahliae* showed that the higher OSE concentration, the higher the RGI. Marked differences between the three olive cultivars were detected from OSE at 10%, whereas that from 'Frantoio' showed higher RGI values than that from 'Arbequina' and 'Picual'; however, in these last two cases, the inhibition did not differ significantly. At the highest concentration evaluated (OSE at 50%), RGI values ranged from 68.9 ± 2.9 to $80.5 \pm 1.9\%$ for 'Picual' and 'Frantoio', respectively (Figure 1).

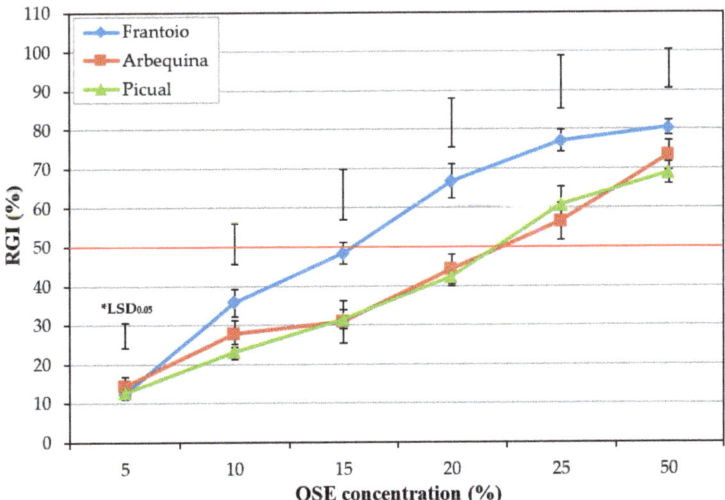

Figure 1. Effect of olive stem extract (OSE) from healthy olive plants of cvs. Frantoio (tolerant), Arbequina (moderately susceptible), and Picual (highly susceptible) on relative conidial germination inhibition (RGI) of *Verticillium dahliae* isolate V180. Data are the means of 240 conidia (six replicates) per combination of olive cultivar and OSE concentration. Vertical bars represent the standard error of the means. *LSD bars represent the critical values for comparison at $P = 0.05$.

In addition, significant differences were observed for EC_{50} between cultivars ($P = 0.0239$). The OSE from 'Frantoio' showed the lowest EC_{50} value ($EC_{50} = 16.2 \pm 0.88$) compared to the OSE from the rest of cultivars, whereas OSE from 'Arbequina' ($EC_{50} = 21.7 \pm 1.15$) and 'Picual' ($EC_{50} = 20.4 \pm 1.08$) did not differ significantly from one another (Table 1).

2.1.2. Experiment II: Influence of Treatments

The influence of the OSE from olive plants of cv. Picual treated with different treatments on conidial viability showed great differences between treatments (Figure 2). In general, the OSE from plants treated with commercial products showed higher RGI values than those from OSE obtained from plants treated with BCAs. Only the OSE from plants treated by root applications with *A. pullulans* AP08, aluminium lignosulfonate, and copper phosphite showed RGI values higher than 80% when they were tested at the highest concentration.

Table 1. Effective concentrations of olive stem extract (OSE) from cvs. Frantoio, Arbequina, and Picual to inhibit 50% of conidial germination (EC$_{50}$; μL mL^{-1}) of *Verticillium dahliae* isolate V180.

Cultivar	EC$_{50}$ (μL mL^{-1}) [a]
Frantoio	16.2 ± 0.88 b
Arbequina	21.7 ± 1.15 a
Picual	20.4 ± 1.08 a

[a] EC$_{50}$ of conidial germination was calculated as predicted value of the linear regression of the relative germination inhibition (%) over OSE concentration. Values represent the average of 240 conidia (six replicates) per extract (cultivar) and OSE concentration. Means followed by a common letter do not differ significantly according to Fisher's protected LSD test at $P = 0.05$.

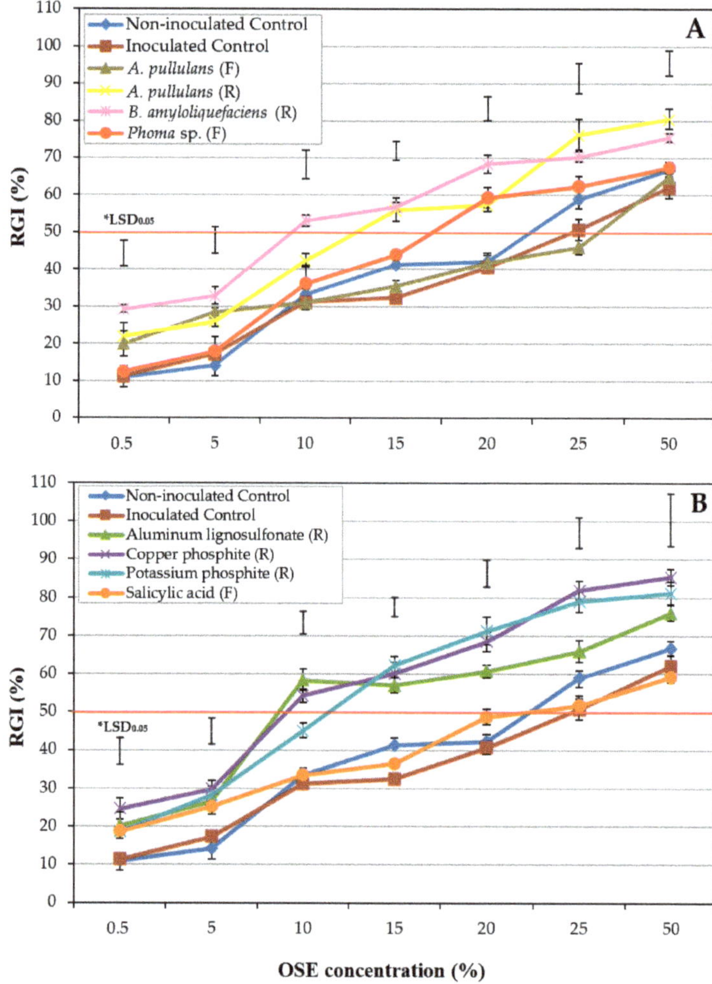

Figure 2. Effect of olive stem extract (OSE) from healthy olive plants of cv. Picual treated by foliar (F) or root (R) applications with (**A**) microorganisms (biological control agents) and (**B**) commercial products on the relative conidial germination inhibition (RGI) of *Verticillium dahliae* isolate V180. In each graph, data are the means of 240 conidia (six replicates) per combination of compound and OSE concentration. Vertical bars represent the standard error of the means. *LSD bars represent the critical values for comparison at $P = 0.05$.

There were significant differences for EC_{50} between treatments ($P < 0.0001$). In general, OSE from treated plants showed significantly lower EC_{50} values than that from both negative and positive control ($EC_{50} = 21.5 \pm 1.43$ and 24.9 ± 0.69, respectively), with the exception of OSE from plants treated by spraying with *A. pullulans* AP08 ($EC_{50} = 24.7 \pm 1.61$), *Phoma* sp. ($EC_{50} = 17.3 \pm 0.55$), and salicylic acid ($EC_{50} = 22.1 \pm 2.05$), which did not differ from the controls (Table 2).

Table 2. Effective concentrations of olive stem extract (OSE) from cv. Picual treated with several compounds to inhibit 50% of conidial germination (EC_{50}; $\mu L\ mL^{-1}$) of *Verticillium dahliae* isolate V180.

Treatment [a]	Application	EC_{50} ($\mu L\ mL^{-1}$) [b]
Control ($-$)		21.5 ± 1.43 ab
Control ($+$)		24.6 ± 0.69 a
Aluminum lignosulfonate	Root	12.1 ± 1.38 cd
Aureobasidium pullulans	Foliar	24.7 ± 1.61 a
A. pullulans	Root	13.1 ± 1.01 cd
Bacillus amyloliquefaciens	Root	10.3 ± 1.26 d
Copper phosphite	Root	10.1 ± 1.34 d
Phoma sp.	Foliar	17.3 ± 0.55 bc
Potassium phosphite	Root	11.8 ± 1.52 cd
Salicylic acid	Foliar	22.1 ± 2.05 ab

[a] All compounds were applied by foliar or root applications in non-inoculated plants. Control (-): non-treated and non-inoculated plants; Control (+): non-treated and inoculated plants with *V. dahliae* isolate V-180. [b] EC_{50} of conidial germination was calculated as predicted value of the linear regression of the relative germination inhibition (%) over OSE concentration. Values represent the average of 240 conidia (six replicates) per extract (cultivar) and OSE concentration. Means followed by a common letter do not differ significantly according to Tukey's HSD test ($P = 0.05$).

2.1.3. Experiment III: Interaction between Olive Cultivar and Treatments

OSE from plants of cv. Frantoio did not show marked differences in RGI between treatments. A moderate effect on RGI was observed for OSE from treated plants of cv. Arbequina, with RGI slightly increasing for OSE from plants treated with *A. pullulans* and copper phosphite. On the other hand, significant differences in RGI were observed in 'Picual' between the OSE from treated and control plants, as well as between the OSE from plants treated with the different BCAs and chemical products (Figure 3).

ANOVA confirmed that there were no significant differences for EC_{50} between treated and non-treated plants of cv. Frantoio ($P = 0.9510$). In 'Arbequina', significant differences for EC_{50} were only observed between the OSE from plants treated with *B. amyloliquefaciens* ($EC_{50} = 13.0 \pm 1.57$) and copper phosphite ($EC_{50} = 14.2 \pm 1.19$) compared to that obtained for OSE of the controls. Finally, in 'Picual', the OSE from treated plants showed significantly lower EC_{50} values than those of the OSE from both negative and positive controls; however, EC_{50} did not differ significantly between the different products evaluated (Table 3).

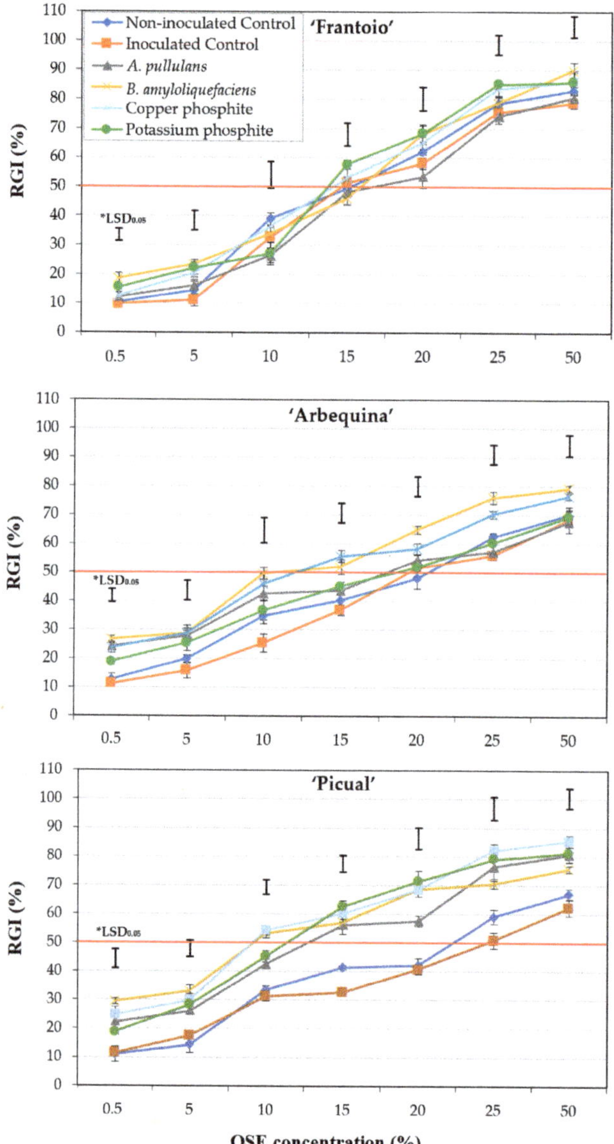

Figure 3. Effect of olive stem extract (OSE) from healthy olive plants of cvs. Frantoio, Arbequina, and Picual treated by root applications with the most effective compounds from *Experiment II* (*Aureobasidium pullulans*, *Bacillus amyloliquefaciens*, copper phosphite, or potassium phosphite) and from non-treated control plants, both non-inoculated and inoculated with *Verticillium dahliae* isolate V-180 on the relative conidial germination inhibition (RGI) of *V. dahliae* isolate V180. In each graph, data are the means of 240 conidia (six replicates) per combination of olive cultivar and OSE concentration. Vertical bars represent the standard error of the means. *LSD bars represent the critical values for comparison at $P = 0.05$.

Table 3. Effective concentrations of olive stem extract (OSE) from cvs. Frantoio, Arbequina, and Picual and treated with the most effective compounds to inhibit 50% of conidial germination (EC_{50}; $\mu L\ mL^{-1}$) of *Verticillium dahliae* isolate V180.

Treatment [a]	EC_{50} ($\mu L\ mL^{-1}$) [b]		
	Frantoio	Arbequina	Picual
Control (−)	15.2 ± 2.07 a	20.1 ± 0.90 a	21.5 ± 0.85 a
Control (+)	16.1 ± 1.39 a	20.0 ± 2.44 a	24.6 ± 0.69 a
Aureobasidium pullulans	16.4 ± 1.36 a	17.9 ± 0.88 ab	13.1 ± 1.01 b
Bacillus amyloliquefaciens	14.8 ± 1.87 a	13.0 ± 1.57 c	10.3 ± 1.26 b
Copper phosphite	14.9 ± 0.83 a	14.2 ± 1.19 bc	10.1 ± 1.34 b
Potassium phosphite	14.7 ± 1.23 a	18.3 ± 0.92 ab	11.8 ± 1.52 b

[a] All evaluated treatments were applied by root applications in non-inoculated plants. Control (−): non-treated and non-inoculated plants; Control (+): non-treated and inoculated plants with *V. dahliae* isolate V-180. [b] EC_{50} of conidial germination was calculated as predicted value of the regression: relative germination inhibition (%) over OSE concentration. Values represent the average of 240 conidia (six replicates) per extract (cultivar) and OSE concentration. Means followed by a common letter do not differ significantly according to Fisher's protected LSD test at $P = 0.05$.

3. Discussion

Although recent studies have revealed that *V. dahliae* modulates the xylem microbiome in olive plants by increasing the diversity of bacterial communities when the pathogen is present in the soil [27], the influence that olive cultivar, biological treatments, and their interaction could have on the endosphere against the viability of the infectious structures of *V. dahliae* has not yet been reported. Thus, this work represents a first approach to elucidate the role of cultivar and biological treatments in modifying the effect of the endosphere contents on the pathogen in olive plants.

In this study, 6-month-old potted olive plants were used because, based on our extensive experience working with the olive tree and *V. dahliae* pathosystem, it is not only the age limit to reproduce the symptoms of the disease by artificial inoculation of plants in small pots [3,14], but also considering the limitations that the size and physiological structure of plant tissues could have in the extraction of contents from the endosphere by Cadahía's method [28]. Regarding this last aspect, we have to consider that Cadahía calls 'sap' to the total liquid extract from the endosphere (called OSE in this study) that comes from both the xylem and phloem of the plant. The ideal would probably be to obtain raw sap (xylem) and elaborated sap (phloem) separately to evaluate the effect of the raw sap alone, but despite this limitation, we chose this method instead of the sap-extraction method using a Scholander chamber [29] for the following reasons: (i) because it has been demonstrated that the extract of pure sap from the xylem of olive plants contains a wide diversity of bacterial communities [27], which could mask the effect of OSE on the germination of conidia in *in vitro* sensitivity tests; (ii) consequently to this first reason, the extraction of OSE using Cadahía's method guaranties the absence of living microorganisms, as well as cellular debris in the extract, since the plant material is subjected to consecutive immersion and freezing treatments in diethyl ether during the extraction process; and (iii) because it has been recognized as reliable method to determine the nutritional levels of plants, since results using 'sap-like' extracts (OSE) contrast well with those of the yield and quality of the harvest of several fruit crops, including the olive tree, in more than 40 years of experience [28].

Only the effect of OSE on the germination of conidia was evaluated in this study because conidia are the infectious structures of *V. dahliae* directly affected by the sap, considering the life cycle of the pathogen, i.e., the pathogen infects plants through the root by the germinated MS and then systemically colonizes the infected plants by producing conidia in the xylem vessels [1,6]. In addition, the treatment combinations (treatments and/or mode of application) evaluated in this study were selected because they all resulted in high effectiveness in inhibiting the viability of the infectious structures of *V. dahliae*, as well as reducing the progress of the disease in previous studies by López-Moral et al. [3].

Our results revealed that OSE from 'Frantoio' (tolerant) showed higher RGI values than that from 'Arbequina' (moderately susceptible) and 'Picual' (highly susceptible), whereas the inhibition did not differ markedly between the last two cultivars (*Experiment I*). In 'Picual', the influence of OSE from the treated plants on the inhibition of conidia germination varied significantly between the evaluated treatments. In this cultivar, the BCA *B. amyloliquefaciens* PAB-024, and a phosphonate salt [copper phosphite (Phoscuprico®)] were the most effective in inhibiting conidia germination by the OSE (*Experiment II*). In addition, when the four selected treatments were applied to the three olive cultivars, their influence on the effect of OSE on the inhibition of conidia germination was not significant between treatments for 'Frantoio', whereas moderate and markedly significant differences between treatments were observed for 'Arbequina' and 'Picual', respectively (*Experiment III*). Finally, although some differences can be observed between both positive and negative controls in RGI as the OSE concentration increases (Figures 2 and 3), the EC$_{50}$ data did not show significant differences between either control in any case. Therefore, these data suggest that the biotic stress caused by the infection of the pathogen in the plant does not influence the effect of OSE on the conidial germination of *V. dahliae*.

As a first conclusion, our results indicate that the influence of biological treatments against the pathogen depends on the genotype, since the greater the resistance of the cultivar, the lower the influence of the treatments on the ability of OSE to inhibit conidia germination. Thus, the results suggest that the high tolerance to *V. dahliae* conferred by the 'Frantoio' genotype prevails over the treatments, even those that were more effective against the pathogen in susceptible cultivars. These results are in agreement with those obtained recently by López-Moral et al. [7], who determined the influence of cultivars and biological treatments on the effect of root exudates from olive plants on the viability of MS and conidia of *V. dahliae*. These authors demonstrated that root exudates induced germination of conidia and MS of *V. dahliae* and that the genotype significantly affected this ability; the root exudates from 'Frantoio' did not show a significant effect on the induction of MS and conidia germination compared to the control, whereas those from 'Arbequina' and 'Picual' showed a moderate and marked effect, inducing the viability of both MS and conidia of *V. dahliae*, respectively [7].

Regarding genetic resistance, our study reveals new knowledge about the relationship that the sap and the olive cultivar could have, favouring or interfering with the colonization of the xylem by the pathogen. However, in order to elucidate how the cultivar factor could influence the effect of its sap on the colonization of the pathogen, further interaction studies with the xylem anatomy of each cultivar should be conducted in the future. In fact, previous studies that evaluated the anatomy of the xylem of healthy olive trees of cvs. Frantoio and Picual showed significant differences not only in the parameters related to water transport but also in the density of vessels associated with a larger or smaller conduction area in the xylem tissue, both parameters being significantly higher in 'Frantoio' than in 'Picual' [30].

Regarding the influence of the treatments on the effect of OSE on the viability of the conidia, our results are also in agreement with those obtained by López-Moral et al. [7]. These authors revealed that the root exudates from plants of the three cultivars (Frantoio, Arbequina, and Picual) treated with the same four treatments evaluated in *Experiment III* of the present study showed significant differences in their effect on MS and conidia germination, and the genotype also significantly affected this ability. In this case, the treatment with *A. pullulans* AP08 was most effective, showing a significant effect inhibiting conidia germination in 'Arbequina' and MS germination in 'Arbequina' and 'Picual', but non-significant effect in these two parameters was observed in 'Frantoio'. Regarding the rest of the evaluated treatments, the root exudates from plants treated with copper phosphite, potassium phosphite, and *B. amyloliquefaciens* PAB-24 gave rise to a significant inhibition in the germination of conidia or MS but only in cv. Arbequina [7]. In addition, previous studies conducted by López-Moral et al. [19] to evaluate these treatments as potential inducers of host resistance against VWO showed that the two BCAs *A. pullulans* AP08 and *B. amyloliquefaciens* PAB-024, as well as the phosphonate salt Phoscuprico®, had the ability to accumulate jasmonic acid (JA) and JA-isoleucine in leaves, stem, or roots of treated olive plants of cv. Picual. These results also suggest an implication of JA in the host resistance induced by these treatments. This last aspect could be directly related to our results, since it is well known that xylem infections by vascular pathogens cause drastic metabolic changes in the cells of the xylem parenchyma adjacent to the infected vessels. These metabolic changes lead to the accumulation of different proteins and secondary metabolites in xylem sap during pathogen colonization, including pathogenesis-related proteins (PR proteins), enzymes (e.g., peroxidases, proteases, xyloglucan-endotransglycosylase, xyloglucan-specific endoglucanase protein inhibitor), phenols, phytoalexins, and lignin, which help to enhance the natural defence mechanisms of the plant [31–35]. On the other hand, due to the fact that the OSE did not have a live microbiome as a consequence of the extraction method used, studies to determine whether the treatments applied in this work could influence the modification of the xylem microbiome of olive plants must be conducted. In this way, recent studies performed by Anguita-Maeso et al. [27] determined that the xylem microbiome of olive plants inoculated with *V. dahliae* increases the diversity of bacterial communities compared to non-inoculated plants. However, how the xylem microbiome could be modified by biological treatments favouring *V. dahliae* inhibition is still uncertain.

In summary, the method used in this study to obtain endosphere contents of olive plants, called OSE or 'sap-like', for further analysis in the laboratory against *V. dahliae* can be considered valid and useful, since all our results agree with those obtained in previous studies [3,7,19]. The knowledge generated here represents a first approach in the study of how genotype and/or biological treatments can influence the extracts of olive plants by inhibiting the germination of conidia or of MS of *V. dahliae*. This knowledge could be useful in the future to prevent infections or mitigate the progression of disease within the framework of the current 'from-farm-to-fork' strategy to obtain safe and healthy fruits.

4. Materials and Methods

4.1. Plant Material

Healthy 6-month-old rooted cuttings of three olive cultivars representative of different degrees of susceptibility to *V. dahliae* were used: 'Frantoio' (tolerant), 'Arbequina' (moderately susceptible), and 'Picual' (highly susceptible) [11,12]. The plants were obtained from a commercial nursery and were grown in peat moss in plastic pots (0.5 L). They were pre-conditioned in a controlled growth chamber (22 $\pm$ 2 °C, with a 14:10 h (light:dark) photoperiod of white fluorescent light (10,000 lux) and 60% relative humidity (RH)) for 1 month to induce active growth. During this month, plants were irrigated three times per week with 350 mL of water per plant.

4.2. Fungal Strain and Inoculum Preparation

The *V. dahliae* isolate V180 was used in all the experiments of this study [3]. It was stored as single-spore isolate on potato dextrose agar (PDA; Difco® Laboratories, MD, USA) slants filled with sterile paraffin oil at 4 °C in the dark in the collection of the Department of Agronomy at the University of Córdoba (DAUCO, Spain). Before conducting each experiment, small mycelial fragments of the colonized agar from the tube were plated onto PDA acidified with 2.5 mL L^{-1} lactic acid (APDA) and incubated at 24 °C in the dark for 10 days in order to obtain fresh colonies. Then, they were transferred to PDA and incubated as previously described.

4.3. Effect of Olive Stem Extract (OSE) on Conidial Viability of Verticillium dahliae

4.3.1. Obtaining Stem Extract

For obtain OSE, the main stem of the plants was cut at its base, and the entire main stem and shoots were used. Leaves and roots were discarded. Subsequently, most of the cortical tissue of stems and shoots was removed manually using sandpaper, and the peeled stems were sprayed with distilled water and kept at 4 °C in the dark to avoid desiccation until further processing. OSE was obtained by the analytical laboratory 'C+E Analítica' (San José de la Rinconada, Seville, Spain) following the protocol described by Cadahía [28]. Specifically, once in the analytical laboratory, the shoots and stems were cut into 0.5 cm fractions, immersed in ethyl ether, and kept at −20 °C for 2 h. As a consequence of the freezer step, the water contained in the plant tissues crystallised, breaking the cell walls, which later allowed a sap-like extract to be obtained. At the same time, the ether is able to extract the chlorophyll that could interfere with the analytical process. After this step, the plant material was defrosted, and the aqueous phase (endosphere contents) was separated from the ether to obtain OSE by means of a hydraulic press [28].

4.3.2. Experiment I: Effect of Olive Cultivars

To evaluate the effect of OSE from different olive cultivars on conidia viability, plants of cvs. Frantoio, Arbequina, and Picual were used. They were maintained for one month in growth chambers, as described above. For this period, they were arranged in a randomized complete block design with three blocks and four replicated plants per cultivar (3 × 4 = 12 plants per cultivar; 36 plants in total). The OSE was obtained by joining all the plants of each block, so there were three experimental units of OSE (≈20 mL) per cultivar.

4.3.3. Experiment II: Influence of Treatments

To evaluate the influence of different treatments on the effect of OSE on *V. dahliae*, olive plants of cv. Picual were treated with seven treatments, including three BCAs and four commercial products (Table 4). These treatments and the type of application (foliar and/or irrigation) were selected for this study because of their significant efficacy against *V. dahliae in vitro*, as well as against the progress of VWO in planta demonstrated in previous studies [3].

Plants were treated with the water solution or suspension of each commercial product or BCA (Table 4), respectively, at 6-, 5-, and 3 weeks before obtaining OSE. Additionally, non-treated but inoculated plants by cornmeal–sand mixture (CSM; sand, cornmeal and distilled water; 9:1:2, weight: weight:volume) colonized by *V. dahliae* isolate V180 (theoretical inoculum density of the final substrate = 10^7 CFU g^{-1}) were included as a positive control; and non-treated and non-inoculated plants as negative control. Plant inoculation was conducted 4 weeks before obtaining the OSE. Treatments or plant inoculation were conducted following the protocols described by López-Moral et al. [3]. A randomized complete block design with three blocks and four replicated olive plants per treatment ($n = 10$; eight treatments and two controls) was used (120 plants in total).

This experiment was maintained for six weeks after the first treatment was applied. Subsequently, OSE was obtained from plants of each treatment, as well as from plants of both positive and negative control, joining all the plants of each block, i.e., there were three experimental units of OSE (≈20 mL) per treatment or control.

4.3.4. Experiment III: Interaction between Olive Cultivars and Treatments

To evaluate the influence of the interaction between olive cultivars and treatments on the effect of OSE on *V. dahliae*, olive plants of the three cvs. described above were treated with two BCAs (fungus *Aureobasidium pullulans* strain AP08 and bacterium *Bacillus amyloliquefaciens* strain PAB-24) and two commercial products [copper phosphite (Phoscuprico®) and potassium phosphite (Naturfos®)] (Table 4). These treatments were selected for this experiment for their efficacy in inhibiting conidial germination of *V. dahliae* in *Experiment II*.

Plants were treated 6-, 5-, and 3 weeks before obtaining the OSE and inoculated 4 weeks before obtaining the OSE. The treatments, OSE extraction, and plant inoculation were conducted as described above. For each olive cultivar, a positive and a negative control were included, as described in *Experiment II*. A randomized complete block design with three blocks and four replicated olive plants per treatment ($n = 6$, four treatments and two controls) and cultivar ($n = 3$) combination was used (72 plants per olive cultivar; 216 plants in total).

Table 4. Biological and chemical products evaluated in this study [a].

Active Ingredient(s)	Trade Name/Formulation [b]	Manufacturer	Class (FRAC Code) [c]	Dose [d]	
				Foliar	Root
Biological Control Agents (BCAs) [e]					
Aureobasidium pullulans	AP08	DAUCO [d]	Fungal (NC)	10^6 conidia mL^{-1}	10^6 conidia mL^{-1}
Bacillus amyloliquefaciens	PAB-024	DAUCO	Bacterial (NC)	n/e	10^8 CFU mL^{-1}
Phoma sp.	ColPat-375	DAUCO	Fungal (NC)	10^6 conidia mL^{-1}	n/e
Chemical Products					
Aluminum lignosulfonate	IDAI Brotaverd®-EW	IDAI Nature	Inorganic salt (NC)	n/e [f]	5 mL L^{-1}
Copper phosphite	Phoscuprico®-EW	Agri nova Science	Phosphorous acid and salts (P07)	n/e	10 mL L^{-1}
Potassium phosphite	Naturfos®-EW	Daymsa	Phosphorous acid and salts (P07)	n/e	8 mL L^{-1}
Salicylic acid	Salicylic acid-SL	Sigma-Aldrich	Organic acid (NC)	5 mM (0.69 g L^{-1})	n/e

[a] Products and type of application evaluated in the present study were selected for their efficacy against *V. dahliae* observed in the previous study conducted by López-Moral et al. [3]. [b] EW: emulsion, oil in water; SL: soluble concentrate. [c] Group and code numbers are assigned by the Fungicide Resistance Action Committee (FRAC) according to different modes of actions (NC: not classified; for more information, see http://www.frac.info/, accessed on 24 November 2021). [d] Maximum dose for foliar or root applications recommended for the manufacturers of the commercial compounds evaluated in this study. Fungal and bacterial inocula from the BCAs (AP08, PAB-024, and ColPat-375) were prepared and adjusted according to Varo et al. [14]. [e] All the BCAs used in this study are maintained in the collection of the Agroforestry Pathology Research Group at the Department of Agronomy, University of Córdoba (DAUCO), Spain. [f] n/e: non-evaluated products and dose combinations in this study.

This experiment was maintained for six weeks after the first treatment was applied. Subsequently, OSE was obtained from plants of each treatment and olive cultivar combination, as well as from plants of the positive and negative control, joining all the plants of each block, so there were three experimental units of OSE ($\approx$20 mL) per treatment or control and cultivar combination.

4.3.5. Conidia Viability *In Vitro*

For each set of experiments, conidial suspensions were obtained from 14-day-old colonies of *V. dahliae* isolate V180 growing on PDA, as described previously, and adjusted to 8×10^5 conidia mL^{-1} using a haematocytometer. In parallel, OSE solutions were adjusted to 0, 1, 10, 20, 30, 40, 50, and 100% in sterile deionized distilled water (SDDW). Subsequently, a 5 µL drop of the conidial suspension was placed in the centre of a microscope coverslip (20×20 mm); then, a 5-µL drop of the OSE solution was mixed. Thus, the OSE was evaluated at the following final concentrations: 0, 0.5, 5, 10, 15, 20, 25, and 50%, with a concentration of 0% consisting of a 5 µL drop of the conidial suspension mixed with a 5 µL drop of sterile SDDW as a control. The coverslips were placed inside Petri dishes containing water agar, which were used as humid chambers, and were incubated at 23 ± 2 °C in the dark for 24 h. After the incubation period, a 5 µL drop of 0.01% acid fuchsine in lactoglycerol (1:2:1 lactic acid:glycerol:water) was added to each coverslip to stop conidial germination, and they were mounted on a slide. For each experiment (*I*, *II*, and *III*), there were three replicated coverslips per concentration of OSE obtained from each block and from each treatment or control (OSE at 0%, i.e., only SDDW). All the experiments were conducted twice.

In all cases, a total of 120 randomly selected conidia per replicated coverslip were observed at a $\times 400$ magnification by means of a Nikon Eclipse 80i microscope (Nikon Corp., Tokyo, Japan), and the germinated and non-germinated conidia were counted. Conidia were considered germinated when the germ tube was at least one-half of the longitudinal axis of the conidia. Conidial viability was estimated as percentage (%) of conidial germination for each OSE concentration, and then, the inhibition of conidial germination (RGI; %) was estimated with respect to the control according to the following formula:

$$\mathrm{RGI}\ (\%) = [(\mathrm{Ge}_{\mathrm{control}} \times \mathrm{Ge}_{\mathrm{OSEsolution}}) / \mathrm{Ge}_{\mathrm{control}}]$$

where $\mathrm{Ge}_{\mathrm{control}}$ = percentage of germinated conidia after incubation in the SDDW, and $\mathrm{Ge}_{\mathrm{OSEsolution}}$ = percentage of germinated conidia after incubation in the OSE solution from treated plants [36]. The RGI data were linearly regressed over the OSE concentration, and the predicted values of the effective OSE concentrations (µg mL^{-1}) inhibiting 50% (EC$_{50}$) of conidial germination were obtained from the fitted regressions.

4.3.6. Data Analysis

Data of EC$_{50}$ from the two repetitions of each experiment were combined after checking for homogeneity of the experimental error variances by the F test ($P \geq 0.05$)p. Subsequently, data were tested for normality, homogeneity of variances, and residual patterns. For *Experiments I* and *II*, a one-way ANOVA was conducted, with the EC$_{50}$ as dependent variable and 'cultivar' or 'treatment' as independent variables. For *Experiment III*, a factorial ANOVA was conducted with EC$_{50}$ as dependent variable, and 'cultivar', 'treatment', and their interaction as independent variables. Since interaction was significant ($P = 0.0001$), one-way ANOVAs were conducted for each olive cultivar. Treatment means were compared according to Fisher's protected LSD test or Tukey´s HSD test (both at $P = 0.05$) for the *Experiment I* ($n = 3$) and *III* ($n = 6$), or for *Experiment II* ($n = 10$), respectively [37]. Study data were analyzed using Statistix 10.0 software [38].

5. Conclusions

Our results indicate that the influence of biological treatments against the pathogen depends on the genotype, since the greater the resistance of the cultivar, the lower the influence of the treatments on the ability of OSE to inhibit conidia germination. In 'Picual', the most susceptible cultivar to VWO, the BCA *B. amyloliquefaciens* PAB-024 and a phosphonate salt [copper phosphite (Phoscuprico®)] were the most effective treatments in inhibiting conidia germination by OSE. Thus, the results suggest that the high tolerance to *V. dahliae* conferred by the 'Frantoio' genotype prevails over the treatments, even those that were more effective against the pathogen in susceptible cultivars. The method used in this study to obtain endosphere contents of olive plants, called OSE or 'sap-like' for further analysis in the laboratory against *V. dahliae* can be considered valid and useful since all our results agree with those obtained in previous studies. Thus, this work represents a first approach to elucidate the role of cultivar and biological treatments in modifying the effect on the pathogen of the endosphere content of olive plants.

Author Contributions: Laboratory tasks, data analyses, writing and editing: A.L.-M.; data analyses, writing, review, and editing: C.A.-B.; laboratory tasks: F.M.L.-E.; study conception and design, funding acquisition, supervision, and manuscript revision: A.T. All authors have read and agreed to the published version of the manuscript.

Funding: This research was funded by the Spanish Ministry of Science, Innovation and Universities (MICINN; projects GIVEROLI Ref. AGL2016-76240-R and VERTOLEA Ref. AGL2016-76240-R), co-financed by the European Union FEDER Funds. A. López-Moral is holder of "Formación de Personal Investigador" fellowship (FPI; contract number BES-2017-081839) from MICINN. We acknowledge financial support of MICINN, the Spanish State Research Agency, through the Severo Ochoa and María de Maeztu Program for Centres and Units of Excellence in R&D (Ref. CEX2019-000968-M).

Institutional Review Board Statement: Not applicable.

Informed Consent Statement: Not applicable.

Data Availability Statement: Not applicable.

Acknowledgments: The authors thank F. González and J.A. Toro for their skilful technical assistance. The authors also thanks AGRI nova Science (Almería, Spain), Daymsa S.A. (Zaragoza, Spain), and IDAI Nature (Valencia, Spain) for their kind collaboration providing experimental samples of Phoscuprico®, Naturfos®, and IDAI Brotaverd®, respectively, to conduct this study.

Conflicts of Interest: The authors declare no conflict of interest. The funders had no role in the design of the study; in the collection, analyses, or interpretation of data; in the writing of the manuscript, or in the decision to publish the results.

References

1. López-Escudero, F.J.; Mercado-Blanco, J. Verticillium wilt of olive: A case study to implement an integrated strategy to control a soil-borne pathogen. *Plant Soil* **2011**, *344*, 1–50. [CrossRef]
2. Montes-Osuna, N.; Mercado-Blanco, J. Verticillium wilt of olive and its control: What did we learn during the last decade? *Plants* **2020**, *9*, 735. [CrossRef] [PubMed]
3. López-Moral, A.; Agustí-Brisach, C.; Trapero, A. Plant biostimulants: New insights into the biological control of Verticillium wilt of olive. *Front. Plant Sci.* **2021**, *12*, 662178. [CrossRef] [PubMed]
4. Jiménez-Díaz, R.M.; Cirulli, M.; Bubici, G.; Jiménez-Gasco, M.; Antoniou, P.P.; Tjamos, E.C. Verticillium wilt, a major threat to olive production: Current status and future prospects for its management. *Plant Dis.* **2012**, *96*, 304–329. [CrossRef]
5. Ayres, P.G. Water relations of diseased plants. In *Water Deficits and Plant Growth*; Kozlowski, T., Ed.; Academic Press: London, UK, 1978; pp. 1–60.
6. Pegg, G.F.; Brady, B.L. *Verticillum Wilts*; CABI Publishing: Wallingford, UK, 2002.
7. López-Moral, A.; Sánchez-Rodríguez, A.R.; Trapero, A.; Agustí-Brisach, C. Elucidating the role of root exudates of olive plants against *Verticillium dahliae*: Effect of the cultivar and biocontrol treatments. *Plant Soil* **2022**. *submitted*.
8. Alström, S. Characteristics of bacteria from oilseed rape in relation to their biocontrol activity against *Verticillium dahliae*. *J. Phytopathol.* **2001**, *149*, 57–64. [CrossRef]
9. Pérez-Rodríguez, M.; Alcántara, E.; Amaro, M.; Serrano, N.; Lorite, I.J.; Arquero, O.; Orgaz, F.; López-Escudero, F.J. The Influence of irrigation frequency on the onset and development of Verticillium wilt of olive. *Plant Dis.* **2015**, *99*, 488–495. [CrossRef]
10. Ostos, E.; García-López, M.T.; Porras, R.; López-Escudero, F.J.; Trapero, A.; Michailides, T.J.; Moral, J. Effect of cultivar resistance and soil management on spatial–temporal development of Verticillium wilt of olive: A long-term study. *Front. Plant Sci.* **2020**, *11*, 584496. [CrossRef]

11. López-Escudero, F.J.; Del Río, C.; Caballero, J.M.; Blanco-López, M.A. Evaluation of olive cultivars for resistance to *Verticillium dahliae*. *Eur. J. Plant Pathol.* **2004**, *110*, 79–85. [CrossRef]

12. Trapero, C.; Serrano, N.; Arquero, O.; Del Río, C.; Trapero, A.; López-Escudero, F.J. Field resistance to Verticillium wilt in selected olive cultivars grown in two naturally infested soils. *Plant Dis.* **2013**, *97*, 668–674. [CrossRef]

13. Poveda, J.; Baptista, P. Filamentous fungi as biocontrol agents in olive (*Olea europaea* L.) diseases: Mycorrhizal and endophytic fungi. *Crop Prot.* **2021**, *146*, 105672. [CrossRef]

14. Varo, A.; Raya-Ortega, M.C.; Trapero, A. Selection and evaluation of micro-organisms for biocontrol of *Verticillium dahliae* in olive. *J. App. Microbiol.* **2016**, *121*, 767–777. [CrossRef] [PubMed]

15. Varo, A.; Mulero-Aparicio, A.; Adem, M.; Roca, L.F.; Raya-Ortega, M.C.; López-Escudero, F.J.; Trapero, A. Screening water extracts and essential oils from Mediterranean plants against *Verticillium dahliae* in olive. *Crop Prot.* **2017**, *92*, 168–175. [CrossRef]

16. Varo, A.; Raya-Ortega, M.C.; Agustí-Brisach, C.; García-Ortiz-Civantos, C.; Fernández-Hernández, A.; Mulero-Aparicio, A.; Trapero, A. Evaluation of organic amendments from agro-industry waste for the control of Verticillium wilt of olive. *Plant Pathol.* **2018**, *67*, 860–870. [CrossRef]

17. Mulero-Aparicio, A.; Agustí-Brisach, C.; Varo, A.; López-Escudero, F.J.; Trapero, A. A non-pathogenic strain of *Fusarium oxysporum* as a potential biocontrol agent against Verticillium wilt of olive. *Biol. Control* **2019**, *139*, 104045. [CrossRef]

18. Mulero-Aparicio, A.; Varo, A.; Agustí-Brisach, C.; López-Escudero, F.J.; Trapero, A. Biological control of Verticillium wilt of olive in the field. *Crop Prot.* **2020**, *128*, 104993. [CrossRef]

19. López-Moral, A.; Llorens, E.; Scalschi, L.; García-Agustín, P.; Trapero, A.; Agustí-Brisach, C. Resistance induction against Verticilium wilt of olive (*Olea europea*) by means of beneficial microorganism and plant biostimulants. *Front. Plant Sci.* **2022**, *13*, 831794.

20. McCully, M.E. Niches for bacterial endophytes in crop plants: A plant biologist's view. *Funct. Plant Biol.* **2001**, *28*, 983–990. [CrossRef]

21. Agrios, G. *Fitopatología*, 5th ed.; Elsevier Academic Press: New York, NY, USA, 2005.

22. Sauter, J.J. Seasonal variation of amino acids and amides in the xylem sap of salix. *Z. Pflanzenphysiol.* **1981**, *101*, 399–411. [CrossRef]

23. Tatar, E.; Mihucz, V.; Varga, A.; Zaray, G.; Fodor, F. Determination of organic acids in xylem sap of cucumber: Effect of lead contamination. *Microchem. J.* **1998**, *58*, 306–314. [CrossRef]

24. García, N.F.; Hernandez, M.; Casado-Vela, J.; Bru, R.; Elortza, F.; Hedden, P.; Olmos, E. Changes to the proteome and targeted metabolites of xylem sap in *Brassica oleracea* in response to salt stress. *Plant Cell Environ.* **2011**, *34*, 821–836. [CrossRef]

25. Anguita-Maeso, M.; Haro, C.; Montes-Borrego, M.; De La Fuente, L.; Navas-Cortés, J.A.; Landa, B.B. Metabolomic, ionomic and microbial characterization of olive xylem sap reveals differences according to plant age and genotype. *Agronomy* **2021**, *11*, 1179. [CrossRef]

26. Álvarez, S.; Marsh, E.L.; Schroeder, S.G.; Schachtman, D.P. Metabolomic and proteomic changes in the xylem sap of maize under drought. *Plant Cell Environ.* **2008**, *31*, 325–340. [CrossRef] [PubMed]

27. Anguita-Maeso, M.; Trapero-Casas, J.L.; Olivares-García, C.; Ruano-Rosa, D.; Palomo-Ríos, E.; Jiménez-Díaz, R.M.; Navas-Cortés, J.A.; Landa, B.B. *Verticillium dahliae* inoculation and *in vitro* propagation modify the xylem microbiome and disease reaction to Verticillium wilt in a wild olive genotype. *Front. Plant Sci.* **2021**, *12*, 632689. [CrossRef]

28. Cadahía, C. *La Savia Como Índice de Fertilización. Cultivos Agroenergéticos, Hortícolas, Frutales y Ornamentales*; Mundi-Prensa: Madrid, Spain, 2008; p. 256.

29. Alexou, M.; Peuke, A. Methods for xylem sap collection. *Methods Mol. Biol.* **2013**, *953*, 195–207.

30. Molina, M. Estudio Anatómico Comparado del Xilema de dos Variedades de Olivo con Distinto Nivel de Resistencia Frente a *Verticillium dahliae*. Bachelor's Thesis, University of Cordoba, Cordoba, Spain, 2010.

31. Hilaire, E.; Young, S.A.; Willard, L.H.; McGee, J.D.; Sweat, T.; Chittoor, J.M.; Guikema, J.A.; Leach, J.E. Vascular defense responses in rice: Peroxidase accumulation in xylem parenchyma cells and xylem wall thickening. *Mol. Plant Microbe Interact.* **2001**, *14*, 1411–1419. [CrossRef]

32. Rep, M.; Dekker, H.L.; Vossen, J.H.; De Boer, A.D.; Houterman, P.M.; Speijer, D.; Back, J.W.; De Koster, C.G.; Cornelissen, B.J. Mass spectrometric identification of isoforms of PR proteins in xylem sap of fungus-infected tomato. *Plant Physiol.* **2002**, *130*, 904–917. [CrossRef] [PubMed]

33. Basha, S.; Mazhar, H.; Vasanthaiah, H. Proteomics approach to identify unique xylem sap proteins in Pierce's disease-tolerant *Vitis* species. *Appl. Biochem. Biotechnol.* **2010**, *160*, 932–944. [CrossRef]

34. Gayoso, C.; Pomar, F.; Novo-Uzal, E.; Merino, F.; Martinez de Ilarduya, O. The Ve-mediated resistance response of the tomato to *Verticillium dahliae* involves H_2O_2, peroxidase and lignins and drives PAL gene expression. *BMC Plant Biol.* **2010**, *10*, 232. [CrossRef]

35. Llorens, E.; García-Agustín, P.; Lapeña, L. Advances in induced resistance by natural compounds: Towards new options for woody crop protection. *Sci. Agric.* **2017**, *74*, 90–100. [CrossRef]

36. Moral, J.; Agustí-Brisach, C.; Agalliu, G.; de Oliveira, R.; Pérez-Rodríguez, M.; Roca, L.F.; Romero, J.; Trapero, A. Preliminary selection and evaluation of fungicides and natural compounds to control olive anthracnose caused by *Colletotrichum* species. *Crop Prot.* **2018**, *114*, 167–176. [CrossRef]

37. Steel, R.G.D.; Torrie, J.H. *Bioestadística*, 2nd ed.; McGraw-Hill: Bogotá, Colombia, 1985.

38. Anonimous. *Analytical Software, Statistix10; User's Manual*; Anonimous: Tallahassee, FL, USA, 2013.

Article

Characterization of Actinobacterial Strains as Potential Biocontrol Agents against *Macrophomina phaseolina* and *Rhizoctonia solani*, the Main Soil-Borne Pathogens of *Phaseolus vulgaris* in Cuba

Miriam Díaz-Díaz [1,2,*], Alexander Bernal-Cabrera [3,4], Antonio Trapero [2], Ricardo Medina-Marrero [1], Sergio Sifontes-Rodríguez [1], René Dionisio Cupull-Santana [1], Milagro García-Bernal [1] and Carlos Agustí-Brisach [2,*]

1   Centro de Bioactivos Químicos (CBQ), Universidad Central "Marta Abreu" de Las Villas (UCLV), Carretera Camajuaní km 5 1/2, Santa Clara 54830, Villa Clara, Cuba; rpmedina@uclv.edu.cu (R.M.-M.); sifontes@uclv.edu.cu (S.S.-R.); rcupull@uclv.cu (R.D.C.-S.); mrgarcia@uclv.edu.cu (M.G.-B.)
2   Departamento de Agronomía, Unit of Excellence María de Maeztu 2020-23, Campus de Rabanales, Universidad de Córdoba, Edif. C4, 14071 Córdoba, Spain; ag1trcaa@uco.es
3   Departamento de Agronomía, Facultad de Ciencias Agropecuarias, Universidad Central "Marta Abreu" de las Villas (UCLV), Carretera Camajuaní km 5 1/2, Santa Clara 54830, Villa Clara, Cuba; alexanderbc@uclv.edu.cu
4   Centro de Investigaciones Agropecuarias (CIAP), Facultad de Ciencias Agropecuarias, Universidad Central "Marta Abreu" de Las Villas (UCLV), Carretera Camajuaní km 5 1/2, Santa Clara 54830, Villa Clara, Cuba
*   Correspondence: miriamdd@uclv.cu (M.D.-D.); cagusti@uco.es (C.A.-B.)

Citation: Díaz-Díaz, M.; Bernal-Cabrera, A.; Trapero, A.; Medina-Marrero, R.; Sifontes-Rodríguez, S.; Cupull-Santana, R.D.; García-Bernal, M.; Agustí-Brisach, C. Characterization of Actinobacterial Strains as Potential Biocontrol Agents against *Macrophomina phaseolina* and *Rhizoctonia solani*, the Main Soil-Borne Pathogens of *Phaseolus vulgaris* in Cuba. *Plants* **2022**, *11*, 645. https://doi.org/10.3390/plants11050645

Academic Editor: Shuji Yokoi

Received: 6 January 2022
Accepted: 24 February 2022
Published: 26 February 2022

Publisher's Note: MDPI stays neutral with regard to jurisdictional claims in published maps and institutional affiliations.

Abstract: *Macrophomina phaseolina* and *Rhizoctonia solani* are considered two major soil-borne pathogens of *Phaseolus vulgaris* in Cuba. Their management is difficult, not only due to their intrinsic biology as soil-borne pathogens, but also because the lack of active ingredients available against these pathogens. Actinobacteria, a heterogeneous bacterial group traditionally known as actinomycetes have been reported as promising biological control agents (BCAs) in crop protection. Thus, the main objective of this study was to evaluate the effectiveness of 60 actinobacterial strains as BCAs against *M. phaseolina* and *R. solani* in vitro by dual culture assays. The most effective strains were characterized according to their cellulolytic, chitinolytic and proteolytic extracellular enzymatic activity, as well as by their morphological and biochemical characters in vitro. Forty and 25 out of the 60 actinobacteria strains inhibited the mycelial growth of *M. phaseolina* and *R. solani*, respectively, and 18 of them showed a common effect against both pathogens. Significant differences were observed on their enzymatic and biochemical activity. The morphological and biochemical characters allow us to identify all our strains as species belonging to the genus *Streptomyces*. *Streptomyces* strains CBQ-EA-2 and CBQ-B-8 showed the highest effectiveness in vitro. Finally, the effect of seed treatments by both strains was also evaluated against *M. phaseolina* and *R. solani* infections in *P. vulgaris* cv. Quivicán seedlings. Treatments combining the two *Streptomyces* strains (CBQ-EA-2 + CBQ-B-8) were able to reduce significantly the disease severity for both pathogen infections in comparison with the non-treated and inoculated control. Moreover, they showed similar effect than that observed for *Trichoderma harzianum* A-34 and with Celest® Top 312 FS (Syngenta®; Basilea, Switzerland) treatments, which were included for comparative purposes.

Keywords: ashy stem blight; biological control; common bean; rhizoctonia blight; *Streptomyces* spp.

1. Introduction

The common bean (*Phaseolus vulgaris* L.) is one of the most important grain legumes in many areas of the world, providing a diet rich in protein, dietary fiber, essential micronutrients and phytochemicals for more than 500 million people [1]. The global cultivated surface of *P. vulgaris* reached 33.1 million hectares in the season 2019/2020, with an annual production of 28.9 million metric tons [2]. Common bean is the most important plant species for Cuba population within the group of edible legumes with an annual production of 169,900 tons. Together with rice (*Oryza sativa* L.), they form the basis of the daily diet in this geographic area [3].

In countries with a subtropical climate, the environmental conditions are favorable for the development and proliferation of a vast and heterogeneous soil microflora, including complexes of fungal species associated with root rot diseases such as *Alternaria alternata* (Fr.) Keissl., *Colletotrichum truncatum* (Schwein.) Andrus & W.D. Moore, *Fusarium oxysporum* Schltdl., *Macrophomina phaseolina* (Tassi) Goid., *Rhizoctonia solani* J.G. Kühn, and *Sclerotium rolfsii* Sacc. [4]. In addition, [5] pointed out that the incidence and severity of root rot diseases caused by these fungi depend on the climatic factors prevailing at each sowing time, as well as the characteristics of the microclimates existing in each region of the country where common beans are grown. Among them, *M. phaseolina* and *R. solani* are considered the most prevalent fungal pathogens associated with root rot diseases of common bean worldwide [6,7].

Macrophomina phaseolina (Ascomycota), causal agent of ashy stem blight, also affects roots and stems of host species via pycnidiospores and microsclerotia that persist in the soil, where the pathogen establishes the primary inoculum [8]. Typical symptoms in common bean include dark, irregular lesions on cotyledons, wilting, systemic chlorosis, premature defoliation, epinasty and early maturity or death in adult plants [6]. Late infections cause the appearance of grey areas on the stems, where microsclerotia and pycnidia of the fungus are produced. The occurrence of *M. phaseolina* in the seeds has major consequences since it causes the disqualification of legumes as propagation material [9]. For instance, six tons of common bean and three tons of broad bean to be used as planting material in the province of Villa Clara were disqualified between 2006–2007 because they were affected by *M. phaseolina* [10].

On the other hand, *R. solani* (Basidiomycota) is the causal agent of rhizoctonia blight, also commonly known as damping off [11]. This soil-borne pathogen can affect more than 500 plant species, including cultivated and wild plants, and causes damping off in stands, necrotic lesions in roots, seeds and stems, as well as foliar lesions with a worldwide distribution [7,12]. This fungus affects young seedlings much more than adult plant tissues. On the stem and hypocotyl of affected plants, reddish-brown cankers of various sizes appear, usually delimited by a dark border, which later become rough, dry up and destroy plant tissues [12]. It also attacks the roots causing foot rot of the plants [5]. The management of soil-borne pathogens including both *R. solani* and *M. phaseolina* is usually difficult, not only due to their intrinsic biology, but also because the lack of effective active ingredients. Thus, the use and extension of eco-friendly control methods such as biological control is required, not only to prevent plant diseases, but also contributing markedly to soil preservation and conservation [12].

Microorganisms belonging to genera *Bacillus* (bacteria) and *Trichoderma* (fungi) are the most commonly used biological control agents (BCAs) against soil-borne plant pathogenic fungi ([13,14]. Within this context, species belonging to *Trichoderma* fungal genus have been studied since 1930, and their use has been successfully applied directly to the soil or by seed treatments [15]. On the other hand, since the last century, bacteria belonging to the genus *Bacillus* have been used as BCAs due to their ability to colonize the rhizosphere of plants and inhibit the growth and development of plant pathogens. In addition, they are used as plant growth promoters. [14]. At the same time, the ability of these bacteria to form endospores gives them resistance to climatic changes, which is an important characteristic for inoculum production [13]. In addition to these well-known BCAs, research in the last decades highlights the benefits of the actinobacteria (*Streptomyces* spp. mainly) and their potential as BCAs (e.g., *Streptomyces griseoviridis*, *S. lydicus*) against soil-borne pathogens, such as species of *Rhizoctonia*, *Phytophthora*, *Fusarium*, and *Pythium* in legumes and other crops [16]. Actinobacteria, which have been traditionally known as actinomycetes, are a heterogeneous group of aerobic, filamentous and Gram-positive bacteria. Traditionally, the main genera isolated from soil samples are *Micromonospora*, *Nocardia*, and *Streptomyces*. The genus *Streptomyces* is represented in nature by the largest number of species among the family Actinomycetaceae [17]. This genus, as a colonizer of the rhizosphere, is able to: (i) act as BCA of plant pathogenic fungi, (ii) produce siderophores, (iii) produce plant growth promoting substances, (iv) promote nodulation, (v) produce biodegradative enzymes such as chitinases, cellulases, glucanases, peroxidases, and (vi) assist *Rhizobium* bacteria in iron assimilation, or in nitrogen fixation in legumes, which indirectly contributes to the promotion of plant growth [16].

As we mentioned above, ashy stem blight and rhizoctonia blight are considered the main diseases of *P. vulgaris* in Cuba since they are associated in a complex disease of this crop that causes root rot and plant death. The control management strategies already available against this complex disease are not enough for its optimum control in the frame of the sustainable agriculture. Thus, it is necessary to explore new alternatives towards biological control of these diseases. Therefore, actinobacteria could play an important role as BCAs against the main causal agents of the disease, *M. phaseolina* and *R. solani*. However, the effect of actinobacteria as BCAs against plant pathogenic fungi is still uncertain. Consequently, no biological based compounds on actinobacteria have been developed so far. Likewise, the 'Centro de Bioactivos Químicos' Universidad Central "Marta Abreu" de Las Villas (Cuba) has a wide collection of actinobacterial strains isolated in the central region of the country, which may be explored as a new biological alternative to be included in the integrated disease management program against soil-borne plant pathogens in the common bean crop. Therefore, the main goal of this study was to evaluate 60 actinobacterial strains for their effectiveness as BCAs against *M. phaseolina* and *R. solani* by in vitro dual-cultures assays and finally to select several actinobacterial strains with high efficiency of reduction the viability of both pathogens in vitro, and the disease progress *in planta*. We expect to select several actinobacterial strains with high efficacy on reducing the viability of *M. phaseolina* and *R. solani* in vitro, and the disease progress *in planta*.

2. Results

2.1. In Vitro Effect of Actinobacterial Strains against Macrophomina phaseolina and Rhizoctonia solani: Dual Culture Assays

For both fungal pathogens *M. phaseolina* and *R. solani*, significant differences between actinobacterial strains were observed on their effectiveness in the Mycelial Growth Inhibition (MGI; %) ($p < 0.001$ in both cases). Regarding their effect against *M. phaseolina* isolate CCIBP-Mp1, 40 out of the 60 strains tested showed antagonistic activity against the pathogen. For this group of 40 strains, the MGI ranged from 70.4 ± 1.23 to $3.24 \pm 1.01\%$ for CBQ-EA-2 and CBQ-ESFe-11, respectively. The most effective strains against *M. phaseolina* were CBQ-EA-2, -Plat-2 and -CD-24 with mean MGI values of 70.4 ± 1.23, 66.6 ± 0.78 and $64.6 \pm 1.48\%$, respectively. On the other hand, 25 out of the 60 actinobacterial strains tested showed antagonistic effect against *R. solani* isolate CCIBP-Rh1. For this group of 25 strains, the MGI ranged from 78.3 ± 0.37 to $5.6 \pm 0.47\%$ for CBQ-EA-12 and CBQ-C-5, respectively. The most effective strains against *R. solani* were CBQ-EA-12, -EA-2 and -CD-24 with mean MGI values of 78.3 ± 0.37, 77.4 ± 1.20 and $75.4 \pm 1.22\%$, respectively. In addition, 19 out of the 60 actinobacteria strains evaluated showed a MGI efficacy higher than 50% for both phytopathogenic fungi, with the strains CBQ-EA-2 (MGI = 70.4 ± 1.23 and $77.4 \pm 1.20\%$ for *M. phaseolina* and *R. solani*, respectively) and CBQ-CD-24 (MGI = 64.6 ± 1.48 and $75.4 \pm 1.22\%$ for *M. phaseolina* and *R. solani*, respectively) showing the highest common effectiveness for the two pathogens. At the same time, 5% of the total actinobacteria tested did not show any effect on MGI for any of the two pathogens evaluated (Table 1; Figure 1).

Table 1. Antagonistic effect of the 60 actinobacterial strains on mycelial growth of *Macrophomina phaseolina* and *Rhizoctonia solani* in dual cultures.

Actinobacterial Strain	(MGI; %) [a,b]	
	Macrophomina phaseolina	*Rhizoctonia solani*
CBQ-CB-14	60.9 ± 1.63	69.8 ± 1.40
CBQ-EA-2	70.4 ± 1.23	77.4 ± 1.20
CBQ-EA-12	46.2 ± 1.32	78.3 ± 0.37
CBQ-OSS-3	62.8 ± 1.55	61.2 ± 0.53
CBQ-ESFe-4	52.0 ± 2.00	63.7 ± 1.19
CBQ-CD-24	64.6 ± 1.48	75.4 ± 1.22
CBQ-EBa-5	60.5 ± 1.65	45.9 ± 1.32
CBQ-EBa-21	54.0 ± 1.11	55.7 ± 2.01
CBQ-Plat-2	66.6 ± 0.78	37.3 ± 1.91
CBQ-WP-14	56.7 ± 1.81	37.1 ± 0.40
CBQ-B-8	63.1 ± 1.54	69.0 ± 1.63
CBQ-J-4	28.1 ± 1.71	35.4 ± 2.70
CBQ-CB-3	40.5 ± 0.91	40.1 ± 1.35
CBQ-EA-3	44.4 ± 0.65	0.0 ± 0.00
CBQ-EB-27	0.0 ± 0.00	0.0 ± 0.00
CBQ-EC-3	32.2 ± 0.60	0.0 ± 0.00
CBQ-EC-18	36.0 ± 1.30	54.6 ± 0.91
CBQ-ECa-24	29.9 ± 1.51	0.0 ± 0.00
CBQ-ESFe-5	31.6 ± 1.31	0.0 ± 0.00
CBQ-ESFe-10	35.8 ± 1.25	0.0 ± 0.00
CBQ-ESFe-11	3.24 ± 1.01	0.0 ± 0.00
CBQ-ESFe-12	38.9 ± 1.32	43.5 ± 1.92
CBQ-Mg-6	8.7 ± 0.42	0.0 ± 0.00
CBQ-Ni-24	21.2 ± 0.58	0.0 ± 0.00
CBQ-Ni-32	5.9 ± 0.34	0.0 ± 0.00
CBQ-OSS-4	8.9 ± 0.30	14.3 ± 2.38
CBQ-Plat-3	20.4 ± 1.52	0.0 ± 0.00
CBQ-Plat-4	24.2 ± 1.43	0.0 ± 0.00
CBQ-SFe-5	4.0 ± 0.11	0.0 ± 0.00
CBQ-Wni-21	28.3 ± 1.55	0.0 ± 0.00
CBQ-RS-3	4.7 ± 0.19	0.0 ± 0.00
CBQ-A-2	0.0 ± 0.00	0.0 ± 0.00
CBQ-A-9	0.0 ± 0.00	52.7 ± 2.98
CBQ-A-17	9.7 ± 0.33	32.6 ± 2.30
CBQ-Amb-3	0.0 ± 0.00	0.0 ± 0.00
CBQ-B-1	0.0 ± 0.00	35.0 ± 1.73
CBQ-B-41	0.0 ± 0.00	21.5 ± 0.63
CBQ-B-44	0.0 ± 0.00	41.5 ± 0.43
CBQ-Be-29	0.0 ± 0.00	0.0 ± 0.00
CBQ-Be-36	0.0 ± 0.00	0.0 ± 0.00
CBQ-C-5	0.0 ± 0.00	5.6 ± 0.47
CBQ-C-7	0.0 ± 0.00	17.8 ± 1.04
CBQ-CB-6	17.8 ± 1.78	0.0 ± 0.00
CBQ-CD-12	0.0 ± 0.00	0.0 ± 0.00
CBQ-CD-19	3.4 ± 0.32	0.0 ± 0.00
CBQ-CD-21	38.2 ± 1.26	0.0 ± 0.00
CBQ-CD-23	26.6 ± 1.56	0.0 ± 0.00
CBQ-CD-25	3.5 ± 0.36	0.0 ± 0.00
CBQ-Cy-5	0.0 ± 0.00	0.0 ± 0.00
CBQ-CYM-2	0.0 ± 0.00	0.0 ± 0.00
CBQ-E-5	0.0 ± 0.00	0.0 ± 0.00
CBQ-EA-29	17.7 ± 0.31	0.0 ± 0.00
CBQ-EBa-1	0.0 ± 0.00	0.0 ± 0.00

Table 1. *Cont.*

Actinobacterial Strain	(MGI; %) [a,b]	
	Macrophomina phaseolina	*Rhizoctonia solani*
CBQ-EB-5	0.0 ± 0.00	0.0 ± 0.00
CBQ-EBa-22	0.0 ± 0.00	44.3 ± 1.40
CBQ-EBe-3	15.0 ± 0.12	0.0 ± 0.00
CBQ-EBe-15	0.0 ± 0.00	0.0 ± 0.00
CBQ-EBe-16	0.0 ± 0.00	0.0 ± 0.00
CBQ-EBe-19	16.1 ± 0.49	0.0 ± 0.00
CBQ-EBe-20	0.0 ± 0.00	0.0 ± 0.00
CBQ-EC-5	20.0 ± 0.78	53.2 ± 0.98
$HSD_{0.05}$	6.7	8.5

[a] Mycelial Growth Inhibition (MGI; %) for *Macrophomina phaseolina* isolate CCIBP-Mp1 and *Rhizoctonia solani* isolate CCIBP-Rh1 were obtained by dual culture assays on PDA at 30 °C for 7 days in darkness. Data represent the average of twelve Petri dishes for each BCA or control ± the standard error of the means. [b] For each pathogen, significant differences between treatment means of MGI are given by a critical value for means comparison [$HSD_{0.05}$ = 6.7 and 8.5% for *M. phaseolina* and *R. solani*, respectively] according to Tukey's honestly significant difference (HSD) tests at $p = 0.05$.

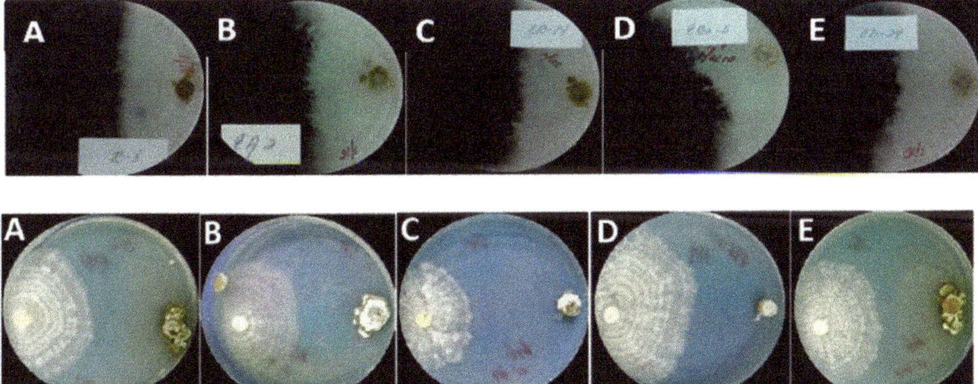

Figure 1. Antagonistic effect of *Streptomyces* strains against *Macrophomina phaseolina* isolate CCIBP-Mp1 (top row photos) and *Rhizoctonia solani* isolate CCIBP-Rh1 (bottom row photos) growing in dual culture on PDA at 7 days after inoculation and incubated at 28 °C in the dark. *Streptomyces* strains evaluated were: (**A**) CBQ-B-8, (**B**) CBQ-EA-2, (**C**) CBQ-CB-14, (**D**) CBQ-EBa-5, and (**E**) CBQ-CD-24 (top row photos); and (**A**) CBQ-B-8, (**B**) CBQ-CB-14, (**C**) CBQ-EA-12, (**D**) CBQ-EBa-21, and (**E**) CBQ-EA-2 (bottom row photos).

2.2. Qualitative Evaluation of Enzyme Activities of Actinobacterial Strains

There were significant differences between the 31 actinobacterial strains evaluated for their chitinolytic, cellulolytic or proteolytic activity ($p < 0.0001$) (Table 2). Twenty out of the 31 strains evaluated showed chitinolytic halo, which ranged from 25.3 ± 0.96 to 33.5 ± 1.91 mm for CBQ-CD-24 to CBQ-EBa-5, respectively. Concerning the cellulolytic activity, the cellulolytic halo ranged between 90.0 ± 0.41 (CBQ-B-8; -CB-14; -ECa-24; -ESFe-12; -Ni-32; -Plat-2; -Plat-3; -Plat-4; and -WP-14) and 36.3 ± 0.75 mm (CBQ-EA-3). Only three out of the 30 strains evaluated did not show cellulolytic halo (CBQ-EB-27; -EC-18; -OSS-4). Finally, 21 out of the 31 strains evaluated showed proteolytic halo, which ranged from 51.5 ± 1.50 to 27.0 ± 0.71 mm for CBQ-EA-12 to CBQ-ECa-24, respectively (Table 2).

Table 2. Chitinolytic, cellulolytic and proteolytic activity of the 31 actinobacterial strains selected for these experiments.

Actinobacterial Strain	Chitinolytic Halo (mm) [a]	Cellulolytic Halo (mm) [b]	Proteolytic Halo (mm) [c]
CBQ-B-8	32.8 ± 0.96 ab	90.0 ± 0.41 a	41.3 ± 0.48 abcd
CBQ-CB-3	0.0 ± 0.0 c	49.3 ± 0.48 ab	0.0 ± 0.0 e
CBQ-CB-14	32.5 ± 2.65 ab	90.0 ± 0.41 a	50.8 ± 2.53 abc
CBQ-CD-24	25.3 ± 0.96 b	63.0 ± 2.38 ab	34.5 ± 0.50 abcd
CBQ-EA-2	34.0 ± 1.41 a	86.3 ± 0.48 ab	44.8 ± 1.65 abcd
CBQ-EA-3	31.3 ± 0.96 ab	36.3 ± 0.75 b	31.8 ± 0.75 bcd
CBQ-CB-4	0.0 ± 0.0 c	86.3 ± 0.48 ab	0.0 ± 0.0 e
CBQ-EA-12	31.3 ± 3.20 ab	80.0 ± 0.71 ab	51.5 ± 1.50 a
CBQ-EBa-5	33.5 ± 1.91 a	85.3 ± 1.18 ab	42.3 ± 0.25 abcd
CBQ-EB-27	0.0 ± 0.0 c	0.0 ± 0.0 c	0.0 ± 0.0 e
CBQ-EBa-21	29.8 ± 1.26 ab	68.5 ± 1.19 ab	47.8 ± 1.4 abc
CBQ-EC-3	0.0 ± 0.0 c	40.0 ± 0.71 ab	0.0 ± 0.0 e
CBQ-EC-18	29.3 ± 0.96 ab	0.0 ± 0.0 c	0.0 ± 0.0 e
CBQ-ECa-24	0.0 ± 0.0 c	90.0 ± 0.41 a	27.0 ± 0.71 d
CBQ-ESFe-4	24.8 ± 0.50 b	82.3 ± 2.59 ab	0.0 ± 0.0 e
CBQ-ESFe-5	0.0 ± 0.0 c	88.8 ± 0.75 ab	41.3 ± 2.39 abcd
CBQ-ESFe-10	31.3 ± 0.96 ab	81.8 ± 1.80 ab	0.0 ± 0.0 e
CBQ-ESFe-11	32.3 ±2.90 ab	84.5 ± 1.55 ab	0.0 ± 0.0 e
CBQ-ESFe-12	29.0 ± 1.41 ab	90.0 ± 0.41 a	49.0 ± 0.71 ab
CBQ-J-4	27.0 ± 1.41 ab	45.0 ± 1.41 ab	27.3 ± 0.48 d
CBQ-Mg-6	0.0 ± 0.0 c	62.0 ± 0.71 ab	44.8 ± 0.48 abcd
CBQ-Ni-24	27.3 ± 0.96 ab	88.8 ± 0.75 ab	31.8 ± 025 cd
CBQ-Ni-32	29.8 ± 0.50 ab	90.0 ± 0.41 a	37.3 ± 0.48 abcd
CBQ-OSS-3	33.0 ± 4.8 ab	67.0 ± 0.71 ab	39.3 ± 0.48 abcd
CBQ-OSS-4	0.0 ± 0.0 c	0.0 ± 0.0 c	44.0 ± 0.71 abcd
CBQ-Plat-2	29.8 ± 0.50 ab	90.0 ± 0.41 a	44.3 ± 0.75 abcd
CBQ-Plat-3	0.0 ± 0.0 c	90.0 ± 0.41 a	32.8 ± 0.25 abcd
CBQ-Plat-4	31.3 ± 0.96 ab	90.0 ± 0.41 a	33.5 ± 0.50 abcd
CBQ-SFe-5	28.3 ± 3.36 ab	47.3 ± 0.75 ab	0.0 ± 0.0 e
CBQ-Wni-21	0.0 ± 0.0 c	42.3 ± 0.48 ab	36.5 ± 0.95 abcd
CBQ-WP-14	29.8 ± 0.50 ab	90.0 ± 0.41 a	44.0 ± 0.71 abc

[a,b,c] Halo develop (mm) for each actinobacterial strain grown onto chitin agar medium Colloidal, Yeast Extract-Malt Extract Agar (ISP2) plates with cellulose (1%, w/v), and ISP2 agar with 1% skimmed milk, respectively, at 28–30 °C in darkness for seven days. For each strain, data represent the average of twelve Petri dishes ± the standard error of the means. In each column, means followed by a common letter do not differ significantly according to Dunn's multiple comparisons for proportions test at $p = 0.05$.

2.3. Phenotypic Characterization

The macroscopic features of the 11 representative actinobacterial strains selected for this experiment are show in Table 3. In general, the colonies were mostly white in color, circular in shape, convex in elevation, with an entire edge, hard consistency and variable pigment production (Figure 2). Microscopic observation of Gram-stained bacterial cells showed stable branched mycelium bearing aerial hyphae, which differentiate into short or long spore chains. Microscopic characterization using the microculture technique revealed details of aerial and vegetative mycelium, mycelial fragmentation and clustering of spores. A spiral arrangement of spores was observed on most of the microculture slides of each sample. In addition, all the strains were characterized as Gram-positive suggesting that they belong to the genus *Streptomyces*.

Table 3. Macroscopic characteristics of colonies of 11 actinobacterial strains (*Streptomyces* spp.) grown on Casein Starch Agar at 28 °C in darkness for 10 days *.

Actinobacterial Strain Code	Gram Stain [a]	Aerial Mycelium [b]	Color	Shape	Elevation	Edge	Consistency	Pigment
CBQ-B-8	+	+	White	Circular	Convex	Full	Hard	Yellow
CBQ-CB-14	+	+	White	Circular	Convex	Full	Hard	Yellow
CBQ-CD-24	+	+	White	Irregular	Convex	Whole	Hard	Beige
CBQ-EA-2	+	+	White	Circular	Convex	Lobed	Hard	Yellow
CBQ-EA-12	+	+	White	Circular	Pulvini	Whole	Hard	Brown
CBQ-EBa-5	+	+	Yellow	Irregular	Convex	Lobular	Hard	Yellow
CBQ-EBa-21	+	+	White	Irregular	Pulvini	Whole	Hard	Orange
CBQ-ESFe-4	+	+	Yellow	Circular	Convex	Lobed	Hard	Beige
CBQ-J-4	+	+	White	Circular	Convex	Whole	Hard	Beige
CBQ-OSS-3	+	+	Yellow	Irregular	Convex	Lobular	Hard	Yellow
CBQ-Plat-2	+	+	White	Circular	Convex	Whole	Hard	Yellow

[a] (+):actinobacteria G+. [b] (+): Presence of aerial mycelium. * The phenotypic characteristics of the colonies of the actinobacterial strains were selected according with [18,19].

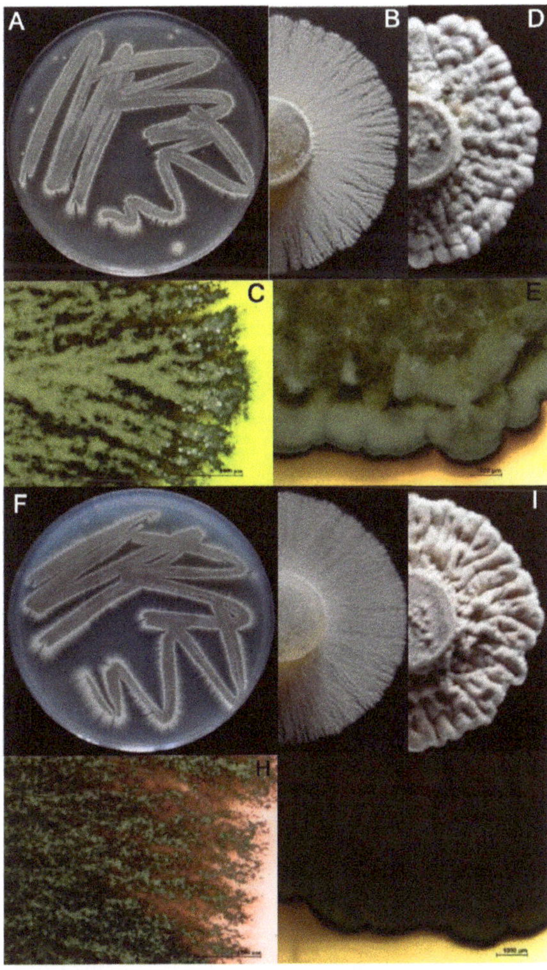

Figure 2. Two-weeks-old colonies of *Streptomyces* strains CBQ-B-8 (**A–E**), and CBQ-EA-2 (**F–J**) growing on ACA medium (**A–C,F–H**) and on PDA medium (**D,E,I,J**) at 28 °C in the dark.

2.4. Biochemical Characterization and Assimilation of Carbon Sources

None of the eleven strains under study were positive for indole production and the Voges Proskauer test. Strains CBQ-J-4, -OSS-3, -EA-2 and -EBa-5, were positive for casein hydrolysis; and the latter two strains were also able to be positive for the methyl red test, in addition to strains CBQ-B-8, -CB-14, -EBa-21 and -Plat-2. Only the strains CBQ-EA-12 and -ESFe-4 did not hydrolyse gelatine. The strains CBQ-OSS-3 and -Plat-2 did not hydrolyse starch (Table 4).

Table 4. Biochemical test results of the 11 actinobacterial strains selected for this experiment.

Biochemical Parameters *	CBQ-B-8	CBQ-CB-14	CBQ-CD-24	CBQ-EA-2	CBQ-EA-12	CBQ-EBa-5	CBQ-EBa-21	CBQ-ESFe-4	CBQ-J-4	CBQ-OSS-3	CBQ-Plat-2
Catalase production	+	+	+	+	+	+	+	+	+	+	+
Lactose Fermentation	+	+	+	+	-	-	-	+	-	+	-
Glucose Fermentation	+	+	+	+	+	D	-	-	+	D	+
Mannitol Fermentation	+	+	-	+	-	+	-	+	-	-	-
Dextrose Fermentation	-	-	+	+	D	-	D	+	-	-	+
Fructose Fermentation	+	+	+	+	+	+	+	+	-	-	+
Maltose Fermentation	+	+	+	+	+	+	+	D	+	+	-
Sucrose Fermentation	+	+	+	+	+	+	+	D	+	-	-
Xylose Fermentation	+	+	-	+	+	-	+	+	-	+	+
Raffinose Fermentation	+	+	+	D	-	-	+	+	+	+	D
Casein Hydrolysis	-	-	-	+	-	+	-	-	+	+	-
Citrate Utilization	+	+	+	+	+	+	+	+	+	+	+
Urea Hydrolysis	+	+	+	+	+	+	+	+	+	+	+
Nitrate reduction	+	+	+	+	+	+	+	+	+	+	+
Indole production	-	-	-	-	-	-	-	-	-	-	-
Methyl red	+	+	-	+	-	+	+	-	-	-	+
Voges Proskauer	-	-	-	-	-	-	-	-	-	-	-
Gelatin hydrolysis	+	+	+	+	-	+	+	-	+	+	+
Starch hydrolysis	+	+	+	+	+	+	+	+	+	-	-

* (+): Positive reaction; (-): Negative reaction; (D): Dubious.

On the other hand, all the evaluated strains were positive for catalase citrate utilization, nitrate reduction and urea hydrolysis. Variability between strains was also observed for the assimilation and utilization of carbohydrates (Table 4).

2.5. Molecular Characterization

BLASTn searches on GenBank showed that the 16S rDNA sequences of the strains CBQ-EA-2 and CBQ-B-8 had 99.71 and 99.93% identity with strains of *Streptomyces* sp. HBUM206419 (MT540570) and MP47-91 (EU263063), respectively. The sequences logged in GenBank and Blast results of the two representative actinobacterial strains selected for their highest effectiveness in vitro in this study are shown in Table 5.

Table 5. Identification by sequencing the 16S rDNA gene of the two actinobacterial strains selected for molecular characterization with their corresponding GenBank accession numbers and data of Blast results obtained from GenBank.

Species	Isolate	Genbank Accession [a]	Blast Accession [b]	Query Length	Gaps [c]	Identities [d]	Maximum Identity (%)
Streptomyces sp.	CBQ-EA-2	OM417233	MT540570	1437	3/1390	1386/1390	99.71
Streptomices sp.	CBQ-B-8	OM417234	EU263063	1491	1/1491	1490/1491	99.93

[a] Corresponding GenBank accession numbers of our isolates. [b] GenBank accession numbers blasted with the isolates obtained in this study. [c] Number of spaces introduced into the alignment to compensate for insertions and deletions in our sequence relative to blasted sequences. [d] Number of nucleotides of our sequences/Number of nucleotides of blasted sequences.

2.6. Effect of Actinobacterial Strains against Macrophomina phaseolina and Rhizoctonia solani Infections in Planta

Because significant differences between sterilized and non-sterilized soils, treatments, and their interaction ($p \leq 0.0001$ in all cases) were observed on their effect on total Disease Severity (DS) (for seedlings inoculated with *M. phaseolina*) and on DSstem and DSroot (for seedlings inoculated with *R. solani*), individual ANOVA per each type of soil was conducted to evaluate the effect of treatment on DS of each tissue.

2.6.1. Effect of Treatments against *Macrophomina phaseolina in Planta*

For the treatments conducted with seedlings grown in non-sterilized soil, significant differences between treatments were observed for their effect on DS ($p \leq 0.0001$). DS ranged from 21.7 ± 2.1 to $6.4 \pm 2.8\%$ for seedlings treated with *Streptomyces* sp. CBQ-EA-2 and Celest® Top 312 FS, respectively, with all treatments showing a significant effect on the disease progress in comparison with the non-treated and inoculated seedlings (positive control; DS = $70.3 \pm 3.1\%$) (Figure 3).

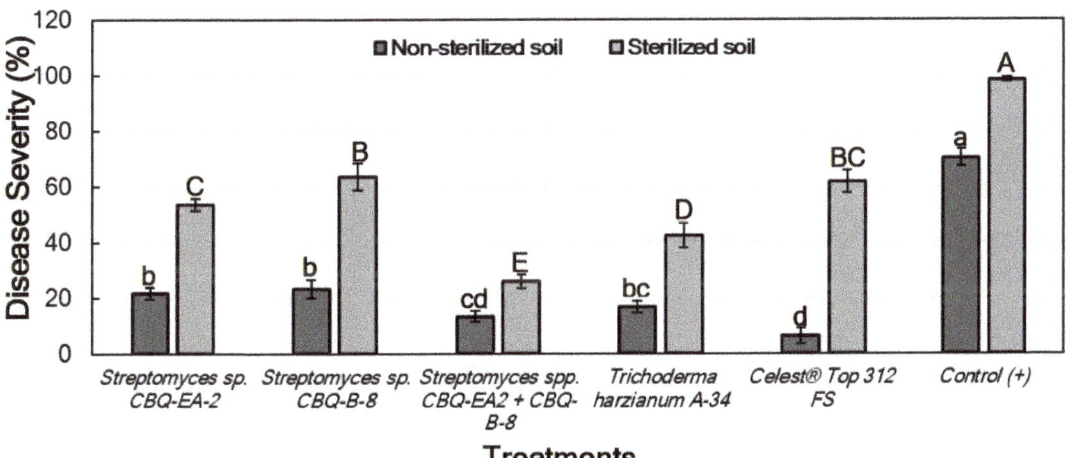

Figure 3. Disease severity (%) in *Phaseolus vulgaris* cv. Quivicán seedlings treated with biological or chemical compounds and inoculated with *Macrophomina phaseolina* isolate CCIBP-Mp1 at 35 days growing on non-sterilized or sterilized soil. Each column represents the mean of 40 seedlings per soil and treatment combination. Columns with a common uppercase or lowercase letter do not differ significantly according to Fisher's protected LSD test ($p = 0.05$) for treatments on non-sterilized or sterilized soil, respectively. Vertical bars are the standard errors of the means.

Concerning the treatments conducted with seedlings grown in sterilized soil, significant differences between treatments were also observed for their effect on DS ($p \leq 0.0001$). In this case, all treatments also resulted in significant effectiveness compared to the positive control (DS = $98.3 \pm 0.7\%$). DS among treated seedlings ranged from 63.6 ± 4.9 to 25.8 ± 2.5 for treatments with *Streptomyces* sp. CBQ-B-8 and *Streptomyces* sp. CBQ-EA-2+ CBQ-B-8, respectively (Figure 3).

The Disease Incidence (DI) was markedly lower in treated seedlings grown in non-sterile soil than those grown in sterile soil. In both cases, not only there were significant differences in DI between treatments, but also significant differences were observed between all the treatments and the positive control ($p \leq 0.0001$ in all cases), the latter always showing the highest DI values. In all cases, the treatments with *Streptomyces* sp. CBQ-EA-2 + CBQ-B-8, *T. harzianum* A-34, or Celest® Top 312 FS showed the lowest DI values (Figure 4). No seedling mortality was not observed in any case, except for the positive control grown in sterile soil which presented 100% mortality.

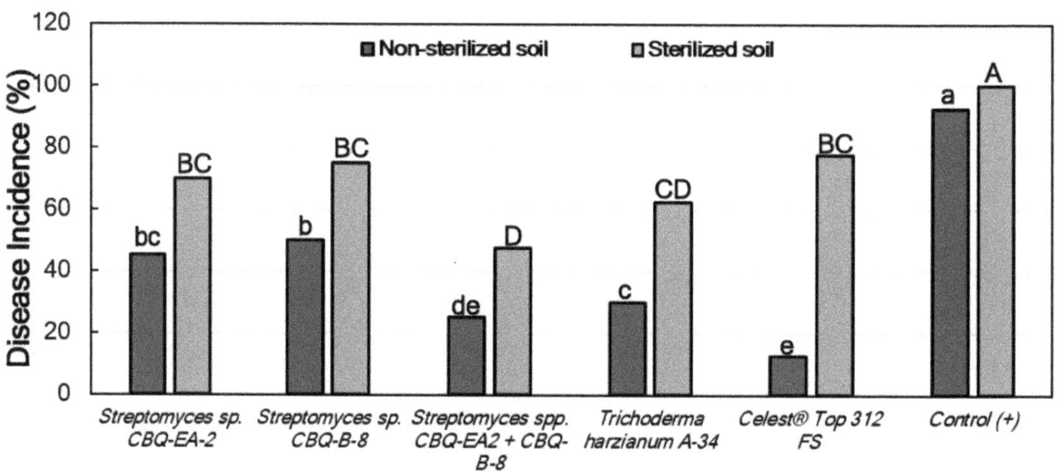

Figure 4. Disease incidence (DI, %) in *Phaseolus vulgaris* cv. Quivicán seedlings treated with biological or chemical compounds and inoculated with *Macrophomina phaseolina* isolate CCIBP-Mp1 at 35 days growing on non-sterilized or sterilized soil. Each column represents the mean of 40 seedlings per soil and treatment combination. Columns with a common uppercase or lowercase letter do not differ significantly according to Dunn's multiple comparisons for proportions test ($p = 0.05$) for treatments on non-sterilized or sterilized soil, respectively.

2.6.2. Effect of Treatments against *Rhizoctonia solani in Planta*

For treatments conducted with seedlings grown in non-sterilized soil, significant differences between treatments were observed for their effect on both DSstem ($p = 0.0015$) and DSroot ($p \leq 0.0001$). In all cases, DSstem was lower than DSroot, ranging from 18.7 ± 2.21 to $4.4 \pm 0.77\%$ for seedlings treated with *Streptomyces* sp. CBQ-EA-2 and Celest® Top 312 FS, respectively. But no important differences were observed for their effect on DSstem compared to the positive control (DSstem = $10.6 \pm 3.78\%$). However, all the treatments showed significantly higher effectiveness on DSroot compared to the positive control (DSroot = $86.8 \pm 4.38\%$). Treatments with *Streptomyces* sp. CBQ-EA-2 (DSroot = $36.9 \pm 1.53\%$) or CBQ-B-8 (DSroot = $40.6 \pm 2.21\%$) were the least effective, while the treatment that combined the two strains was highly effective (DSroot = $21.2 \pm 1.53\%$) showing results similar to those observed for *T. harzianum* A-34 (DSroot = $27.5 \pm 2.07\%$) (Figure 5).

Regarding the treatments conducted with seedlings grown on sterilized soil, significant differences were also observed between treatments for their effect on both DSstem ($p \leq 0.0001$) and DSroot ($p \leq 0.0001$). In this case, all the treatments were highly effective compared to the positive control (DSstem = 100%), but no significant differences in effectiveness between treatments were observed. DSstem ranged from 18.1 ± 2.50 to $15.6 \pm 1.71\%$ for treatments with *Streptomyces* sp. CBQ-EA-2+ CBQ-B-8, and with *Streptomyces* sp. CBQ-EA-2, respectively. On the other hand, all the treatments also showed significantly lower DSroot values compared to the positive control (DSroot = 100%), but significant differences were also observed between treatments for their effect against the disease. The most effective treatment was *Streptomyces* sp. CBQ-EA-2+ CBQ-B-8 (DSroot = $40.6 \pm 2.21\%$), and the least effective were *Streptomyces* sp. CBQ-EA-2 (DSroot = 72.5 ± 1.82%), *Streptomyces* sp. CBQ-B-8 (DSroot = 71.3 ± 1.17%) and *T. harzianum* A-34 (DSroot = 68.7 ± 2.62%) (Figure 5).

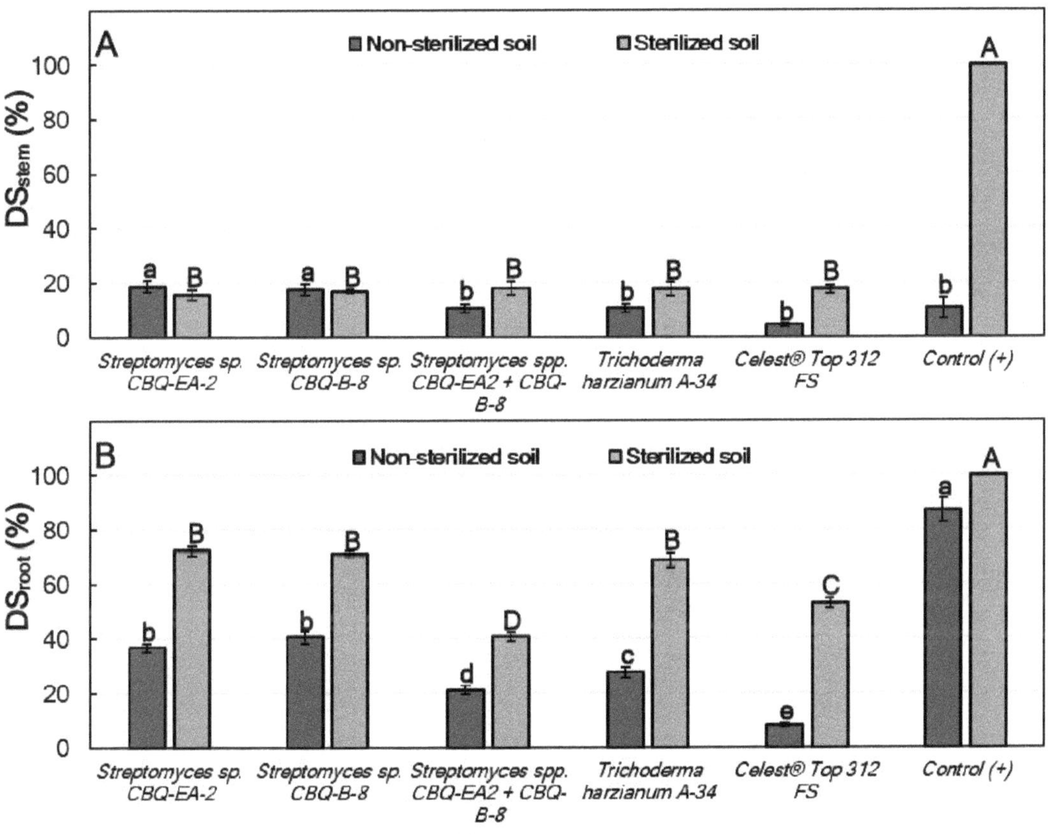

Figure 5. Disease severity (%) in stem (**A**; DSstem) and roots (**B**; DSroot) of *Phaseolus vulgaris* of cv. Quivicán seedlings treated with biological or chemical compounds and inoculated with *Rhizoctonia solani* isolate CCIBP-Rh1 at 28 days growing on non-sterilized or sterilized soil. Each column represents the mean of 40 seedlings per soil and treatment combination. Columns with a common uppercase or lowercase letter do not differ significantly according to Fisher's protected LSD test ($p = 0.05$) for treatments on non-sterilized or sterilized soil, respectively. Vertical bars are the standard errors of the means.

A pattern similar to that observed for seedlings inoculated with *M. phaseolina* was found for the effect of the treatments on the DI of seedlings inoculated with *R. solani*, with significant differences in DI being between all treatments and the positive control ($p \leq 0.0001$ in all cases). In all cases, the treatment with *Streptomyces* sp. CBQ-EA-2 + CBQ-B-8 showed significantly less DI than the treatments with *Streptomyces* sp. CBQ-EA-2 or *Streptomyces* sp. CBQ-B-8, and also showed DI values similar to those observed for *T. harzianum* A-34 and Celest® Top 312 FS (Figure 6).

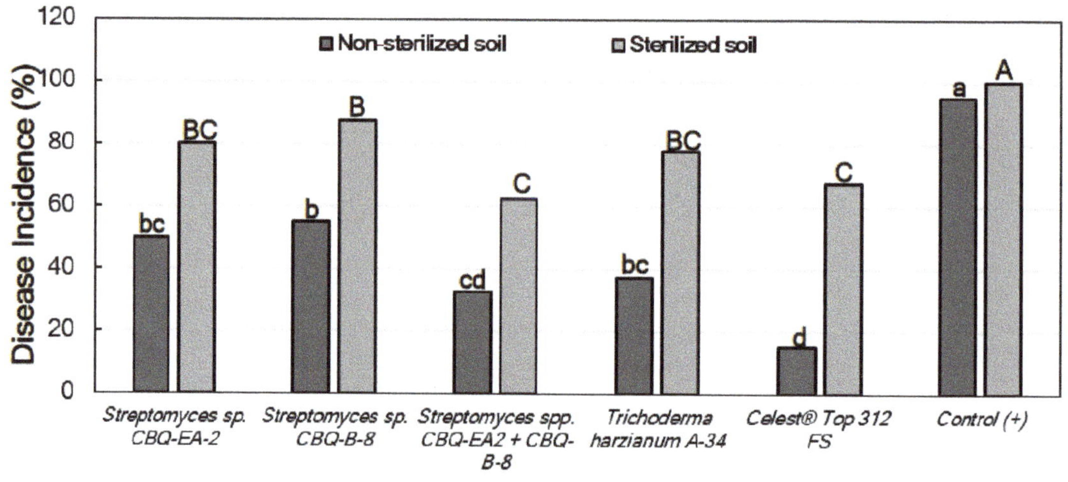

Figure 6. Disease incidence (DI, %) in *Phaseolus vulgaris* cv. Quivicán seedlings treated with biological or chemical compounds and inoculated with *Rhizoctonia solani* isolate CCIBP-Rh1 at 28 days growing on non-sterilized or sterilized soil. Each column represents the mean of 40 seedlings per soil and treatment combination; and columns with a common uppercase or lowercase letter do not differ significantly according to Zar's multiple comparisons for proportions test ($p = 0.05$) for treatments on non-sterilized or sterilized soil, respectively.

Finally, for both lots of plants growing in non-sterilized and sterilized soil, the linear correlation analysis showed that there was not significant linear correlation between DSstem and DSroot (non-sterilized soil: $r = 0.2869$; $p = 0.5815$; sterilized soil: $r = 0.7739$; $p = 0.0709$), and DSstem and DI (non-sterilized soil: $r = 0.3826$; $p = 0.4541$; sterilized soil: $r = 0.7409$; $p = 0.0920$). Nevertheless, a significant positive linear correlation was observed between DSroot and DI in both non-sterilized soil ($r = 0.9944$; $p = 0.0001$) and sterilized soil ($r = 0.9710$; $p = 0.0013$).

3. Discussion

Actinobacteria (*Streptomyces* spp. mainly) have been reported as potential BCAs against soil-borne pathogens of legumes during the last decade [16]. However, the use of actinobacteria as BCAs in the frame of the integrated management of the major diseases of common bean caused by soil-borne pathogens in Cuba has not been explored yet. Therefore, this study aimed to characterize a collection of 60 actinobacterial strains from Cuba based on their in vitro effectiveness against the two main soil-borne pathogens of common bean in Cuba, as well as on their phenotypic and biochemical characteristics.

All the selected actinobacteria formed a smooth surface colony in CAS, becoming white to beige, hard and compact with age, varying in pigmentation, powdery or velvety appearance as a result of the formation of short and long chains of spores, with typical smell of wet soil (Figure 2). In the totality of the microcultures a spiral arrangement of the spores was observed. Similar results were obtained by Ayuningrum and Jati [20], whom reported that isolates of actinobacteria forming powdery colonies with well-developed aerial hyphae divided into spore chains were termed *Streptomyces*-like actinomycete bacteria. This fact together with the concordance of the morphological characters of our strains with those described by Bergey [19] for the *Streptomyces* genus, indicate that all of our actinobacteria strains belong to this genus. In addition, our *Streptomyces* strains showed high levels of cellulolytic and proteolytic activity. Our results are also in concordance with those previously obtained by several authors, who reported the ability of *Streptomyces* strains to produce high levels cellulase and protease [21]. For instance, 62% of our *Streptomyces* strains revealed a high cellulolytic capacity with a halo between 80 to 90 mm in diameter, and 90% of them developed a halo with considerable extension around the colony, which denotes an important cellulolytic hydrolysis. Similar results were recently obtained by Rani et al. [22], who reported that the 67.5 and 60.0% of the *Streptomyces* isolates of their collection showed cellulolytic and proteolytic activity, respectively. Furthermore, the 66.7% of our *Streptomyces* strains showed chitinolytic capacity, highlighting the CBQ-EBa-5 strain, with a 35.5 mm clearance halo surrounding the colony. These results are also in agreement with those obtained by Liu et al. [23], who showed that *S. hydrogenans* (SSD60) and *S. spororaveus* (SDL15) had strong chitinolytic activity, and the 24% of the *Streptomyces* strains of their collection ($n = 94$) developed a clear halo surrounding the colony when evaluating their chitinolytic activity. Altogether, it not only confirms that our strains are well identified as *Streptomyces*, but also suggests that the actinobacteria form one of the most important microbial communities in soil rehabilitation and conservation, as they are largely responsible for their ability to produce extracellular cellulolytic, chitinolytic and proteolytic enzymes.

Actinobacteria represent a source of biologically active secondary metabolites, including enzymes [24]. In this study, we achieved specific qualitative metabolic characterization such as enzymatic, biochemical, morphological and antagonistic of at least 11 strains, which is the main criterion for determining their environmental role and their action in biogeochemical cycles. The challenge of our future research has its origins in this study, so evaluating the in vitro antagonistic activity of our strains showed that many of them disseminate secondary metabolites in the same culture medium in which they inhibit the growth of *M. phaseolina* and *R. solani*. After having evaluated the enzymatic activities, we could infer that the production of chitinases has a positive effect in this regard, since chitin is one of the major components of the fungal cell wall. In addition, actinobacteria combine with other soil microorganisms in their natural environment to decompose resistant plant debris, such as cellulose, as well as animal debris to maintain the biotic balance of the soil by cooperating with the nutrient cycle [25]. Although we were able to identify well all of our actinobacterial strains as *Streptomyces* spp. based on their phenotypic and biochemical characters, the identity of the two representative strains that showed that highest effectiveness on MGI in vitro in this study (CBQ-EA-2, and -B-8) was confirmed by sequencing the 16S rRNA gene using the universal primers 27f and 1492r for eubacteria. The consensus sequences obtained were blasted in GenBank and they match with more than 99% of maximum identity with reference sequences from *Streptomyces* spp. According to Law et al. [26], the 16S rRNA gene has been extensively studied with proven sensitivity for taxonomic and phylogenetic identification of most bacteria including actinobacteria such as *Streptomyces* spp.

Regarding the in vitro efficacy of our 60 *Streptomyces* potential strains against *M. phaseolina* and *R. solani*, it varied depending on the soil-borne pathogen tested. It is worth mentioning that 40 and 25 out of the 60 actinobacterial strains inhibited the mycelial growth of *M. phaseolina* and *R. solani*, respectively. Among the most effective strains, 18 of them showed a common effect against both pathogens, with the CQB-EA2, and -CD-24 being among the strains that showed greater efficacy in inhibiting mycelial growth of the two pathogens. Our results are similar than those described by Dalal et al. [27], who evaluated in vitro the antagonistic activity of 15 strains of actinobacteria against various soil-borne soybean pathogens. These authors reported that the 15 strains showed some effectiveness in inhibiting the mycelial growth of *R. solani*, and six of the 15 strains were also able to inhibit mycelial growth of *M. phaseolina* [25]. Similarly, Singh et al. [28] evaluated the antifungal activity of 80 strains of actinobacteria against *C. truncatum, F. oxysporum, M. phaseolina*, and *S. rolfsii*, highlighting the greater efficacy of *Streptomyces* sp. strain ACITM-1 on inhibition of mycelial growth of all pathogens. In addition to these, several *Streptomyces* sp. strains has also been reported for their high efficacy in inhibiting the mycelial growth of soil-borne pathogenic fungi, such as *R. solani* [29], *R. bataticola* [30], *M. phaseolina* [21], *F. oxysporum*, *Alternaria* sp., and *Magnaporthe oryzae* [21].

Finally, seed treatments with *Streptomyces* sp. CBQ-EA-2 and -B-8 were evaluated separately and in combination against infections by *M. phaseolina* and *R. solani* in inoculated seedlings of common bean under semi-controlled conditions. In general, the treatments conducted using a mix of the two *Streptomyces* sp. strains (CBQ-EA-2 + -B-8) showed a significant greater effectiveness against both pathogens compared to treatments performed with the two strains alone. In addition, the effectiveness of the two combined *Streptomyces* strains in controlling the disease was similar to that observed for the other comparative treatments such as *T. harzianum* A-34 or the chemical (Celest® Top 312 FS). Interestingly, the DS was higher in seedlings grown in sterilized soils than in those grown in non-sterilized soils, also varying the effectiveness of the different treatments with the soil used. It suggests that the microbiota of the soil is in active and positive interaction with the plant and the pathogen, making difficult the pathogen infection and development. Further research to evaluate the effect of the microbiota of the soils used in this study on the biology of both *M. phaseolina* and *R. solani* should be conducted to determine the potential plant-soil-pathogen interactions.

Our results are in concordance with those reported by Yadav et al. [31], who showed that *Streptomyces* sp. S160 reduced the incidence of charcoal rot caused by *M. phaseolina* under greenhouse conditions in chickpea by 33.3% relative to the control. Similarly, Alekhya et al. [32] found that *Streptomyces* sp. (BCA-546 and CAI-8) significantly reduced charcoal rot in sorghum caused by *M. phaseolina* under semi-controlled conditions. On the other hand, our results are also in correspondence with those reported by Korayem et al. [33] who evaluated the biological activity of *S. parvulus* strain 10d against *R. solani* on green beans in a semi-controlled trial with sterilized and non-sterilized soil. These authors showed that seedlings plants treated with a spore suspension of *S. parvulus* strain 10d showed the highest survival rate (88%) and the lowest DSroot (28%) in the whole of the experiment, showing much better results than those observed for seedlings treated with specific chemicals such as Rhizolex® [31]. Similarly, Fatmawati et al. [29] evaluated 10 strains of actinobacteria against *R. solani* on soybean seeds under controlled conditions, with *Streptomyces* spp. strain ASR53 showing the best results in suppressing damping-off disease by 68% and 91% in sterile soil and non-sterile soil, respectively.

This study represents the first report evaluating the effect actinobacteria against the main soil-borne pathogenic fungi of common bean in Cuba. It also shows that *Streptomyces* spp. should be considered as possible biocontrol alternatives against soil-borne pathogens, not only for their effectiveness in disease control, but also for their role in soil preservation which is highly recommended in the frame of sustainable agriculture. Due to the conclusions of this study are based on experiments under controlled conditions, the most effective *Streptomyces* strains of this study may be evaluated against the disease under natural field conditions in the future. Altogether will help us to develop potential BCAs for the control of *M. phaseolina* and *R. solani* associated with stem and root-rot diseases of common bean in Cuba.

4. Materials and Methods

4.1. Actinobacterial Strains and Growth Conditions

A total of 60 actinobacterial strains isolated from different substrates or geographical areas of west-central Cuba were included in this study. They were recovered from rhizosphere (21), stem (15) or root (9) samples from a wide diversity of hosts, among other sources (Table 6), and stored in the laboratory at 4 °C for no more than 72 h until processing. For isolation of actinobacteria from rhizosphere samples, 1 g of each sample was suspended in 9 mL of sterile distilled water (SDW) by vortexing and incubated in water bath at 55 °C for 6 min. Subsequently, serial dilutions (up to 10−5) were performed. The same procedure was carried out with stem or root samples, but they were previously macerated in a mortar with sterile sand. In all cases, 100 µL aliquots of each dilution were spread in 9.0 cm diameter Petri dishes containing casein-starch agar (CSA) supplemented with filtered cycloheximide (100 µg/mL) and nalidixic acid (30 µg/mL) [34]. The inoculated Petri dishes were incubated at 28 °C for 28 days in darkness. Based on macroscopic characters i.e., texture, appearance, surface with or without aerial mycelium, colonies of actinobacteria were selected, transferred to CSA, and incubated as described before. Subsequently, spore suspensions were obtained from the pure cultures of each selected strain, and they were kept in 2 mL translucent screw-capped microtubes (Zhejiang Runlab Technology Co., Taizhou, China) at −20 °C in 20% glycerol for further studies [35]. The collection belongs to the Microbiology Laboratory of the CBQ of the Universidad Central "Marta Abreu" de Las Villas (Cuba).

Table 6. Origen of actinobacterial strains used in this study.

Strain *	Isolation Substrate	Origin (Location, State)	Year of Collection
CBQ-RS-3	Sediment	River Seibabo, Villa Clara	2007
CBQ-A-2	Rhizosphere	Arco Iris, V. Clara	2007
CBQ-EA-2 [a–c]	Endophytic, stem of *Mosiera bullata*	Arco Iris, V. Clara	2008
CBQ-B-8 [a–c]	Rhizosphere, Carbonated brown	Botanical Garden UCLV, V. Clara	2008
CBQ-J-4 [b,c]	Rhizosphere Ferrallitic red	River Seco, Jibacoa, Manicarargua. V. Clara	2008
CBQ-A-9	Rhizosphere	Arco Iris, V. Clara	2008
CBQ-A-17	Rhizosphere	Arco Iris, V. Clara	2008
CBQ-C-5	Rhizosphere	Cienfuegos	2008
CBQ-C-7	Rhizosphere	Cienfuegos	2008
CBQ-B-1	Rhizosphere	Botanical GardenUCLV, V. Clara	2008
CBQ-E-5	'Fangos de Elguea'	Corralillo, V. Clara	2009
CBQ-B-41	Rhizosphere	Botanical Garden UCLV, V. Clara	2009
CBQ-B-44	Rhizosphere	Botanical Garden UCLV, V. Clara	2009
CBQ-Be-29	Rhizosphere	Escambray, Bernal	2010
CBQ-EC-3 [b]	Endophytic	Coge Finca, Camajuaní, V. Clara	2010
CBQ-EC-5	Endophytic, stem of *Petiveria alliacea.*	V. Clara	2010
CBQ-Be-36	Rhizosphere	Escambray, Bernal	2010
CBQ-EBe-3	Endophytic, root of *Hibiscus elatus*	Bernal, Herradura, Manicarargua, V. Clara	2010
CBQ-Cy-5	Rhizosphere	Key I, V. Clara	2010
CBQ-CYM-2	Rhizosphere	Salinas, V. Clara	2010
CBQ-EA-29	Endophytic, stem	Arco Iris, V. Clara	2011
CBQ-EBa-1	Endophytic, root	Banao, S. Spíritus	2011
CBQ-EB-5	Endophytic	Botanical Garden UCLV, V. Clara	2011
CBQ-EBa-22	Endophytic, stem	Banao, S. Spíritus	2011
CBQ-EBe-15	Endophytic, root	Planta Escambray, Bernal	2011
CBQ-EBe-16	Endophytic, root	Planta Escambray, Bernal	2011
CBQ-EBe-20	Endophytic, root	Planta Escambray, Bernal	2011
CBQ-EA-3 [b]	Endophytic	Arco Iris. V. Clara	2011
CBQ-EB-27 [b]	Endophytic, Stem	Jandín Botánico UCLV, V. Clara	2011
CBQ-EA-12 [a,b]	Endophytic, leaf of *Mosiera bullata,*	Arco Iris. V. Clara	2011
CBQ-EC-18 [b]	Endophytic, stem of *Petiveria alliacea.*	Coge Finca, Camajuaní, V. Clara	2011
CBQ-ECa-24 [b]	Endophytic, root	Caguanes, S. Spíritus	2011
CBQ-EBa-5 [a,b]	Endophytic, root	Banao. S. Spíritus	2011
CBQ-EBa-21 [a,b]	Endophytic, root of *Piper aducum*	Banao. S. Spíritus	2011
CBQ-Ni-24 [b]	Endophytic, stem	Nicho, V. Clara	2011
CBQ-Ni-32 [b]	Endophytic, stem	Nicho, V. Clara	2011
CBQ-ESFe-12 [b]	Endophytic, stem of *Fleurya cuneata,*	Loma Sta Fé, V. Clara	2012
CBQ-ESFe-5 [b]	Endophytic, stem of *Fleurya cuneata,*	Loma Sta Fé, V Clara	2012
CBQ-ESFe-10 [b]	Endophytic, stem of *Fleurya cuneata,*	Loma Sta Fé, V. Clara	2012
CBQ-ESFe-11 [b]	Endophytic, stem of *Fleurya cuneata,*	Loma Sta Fé, V. Clara	2012
CBQ-WP-14 [b]	Sediment, *Clarias batrachus,*	V. Clara	2012
CBQ-ESFe-4 [a,b]	Endophytic, leaf of *Piper aduncum*	Loma Santa Fé. V. Clara	2012

Table 6. *Cont.*

Strain *	Isolation Substrate	Origin (Location, State)	Year of Collection
CBQ-Wni-21 [b]	Sediment	River Nicho, V. Clara	2012
CBQ-Amb-3	Endophytic, Stem of *Cecropia adenopu,*	V. Clara	2013
CBQ-CB-14 [a,b]	Sediment	Caves de Bellamar, Matanzas	2013
CBQ-OSS-4 [b]	Endophytic	Topes de Collantes, S. Spíritus	2013
CBQ-Plat-3 [b]	Endophytic, Stem of *Comocladia platyphylla.*	V. Clara	2013
CBQ-Plat-4 [b]	Endophytic, Root of *Comocladia platyphylla,*	V. Clara	2013
CBQ-CB-3 [b]	Sediment	Caves de Bellamar, Matanzas	2013
CBQ-Plat-2 [a,b]	Endophytic, stem of *Comocladia platyphylla.*	V. Clara	2013
CBQ-OSS-3 [a,b]	Endophytic, leaf of *Ossanum*	Topes de Collantes, S. Spíritus	2013
CBQ-CB-4 [b]	Sediment	Caves de Bellamar, Matanzas	2013
CBQ-CD-12	Rhizosphere	Key Las Dunas, V. Clara	2014
CBQ-CD-19	Rhizosphere	Key Las Dunas, V. Clara	2014
CBQ-CD-21	Rhizosphere	Key Las Dunas, V. Clara	2014
CBQ-CD-23	Rhizosphere	Key Las Dunas, V. Clara	2014
CBQ-CD-25	Rhizosphere	Key Las Dunas, V. Clara	2014
CBQ-CD-24 [a,b]	Rhizosphere	Key Las Dunas. V. Clara	2014
CBQ-Mg-6 [b]	Endophytic, stem of *Rhizophora mangle*	Mégano, La Habana	2014
CBQ-SFe-5 [b]	Rhizosphere	Sta Fé, V. Clara	2015

[a] Strains selected for biochemical characterization. [b] Strains selected for qualitative determination of their chitinolytic, cellulolytic and proteolytic activity. [c] Strains selected for *in planta* bioassays. * All actinobacterial strains used in this study were collected in Cuba by Dr. C. R. Medina-Marrero (CBQ: 'Centro de Bioactivos Químicos').

4.2. In Vitro Effect of Actinobacterial Strains against Macrophomina phaseolina and Rhizoctonia solani: Dual Culture Assays

All the 60 actinobacterial strains (Table 6) were evaluated for their effectiveness inhibiting mycelial growth of *M. phaseolina* isolate CCIBP-Mp1 and *R. solani* isolate CCIBP-Rh1 by means in vitro dual culture assays. The two pathogenic fungi were obtained from the collection of plant pathogenic fungi of the Instituto de Biotecnología de las Plantas (IBP) of the Universidad Central Marta Abreu de Las Villas (Cuba), where are maintained growing on PDA at 5 °C in darkness. These isolates were selected due to their high aggressiveness previously tested in the common bean crop [36].

Prior to conduct the dual culture assay, the 60 actinobacterial strains were grown on CSA (pH = 7) at 30 °C for seven days in darkness. The inoculum of *M. phaseolina* and *R. solani* was prepared by seeding suspensions of mycelial fragments of each isolate on Potato Dextrose Agar (PDA; BioCen, Bejucal, Mayabeque, Cuba) at 28 °C for three days in darkness. In vitro dual culture assays were conducted in 9.0 cm in diameter Petri dishes with PDA [37]. To this end, a 7.0 mm in diameter mycelial plug of the pathogen was placed at one end of the plate, and another 7.0 mm in diameter mycelial plug of the actinobacterial strain was plated at 50.0 mm apart at the opposite end. Additionally, 7.0 mm in diameter mycelial plugs of *M. phaseolina* or *R. solani* isolates were seeded in the center of PDA plates without actinobacteria as a positive growth control. All Petri dishes were incubated at 28 °C in total darkness, and the radial mycelial growth of the two plant pathogens was assessed every 24 h, until seven days of incubation [38,39]. There were three replicated Petri dishes per actinobacterial strain ($n = 60$) and plant pathogen ($n = 2$) or control ($n = 2$) combination in a completely randomized design [(60 actinobacterial strains $\times$ 2 fungal pathogens $\times$ 3 Petri dishes) + (2 control $\times$ 3 Petri dishes) = 366 Petri dishes in total]. The experiment was performed three times under similar conditions.

For each fungal pathogen, the percentage of the inhibition of mycelial growth was calculated using the following formula:

$$\text{Mycelial growth inhibition (MGI) (\%)} = [(\text{RGR-rgr})/\text{RGR}] \times 100$$

where 'rgr' is the radial growth of *M. phaseolina* or *R. solani* in dual culture with each actinobacterial strain, and 'RGR' is the radial growth rate of the control treatment (fungal pathogen isolates growing on PDA without actinobacterial strains).

4.3. Qualitative Evaluation of Enzyme Activities of Actinobacterial Strains

Of the 60 strains analyzed in vitro (4.2), the 31 most effective were selected to determine their chitinolytic, cellulolytic, proteolytic activity (Table 6). For chitinolytic activity, all the strains were grown on Colloidal Chitin Agar culture medium (pH = 7) at 28 °C for seven days in darkness [40]. For cellulolytic activity, the strains were grown on ISP2 (International Streptomyces Project) [41] with cellulose (1%, *w/v*) (pH = 7.2) also at 28 °C for seven days in darkness; then a congo red solution (1%) was added as developer for 15 min; and finally, the congo red solution was removed and NaCl solution (1 M) was added for 15 min [42]. For proteolytic activity, the strains were grown on ISP2 with 1% skimmed milk at 30 °C for seven days in darkness [43]. For each parameter evaluated, there were three replicated Petri dishes per strain in a completely randomized design (93 Petri dishes in total), and the experiment was performed three times under similar conditions.

In all cases, after seven days of incubation, the halo surrounding the colonies of the actinobacterial strains was measured (mm) from the center of the inoculated mycelial disc.

4.4. Phenotypic Characterization

Taking into account the macroscopic appearance of the 60 actinobacterial strains evaluated for their effectiveness on MGI of the two pathogens in this study, a total of 11 strains (Table 6) were selected as representative of the main groups with slightly differences on the colony morphology to complete their macro- and microscopic morphological characterization. These strains were grown on CSA as described above, and then, macroscopic colony characters such as presence and color of aerial mycelium, as well as substrate color, shape, elevation, edges and consistency of colonies were recorded [18,19]. Subsequently, microscopic observations were conducted under optical microscope (LABOMED®, Fremont, CA, USA). Bacterial cell observations were carried out on fresh and stained preparations (simple and Gram staining) to define the shape, clustering and response to Gram stain [19]. Additional microscopic features such as aerial and vegetative mycelium, mycelial fragmentation, or clumping of spores were recorded by microcultures with lactophenol blue as a contrast stain [44], and they were compared with those described in Bergey's Manual of Bacteriological Determination [45]. There were three replicated Petri dishes per strain in a completely randomized design (33 Petri dishes in total), and the experiment was performed three times under similar conditions.

4.5. Biochemical Characterization and Assimilation of Carbon Sources

The biochemical characterization using traditional techniques of the same 11 actinobacterial analyzed in the Section 4.4 (Table 6) was evaluated by applying the following tests: catalase, acid production by using different carbohydrate sources (e.g., glucose, mannitol, dextrose, fructose, maltose, raffinose, sucrose and xylose), casein hydrolysis, citrate utilization, indole test, and gelatin hydrolysis [46]. The ability to produce hydrolytic enzymes for the utilization of polysaccharides such as starch was also determined. The hydrolysis of urea to reveal the activity of the enzyme urease [47], methyl red (MR) and Voges Proskauer (VP) tests were carried out according to the ISP [18]. There were three replicated Petri dishes per strain in a completely randomized design (33 Petri dishes in total), and the experiment was performed three times under similar conditions.

4.6. Molecular Characterization

The actinobacterial strains CBQ-EA-2 and CBQ-B-8 were grown in tryptone-soya broth (BioCen) at 30 °C for three days, and centrifuged at 16,000 rpm. DNA was extracted from the resulting pellet using the PureLink™ Genomic DNA Mini Kit reagent (Invitrogen, Waltham, MA, USA), following the manufacturer's instructions. The universal primers 27f and 1492r [48] for eubacteria were used to amplify the 16S rRNA gene via Polymerase Chain Reaction (PCR). Each reaction mixture contained each primer at 20 µM, dNTPs at 10 µM, 5 µL of 10X $MgSO_4$ and buffer, dimethyl sulfoxide (5%), 1 µg of genomic DNA and 1 unit of taq DNA polymerase, for a final volume of 50 µL. PCR steps included an initial denaturation at 94 °C for 3 min, followed by 30 cycles at 94 °C for 30 s, 47 °C for 33 s and 72 °C for 90 s and a final extension step at 72 °C for 7 min. PCR products were run through 1% agarose gel electrophoresis stained with RedSafe™ dye (iNtRONBiotechnology), followed by purification using the PureLink™ kit (Invitrogen, Waltham, MA, USA) and determination of amplicon quality by spectrophotometry (NanoDrop 2000, ThermoScientific; Waltham, MA, USA). Sequencing was carried out on the ABI310 Prism automated sequencer (Applied Biosystems; Waltham, MA, USA), and the resulting sequences were compared with those in the GenBank database using the BLAST (Basic Local Alignment Search Tool) algorithm to identify closely related sequences [49,50]. The consensus sequences were uploaded to GenBank data base (Table 6).

4.7. Effect of Actinobacterial Strains against Macrophomina phaseolina and Rhizoctonia solani Infections in Planta

4.7.1. Plant Material

Seedlings of the common bean (*P. vulgaris* L.) of cv. Quivicán (white testa) were used in this study. The seeds used are registered in the official list of commercial cultivars [51] from the 'UEB Semillas Villa Clara'. Prior to conduct the experiments, the viability of seeds was tested estimating the percentage of germination (%) using a humid chamber at 100% of relative humidity (RH). The seeds were previously disinfected in a serial wash by dipping them first in a 70% ethanol solution for 5 min, then in a 1.5% sodium hypochlorite solution for 15 min, and finally, three times in distilled water for 20 min.

4.7.2. Biological Control Agents and Inoculum Preparation

The actinobacterial strains CBQ-EA-2 and CBQ-B-8 were selected to conduct the experiments *in planta* because they were considered as representative of the strains showing high (CBQ-EA-2; MGI = 70.4 and 77.4% for *M. phaseolina* and *R. solani*, respectively) and moderate (CBQ-B-8; MGI = 63.1 and 69.0% for *M. phaseolina* and *R. solani*, respectively) effectiveness to both pathogens in the dual culture assays. In addition, their morphological, biochemical, and extracellular enzymatic characteristics together with their molecular characterization were also taken into account to ensure that they belong to *Streptomyces* genus together. To prepare the inoculum of the two strains for seed treatments (see below), 20 µL of the original spore suspension preserved at −20 °C in 20% glycerol were firstly added in a 5 mL sterile plastic tubes with tryptone soy broth (BioCen) and incubated at 28 °C for 48 h [52]. Then, they were transferred to 250 mL Erlenmeyer flasks with 100 mL of tryptone soy broth and shaken in a Gerhardt orbital shaker at 28 °C at a speed of 120 G for 3 days. Finally, the inoculum of each actinobacterial strain was adjusted at 1×10^8 spores mL^{-1} using a hemocytometer.

Additionally, *Trichoderma harzianum* strain A-34 belonging to the Plant Health Research Institute (INISAV, La Habana, Cuba) was also included in this experiment as a BCA for comparative purposes. The selected strain is the active ingredient of a bioproduct for the control of phytopathogenic soil fungi, foliar diseases and nematodes commonly used in Cuba [53]. To prepare the inoculum of *T. harzianum* A-34 for seed treatments (see below), sterile 250 mL Erlenmeyer flasks with 100 mL of Potato Dextrose Broth (PDB; BioCen) were inoculated by adding five 10-mm in diameter mycelial plugs of *T. harzianum* A-34 obtained from the active margin of colonies previously grown on PDA at 28 °C in darkness for 72 h. Then, the inoculated Erlenmeyer flask were shaken as described above, and the inoculum was adjusted at 1×10^8 spores mL^{-1}.

4.7.3. Soil Inoculation with *Macrophomina phaseolina* and *Rhizoctonia solani*

The effectiveness of the selected BCAs was evaluated in planta against *M. phaseolina* isolate CCIBP-Mp 2, and *R. solani* isolate CCIBP-Rh1. To prepare the inoculum of both isolates, 1-L Erlenmeyer flasks were filled with 200 g of an artificial substrate (risk husk, part rice grain and distilled water; 3:1:0.5, weight:weight:volume) and sterilized at 120 °C for 1 h. Subsequently, the flasks were seeded with five 1.0-cm in diameter of mycelial plugs of *M. phaseolina* isolate CCIBP-Mp 2 or *R. solani* CCIBP-Rh1 anastomosis groups (AG-4_HGI), taken from the edge of the active growing colonies previously grown on PDA as described before. The inoculated flasks were incubated at 28 °C in darkness for 10 days, and they were manually shaken each 2 days to favor the homogeneous colonization of the substrate [54]. In this study, a medium washed fluffy brown soil [55] non-sterilized and sterilized (120 °C for 20 min in cycles of three consecutive days, and subsequent sterility testing) was used in this study. In all cases, and for each pathogen, the inoculation was carried out at 2% by homogenizing the colonized substrate with the soil [56].

Subsequently, plastic pots were filled with 1.5 Kg of this mix. After 48 h of mix preparation (soil + colonized substrate), four common bean seeds previously treated were sown per plastic pot, and soil moisture was kept at 80% of the field capacity (FC).

4.7.4. Seed Treatments, Growth Conditions and Experimental Design

Seed treatments were conducted by dipping the seeds for 30 min in the following suspensions: (i) actinobacterial strain CBQ-EA-2 at 1×10^8 spores mL^{-1}; (ii) actinobacterial strain CBQ-B-8 at 1×10^8 spores mL^{-1}; (iii) a mix of the actinobacterial strains CBQ-EA-2 and CBQ-B-8 at 1×10^8 spores mL^{-1} global concentration; (iv) *T. harzianum* strain A-34 at 1×10^8 spores mL^{-1}; and (v) Celest® Top 312 FS (Syngenta®; Basilea, Switzerland) prepared in a water suspension of 192 mL of active ingredient per kg of seeds. The latter chemical compound was included for comparison purposes. Additionally, seeds dipped for 30 min in SDW were also included as non-treated control seeds, and lots of non-treated seeds were sowed in plastic pots with inoculated soil (treatment (vi): positive control) as well as in plastic pots with non-inoculated soil (treatment (vii): negative control).

After more than 50% of the seeds emerged, seedlings were treated every three days by wetting the substrate with 1 mL of the respective biological treatment (actinobacterial or *T. harzianum*) adjusted to 1×10^8 spores mL^{-1} until the end of the experiment [28 days after sowing (das)]. Both positive and negative controls and the chemical treatment were wetted every three days with 1 mL of SDW.

For each pathogen, a split-plot design was used with soil ($n = 2$; sterilized and non-sterilized) as the main plot factor and treatments ($n = 7$) as sub-plot factor; with ten pots (replicates) per treatment, and 4 seeds per replicate ($n = 40$). They were maintained in a CBQ greenhouse at 28 °C, 70% RH and 1100 µmol m^{-2} s^{-1} light intensity.

4.7.5. Disease Severity Assessment

For treated seedlings inoculated with *M. phaseolina*, disease severity (DS) was assessed at 35 days after inoculation using the following DS rating scale: 1 = no visible disease symptoms; 3 = wilt restricted to cotyledons, lower stem tissues with small necrotic lesions; 5 = 10% of hypocotyl and lower stem tissues showing lesions, fungal fruiting structures starting the development in the affected tissues, 7 = 25% of hypocotyl and lower stem tissues showing lesions, with development of fungal fruiting structures in the affected tissues; and 9 = ≥50% of hypocotyl and lower stem tissues with lesions, with abundant development of fungal fruiting structures [57]. Subsequently, a DS index was estimated using the following formula:

$$\mathrm{DS}(\%) = \left[\sum_{i=1}^{9} n_i (st_i) / (N \times K) \right] \times 100$$

in which n_i = number of seedlings in the DS development stage i, st_i = value of the DS stage (1–9), N = total number of plants assessed, and K = largest scale level (9) [58].

Regarding treated seedlings inoculated with *R. solani*, DS was evaluated separately on stem and roots tissues at 28 days after inoculation by using the following DS rating scales: (i) DSstem: 1 = absence of lesions on hypocotyl, 2 = superficial lesions (yellow-brown discoloration) on hypocotyl, 3 = deep tissue lesions, and 4 = seedlings dead or wilted [59]; (ii) DSroot: 0 = healthy seedlings, 1 = yellowish-brown discoloration near hypocotyl, 2 = yellowish-brown discoloration plus lesions or brown spots near hypocotyl, 3 = entirely brown surface or lesions covering more than 75% of root surface, and 4 = pre-emergence damping off, seedlings dead or wilted [60]. Subsequently, a DS index was estimated for each tissue using the following formulas:

$$\text{DSstem}(\%) = [\sum_{i=1}^{4} n_i(st_i)/(N \times Kstem)] \times 100$$

$$\text{DSroot}(\%) = [\sum_{i=1}^{5} n_i(st_i)/(N \times Kroot)] \times 100$$

in which n_i = number of stems or roots in the DS development stage i, st_i = value of the DS stage (1–4 and 0–4 for stems and roots, respectively), N = total number of plants assessed, and K = largest scale level (4 in all cases) [58]. Furthermore, for each combination of soil and treatment, the incidence of disease (DI; % of affected plants) and mortality (% of dead plants) were estimated at 28 days after inoculation.

Ungerminated seeds or plants with lesions on the hypocotyl, roots and/or stem were subjected to wet chamber and microscope preparations to confirm the identity of the inoculated pathogens.

4.8. Data Analyses

Data from the repetitions of each experiment were combined after checking for homogeneity of the experimental error variances by the F test ($p \geq 0.05$). Subsequently, data were tested for normality and homogeneity of variances prior to conduct analyses of variance (ANOVA). For the dual culture assay, factorial ANOVA was conducted with MGI as dependent variable, and actinobacterial strains, fungal pathogens and their interaction as independent variables. Significant differences were observed for the two independent variables as well as for their interaction ($p < 0.0001$ in any cases). Thus, independent ANOVA were conducted to determine differences between actinobacterial strains against each fungal pathogen. For each fungal pathogen, mean values were compared using Tukey's honestly significant difference (HSD) tests at $p = 0.05$ [61]. For the enzymatic activity, data of the halo (mm) for each of the three parameters evaluated were analyzed separately by the non-parametric Kruskal-Wallys test due to the assumptions of normality and homogeneity of variances were not fulfilled even though logarithmically, arcsine or square root transformation of the data were conducted. Data from the actinobacterial strains that not develop halo (0.0 mm) were excluded from the statistical analysis in any cases. Mean values were compared using Dunn's comparisons test at $p = 0.05$. In the *in planta* experiment, data of total DS (seedlings inoculated with *M. phaseolina*), and DSstem and DSroot (seedlings inoculated with *R. solani*) were tested for normality and homogeneity of variances prior to conduct analyses of variance (ANOVA). Data from negative control were omitted since no symptoms were observed in all cases. For each dependent variable, a split-plot ANOVA was conducted with soil ($n = 2$) as main-plot factor and treatments ($n = 6$) as the subplot factor. Due to significant differences were observed in all cases for the two independent variables as well as for their interaction ($p < 0.005$), independent ANOVA were conducted to determine differences between treatments for each disease. The treatment means of total DS, or DSstem and DSroot were compared according to Fisher's protected LSD test at $p = 0.05$ [61]. For both inoculated plants with *M. phaseolina* and *R. solani*, data on the final DI (% of affected plants) and mortality (% of dead plants) were analyzed by multiple comparisons for proportions tests at $p = 0.05$ [62]. Additionally, for plants inoculated with *R. solani*, the Pearson correlation coefficients (r) between the DSstem and DSroot were calculated using the average values of the two variables for each of the treatment evaluated in sterilized or non-sterilized soil ($n = 6$ in each type of soil). All data analyses were conducted using Statistix 10 [63].

5. Conclusions

The qualitative characterization of the extracellular enzyme activities, the antagonism of the *Streptomyces* spp. strains, as well as the in vivo studies against *M. phaseolina* and *R. solani* under semi-controlled conditions have allowed us to characterize promising strains as BCAs, and to have a biological alternative in the framework of the integrated management of the main common bean diseases caused by soil pathogens in Cuba. To confirm our laboratory results, the research should and will be evaluated under natural field conditions in further studies.

Author Contributions: Experimental and laboratory tasks, data analysis, wrote and edit the manuscript, M.D.-D.; supervising, review and edit the manuscript, A.B.-C.; review the manuscript, A.T.; sampling, isolation and collection of the actinobacterial strains used in this study, R.M.-M.; experimental design and data analysis, S.S.-R.; experimental and laboratory tasks, R.D.C.-S., isolation of actinobacterial strains, M.G.-B., data analyses, review and edit the manuscript, funding for Open Access publishing, C.A.-B. All authors have read and agreed to the published version of the manuscript.

Funding: This research was financially supported by the Institutional Project 9453: "Search for new strains of actinomycetes that produce antibiotics" of the Center of Chemical Bioactives (CBQ) of the Central University "Marta Abreu" de Las Villas (UCLV), Cuba. M.D-D is holder of scholarships from the Ibero-American University Association for Postgraduate Studies (AUIP) to carry out her pre-doctoral stages (3 stages of 3 months) at the Department of Agronomy at the University of Córdoba (UCO; Spain). We also acknowledge financial support from the Spanish Ministry of Science and Innovation, the Spanish State Research Agency, through the Severo Ochoa and María de Maeztu Program for Centres and Units of Excellence in R&D (Ref. CEX2019-000968-M).

Institutional Review Board Statement: Not applicable.

Informed Consent Statement: Not applicable.

Data Availability Statement: The data that support the findings of this study are available from the corresponding author upon reasonable request.

Acknowledgments: The authors thank the Center of Chemical Bioactives (CBQ) of the Central University "Marta Abreu" de Las Villas (UCLV; Cuba) since all the Streptomyces strains used in this study were donated by the CBQ to develop the present study. The authors thank Edisleidy Águila Jiménez, Yoandry Arencibia, Eliannys Rodríguez Soris and Juan Pablo Mesa Marcillas and Technician Marlén Casanova González for their skillful technical assistance, as well as the colleagues of Department of Agronomy of the University of Córdoba (Spain) for host during the pre-doctoral stages.

Conflicts of Interest: The authors declare no conflict of interest. The funders had no role in the design of the study; in the collection, analyses, or interpretation of data; in the writing of the manuscript, or in the decision to publish the results.

References

1. Singh, G.; Dukariya, G.; Kumar, A. Distribution, Importance and Diseases of Soybean and Common Bean: A Review. *Biotechnol. J. Int.* **2020**, *24*, 86–98. [CrossRef]
2. Food and Agricultural Organisation of the United Nations (FAOSTAT) Website. Available online: http://www.fao.org/faostat/es/#data/QCL/visualize (accessed on 11 August 2021).
3. Ulloa, J.A.; Ulloa, P.R.; Ramírez, R.J.; Ulloa, R.B. El frijol (*Phaseolus vulgaris*): Su importancia nutricional y como fuente de fitoquímicos. *Rev. Fuente.* **2011**, *3*, 5–9.
4. Were, S.A.; Narla, R.; Mutitu, E.W.; Muthomi, J.W.; Munyua, L.M.; Roobroeck, D.; Valauwe, B. Biochar and Vermicompost Soil Amendments Reduce Root Rot Disease of Common Bean (*Phaseolous vulgaris* L.). *Afr. J. Biol. Sci.* **2021**, *3*, 176–196. [CrossRef]
5. González, M. *Enfermedades Fungosas Del Frijol en Cuba*; Técnica, C., Ed.; Científico-Técnica: La Habana, Cuba, 1988; pp. 39–60.
6. Olaya, G.; Abawi, G.S. Effect of water potential on mycelial growth and on production and germination of sclerotia of *Macrophomina phaseolina*. *Plant Dis.* **1996**, *80*, 1347–1350. [CrossRef]

7. Spedaletti, Y.; Cardenas Mercado, G.; Taboada, G.; Aban, C.; Aparicio, M.; Rodriguero, M.; Vizgarra, O.; Sühring, S.; Galíndez, G.; Galván, M. Molecular identification and pathogenicity of *Rhizoctonia* spp. recovered from seed and soil samples of the main bean growing area of Argentina. *Aust. J. Crop Sci.* **2017**, *11*, 952–959. [CrossRef]

8. Zhao, X.; Ni, Y.; Liu, X.; Zhao, H.; Wang, J.; Chen, Y.C.; Liu, H. A simple and effective technique for production of pycnidia and pycnidiospores by *Macrophomina phaseolina*. *Plant Dis.* **2020**, *104*, 1183–1187. [CrossRef] [PubMed]

9. Ministerio de la Agricultura (MINAG). *Guía Técnica Para el Cultivo de Frijol en Cuba*; Instituto de Investigaciones Hortícolas "Liliana Dimitrova": Buenaventura, Cuba, 2008; 18p.

10. González, D. *Micoflora Patogénica en Semillas de Frijol (Phaseolus vulgaris) y Habichuela (Vigna Unguiculata Sesquipedalis), su Efecto en la Germinación y su Control*; Trabajo de Diploma; Facultad de Ciencias Agropecuarias; UCLV: Santa Clara, Cuba, 2007; pp. 18–30.

11. Abu-Tahon, M.A.; Mogazy, A.M.; Isaac, G.S. Resistance assessment and enzymatic responses of common bean (*Phaseolus vulgaris* L.) against *Rhizoctonia solani* damping-off in response to seed presoaking in Vitex agnus-castus L. oils and foliar spray with zinc oxide nanoparticles. *S. Afr. J. Bot.* **2022**, *146*, 77–89. [CrossRef]

12. Muharrem, T.; Kılıçoğlu, M.Ç.; Ismail, E. Characterization and pathogenicity of *Rhizoctonia* isolates collected from *Brassica oleracea* var. *acephala* in Ordu, Turkey. *Phytoparasitica* **2020**, *48*, 273–286.

13. Sabaté, D.C.; Petroselli, G.; Erra-Balsells, R.; Carina Audisio, M.; Brandan, C.P. Beneficial effect of *Bacillus* sp. P12 on soil biological activities and pathogen control in common bean. *Biol. Control* **2019**, *141*, 104–131. [CrossRef]

14. Torres, M.J.; Pérez-Brandan, C.; Sabaté, D.C.; Petroselli, G.; Erra-Balsells, R.; Audisio, M.C. Biological activity of the lipopeptide-producing *Bacillus amyloliquefaciens* PGPBacCA1 on common bean *Phaseolus vulgaris* L. pathogens. *Biol. Control* **2016**, *105*, 93–99. [CrossRef]

15. Mayo-Prieto, S.; Porteous-Álvarez, A.J.; Mezquita-García, S.; Rodríguez-González, Á.; Carro-Huerga, G.; del Ser-Herrero, S.; Casquero, P.A. Influence of Physicochemical Characteristics of Bean Crop Soil in *Trichoderma* spp. Development. *Agronomy* **2021**, *11*, 274. [CrossRef]

16. Vurukonda, S.S.K.P.; Giovanadri, D.; Stefani, E. Growth Promotion and Biocontrol Activity of Endophytic *Streptomyces* spp. In *Prime Archives in Molecular Sciences*, 2nd ed.; Giampietro, L., Ed.; Vide Leaf: Hyderabad, India, 2021; 55p.

17. Olanrewaju, O.S.; Babalola, O.O. *Streptomyces*: Implications an dinteractions in plant growth promotion. *Appl. Microbiol. Biotechnol.* **2019**, *103*, 1179–1188. [CrossRef] [PubMed]

18. Shirling, E.T.; Gottlieb, D. Methods for characterization of *Streptomyces* species. *Int. J. Syst. Evol. Microbiol.* **1966**, *16*, 313–340. [CrossRef]

19. Bergey, D.H.; Holt, J.G.; Hensyl, R. *Bergey's Manual of Determinative Bacteriology*; Lippincott Williams and Wilkins: Baltimore, MD, USA, 2005; Volume 2, pp. 1208–1232.

20. Ayuningrum, D.; Jati, O. Screening of actinobacteria-producing amylolytic enzyme in sediment from Litopenaeus vannamei (Boone, 1931) ponds in Rembang District, Central Java, Indonesia. *Biodiversitas* **2021**, *22*, 1819–1828. [CrossRef]

21. Manigundan, K.; Joseph, J.; Ayswarya, S.; Vignesh, A.; Vijayalakshmi, G.; Soytong, K.; Radhakrishnan, M. Identification of biostimulant and microbicide compounds from *Streptomyces* sp. UC1A-3 for plant growth promotion and disease control. *Int. J. Agric. Technol.* **2020**, *16*, 1125–1144.

22. Rani, K.; Parashar, A.; Wati, L. Estimation of hydrolyzing potential of chickpea actinomycetes for degradation of complex compounds through enzymes and acid production. *Pharma. Innov.* **2021**, *10*, 220–223.

23. Liu, X.; Dou, G.; Ma, Y. Potential of endophytes from medicinal plants for biocontrol and plant growth promotion. *J. Gen. Plant Pathol.* **2016**, *82*, 165–173. [CrossRef]

24. Selim, M.S.M.; Abdelhamid, S.A.; Mohamed, S.S. Secondary metabolites and biodiversity of actinomycetes. *J. Genet. Eng. Biotechnol.* **2021**, *19*, 72. [CrossRef]

25. Bhatti, A.A.; Haq, S.; Bhat, R.A. Actinomycetes benefaction role in soil and plant health. *Microb. Pathog.* **2017**, *111*, 458–467. [CrossRef]

26. Law, J.W.F.; Tan, K.X.; Wong, S.H.; Ab Mutalib, N.S.; Lee, L.H. Taxonomic and characterization methods of *Streptomyces*: A review. *Prog. Microb. Mol. Biol.* **2018**, *1*, a0000009. [CrossRef]

27. Dalal, J.M.; Kulkarni, N.S. Antagonistic and Plant Growth Promoting Potentials of Indigenous Endophytic Actinomycetes of Soybean (*Glycine max* (L.) Merril). *CIBTech J. Microbiol.* **2014**, *3*, 1–12.

28. Singh, C.; Parmar, R.S.; Kumar, A.; Jadon, P. Characterization of actinomycetes against phytopathogenic fungi of *Glycine max* (L.). *Asian J. Pharm. Clin. Res.* **2016**, *9*, 216–219.

29. Fatmawati, U.; Meryandini, A.; Nawangsih, A.A.; Wahyudi, A.T. Damping-off disease reduction using actinomycetes that produce antifungal compounds with beneficial traits. *J. Plant Prot. Res.* **2020**, *60*, 233–243.

30. Khendkar, A.S.; Deshpande, A.R. Isolation and screening of *Streptomyces* for biocontrol potential against *Rhizoctonia bataticola* infection of soybean. *Indian J. Appl. Microbio.* **2018**, *21*, 78–86.

31. Yadav, A.K.; Yandigeri, M.; Vardhan, S.; Sivakumar, G.; Rangeshwaran, R.; Tripathi, C.P.M. *Streptomyces* sp. S160: A potential antagonist against chickpea charcoal root rot caused by *Macrophomina phaseolina* (Tassi) Goid. *Ann. Microbiol.* **2014**, *64*, 1113–1122. [CrossRef]

32. Alekhya, G.; Sharma, R.; Gopalakrishnan, S. *Streptomyces* spp., a potential biocontrol agent of charcoal rot of sorghum caused by *Macrophomina phaseolina*. *Indian J. Plant Prot.* **2016**, *44*, 222–228.

33. Korayem, A.S.; Abdelhafez, A.A.; Zaki, M.M.; Saleh, E.A. Biological control of green bean damping-off disease caused by *Rhizoctonia solani* by *Streptomyces parvulus* strain 10d. *Egypt. J. Microbiol.* **2020**, *55*, 87–94. [CrossRef]

34. El Karkouri, A.; Assou, S.A.; El Hassouni, M. Isolation and screening of actinomycetes producing antimicrobial substances from an extreme Moroccan biotope. *Pan Afr. Med. J.* **2019**, *33*, 329. [CrossRef]

35. Bernal, M.G.; Campa-Córdova, Á.I.; Saucedo, P.E.; González, M.C.; Marrero, R.M.; Mazón-Suástegui, J.M. Isolation and in vitro selection of actinomycetes strains as potential probiotics for aquaculture. *Vet. World* **2015**, *8*, 170–176. [CrossRef]

36. Díaz, C.M. Incidence of *Rhizoctonia* spp., *Sclerotium rolfsii* and *Macrophomina phaseolina* on common bean in Villa Clara. Ph.D. Thesis, Universidad Central "Marta Abreu" de Las Villas, Santa Clara, Cuba, 2011.

37. Sellem, I.; Triki, M.; Elleuch, L.; Cheffi, M.; Chakchouk, A.; Smaoui, S.; Mellouli, L. The use of newly isolated *Streptomyces* strain TN258 as potential biocontrol agent of potato tubers leak caused by *Pythium ultimum*. *J. Basic Microbiol.* **2017**, *57*, 393–401. [CrossRef]

38. Misk, A.; Franco, C. Biocontrol of chickpea root rot using endophytic Actinobacteria. *BioControl* **2011**, *56*, 811–822. [CrossRef]

39. Cuesta, G.; García-de-la-Fuente, R.; Abad, M.; Fornes, F.; Torriani-Gorini, A.; Yagil, E.; Silver, S. Isolation and identification of actinomycetes from a compost-amended soil with potential as biocontrol agents. *J. Environ. Manag.* **2012**, *95*, 281–284. [CrossRef] [PubMed]

40. Kawase, T.; Saito, A.; Sato, T.; Kanai, R.; Fujii, T.; Nikaidou, N.; Miyashita, K.; Watanabe, T. Distribution and phylogenetic analysis of family 19 chitinases in Actinobacteria. *Appl. Environ. Microbiol.* **2004**, *70*, 1135–1144. [CrossRef]

41. Meena, B.; Rajan, L.A.; Vinithkumar, N.V.; Kirubagaran, R. Novel marine actinobacteria from emerald Andaman & Nicobar Islands: A prospective source for industrial and pharmaceutical byproducts. *BMC Microbiol.* **2013**, *13*, 13–45.

42. El-Sersy, N.A.; Abd-Elnaby, H.; Abou-Elela, G.M.; Ibrahim, H.A.; El-Toukhy, N.M. Optimization, economization and characterization of cellulase produced by marine *Streptomyces ruber*. *Afr. J. Biotechnol.* **2010**, *9*, 6355–6364.

43. Ara, I.; Bukhari, N.A.; Wijayanti, D.R.; Bakir, M.A. Proteolytic activity of alkaliphilic, salt-tolerant actinomycetes from various regions in Saudi Arabia. *Afr. J. Biotechnol.* **2012**, *11*, 3849–3857.

44. Franco-Correa, M. Utilización de los actinomicetos en procesos de biofertilización. *Rev. Peru. Biol.* **2009**, *16*, 239–242. [CrossRef]

45. Oskay, A.M.; Üsame, T.; Cem, A. Antibacterial activity of some actinomycetes isolated from farming soils of Turkey. *Afr. J. Biotechnol.* **2004**, *3*, 441–446.

46. Selvakumar, P.; Viveka, S.; Prakash, S.; Jasminebeaula, S.; Uloganathan, R. Antimicrobial activity of extracellularly synthesized silver nanoparticles from marine derived *Streptomyces rochei*. *Int. J. Pharma Bio Sci.* **2012**, *3*, 188–197.

47. Saxena, A.; Upadhyay, R.; Kango, N. Isolation and identification of actinomycetes for production of novel extracellular glutaminase free L-asparaginase. *Indian J. Exp. Biol.* **2015**, *53*, 786–793.

48. Coombs, J.T.; Franco, C.M. Isolation and identification of actinobacteria from surface-sterilized wheat roots. *Appl. Environ. Microbiol.* **2003**, *69*, 5603–5608. [CrossRef] [PubMed]

49. Altschul, S.F.; Gish, W.; Miller, W.; Myers, E.W.; Lipman, D.J. Basic local alignment search tool. *J. Mol. Biol.* **1990**, *215*, 403–410. [CrossRef]

50. Cole, J.R.; Chai, B.; Farris, R.J.; Wang, Q.; Kulam, S.A.; Mcgarrell, D.M.; Garrity, G.M.; Tiedje, J.M. The Ribosomal Database Project (RDP-II): Sequences and tools for high-throughput rRNA analysis. *Nucleic Acids Res.* **2005**, *33*, 294–296. [CrossRef] [PubMed]

51. Ministerio de la Agricultura (MINAG). *Listado oficial de Variedades Comerciales*; Dirección de Semillas y Recursos Filogenéticos; CENSA: La Habana, Cuba, 2016; 41p.

52. Hamdali, H.; Hafidi, M.; Virolle, M.J.; Ouhdouch, Y. Growth promotion and protection against damping-off of wheat by two rock phosphate solubilizing actinomycetes in a P-deficient soil under greenhouse conditions. *Appl. Soil Ecol.* **2008**, *40*, 510–517. [CrossRef]

53. Stefanova, M.; Díaz de Villegas, M.E.; Mena, C.J. Control biológico de enfermedades de plantas en Cuba. In *Control Biológico de Enfermedades de Plantas en América Latina y el Caribe*; Bettiol, W., Rivera, M.C., Mondino, P., Montealegre, J.R., Colmenarez, Y.C., Eds.; Faculdad de Agronomia, Universidad de la Republica: Montevideo, Uruguay, 2014; pp. 201–204, ISBN 978-9974-0-1091-8.

54. Suryawanshi, P.P.; Krishnaraj, P.U.; Suryawanshi, M.P. Evaluation of actinobacteria for biocontrol of sheath blight in rice. *J. Pharmacogn. Phytochem.* **2020**, *9*, 371–376.

55. Hernández, J.A.; Pérez, J.M.; Bosch, I.D.; Rivero, S.N. *Clasificación de los Suelos de Cuba*; Ediciones INCA: Mayabeque, Cuba, 2015; pp. 10–75.

56. Hernández Pérez, D.; Díaz Castellanos, M.; Quiñones Ramos, R.; Santos Bermúdez, R.; Portal González, N.; Herrera Isla, L. Control de *Rhizoctonia solani* en frijol común con rizobacterias y productos naturales. *Cent. Agríc.* **2018**, *45*, 55–60.

57. Van Schoonhoven, A.; Corrales, P. *Sistema Estándar Para la Evaluación de Germoplasma de Frijol*; CIAT: Cali, Colombia, 1987; 56p.

58. Townsend, G.R.; Heuberger, J.W. Methods for estimating losses caused by diseases in fungicide experiments. *Plant Dis. Rep.* **1943**, *27*, 340–343.

59. Bradley, C.A.; Hartman, G.L.; Nelson, R.L.; Müller, D.S.; Pedersen, W.L. Response of ancestral soybean lines and comercial cultivars to *Rhizoctonia solani* root and hypocotyls rot. *Plant Dis.* **2001**, *85*, 1091–1095. [CrossRef]

60. Keijer, J.; Korman, M.G.; Dullemans, A.M.; Houterman, P.M.; De Bree, J.; Van Silfhout, C.H. In vitro analysis of host plant specificity in *Rhizoctonia solani*. *Plant Pathol.* **1997**, *46*, 659–669. [CrossRef]

61. Steel, R.G.D.; Torrie, J.H. *Bioestadística: Principios y Procedimientos, 2nd ed*; McGraw Hill: Bogotá, Colombia, 1985; 613p.

62. Zar, J.H. *Biostatistical Analysis*, 5th ed.; Pte: New Delhi, India; Pearson Education: Singapore, 2010.
63. Analytical Software. StatistixUser's Manual. Analytical Software, Tallahassee, FL, USA. 2013. Available online: https://www.scribd.com/document/381816978/Statistix-10-Manual (accessed on 18 October 2021).

 *plants*

Article

Streptomyces albidoflavus Strain CARA17 as a Biocontrol Agent against Fungal Soil-Borne Pathogens of Fennel Plants

Antonia Carlucci *, Maria Luisa Raimondo *, Donato Colucci and Francesco Lops

Department of Agricultural Sciences, Food, Natural Resources and Engineering, University of Foggia, Via Napoli 25, 71122 Foggia, Italy; donatocolucci.dc@gmail.com (D.C.); francesco.lops@unifg.it (F.L.)
* Correspondence: antonia.carlucci@unifg.it (A.C.); marialuisa.raimondo@unifg.it (M.L.R.)

Abstract: Fennel crop is a horticultural plant susceptible to several soil-borne fungal pathogens responsible for yield losses. The control of fungal diseases occurring on fennel crops is very difficult with conventional and/or integrated means; although several chemical fungicides are able to contain specific fungal diseases, they are not registered for fennel crops. The intensive use of some fungicides causes public concern over the environment and human health. The main aims of this study were to assess the ability of a strain of *Streptomyces albidoflavus* CARA17 to inhibit the growth of fungal soil-borne pathogens, and to protect fennel plants against severe fungal soil-borne pathogens such as *Athelia rolfsii*, *Fusarium oxysporum*, *Plectosphaerella ramiseptata*, *Sclerotinia sclerotiorum* and *Verticillium dahliae*. This study confirmed that the CARA17 strain has been able to inhibit the mycelium growth of pathogens in vitro conditions with significant inhibition degrees, where *S. sclerotiorum* resulted in being the most controlled. The strain CARA17 was also able to significantly protect the fennel seedlings against fungal soil-borne pathogens used in vivo conditions, where the treatment with an antagonist strain by dipping resulted in being more effective at limiting the disease severity of each fungal soil-borne pathogen. Moreover, any treatment with the CARA17 strain, carried out by dipping or after transplanting, produced benefits for the biomass of fennel seedlings, showing significant effects as a promoter of plant growth. Finally, the results obtained showed that CARA17 is a valid strain as a biocontrol agent (BCA) against relevant fungal soil-borne pathogens, although further studies are recommended to confirm these preliminary results. Finally, this study allowed for first time worldwide the association of *Plectosphaerella ramiseptata* with fennel plants as a severe pathogen.

Keywords: microbial antagonist; *Athelia rolfsii*; *Fusarium oxysporum*; *Plectosphaerella ramiseptata*; *Sclerotinia sclerotiorum*; *Verticillium dahliae*; biological control

Citation: Carlucci, A.; Raimondo, M.L.; Colucci, D.; Lops, F. *Streptomyces albidoflavus* Strain CARA17 as a Biocontrol Agent against Fungal Soil-Borne Pathogens of Fennel Plants. *Plants* **2022**, *11*, 1420. https://doi.org/10.3390/plants11111420

Academic Editors: Carlos Agustí-Brisach and Eugenio Llorens

Received: 9 May 2022
Accepted: 23 May 2022
Published: 26 May 2022

Publisher's Note: MDPI stays neutral with regard to jurisdictional claims in published maps and institutional affiliations.

1. Introduction

Fennel (*Phoeniculum vulgare* Mill) is a native horticultural plant of the Southern Europe and Mediterranean area. In particular, in Italy, this crop is of great economic importance, with about 85% of the world production consisting of 507,054 tons [1]. Many soil-borne fungi, reported throughout world, are severe pathogens for fennel crops, causing general symptoms of decline such as root and stem rot (*Rhizoctonia solani*, *Sclerotinia sclerotiorum*, *Athelia rolfsii*) [2–4], vascular diseases (*Verticillium dahliae*, *Fusarium oxysporum*) [2,5] and damping off of seedlings (*Pythium aphanidermatum*) [4]. These fungal pathogens are all notoriously difficult to control with conventional and/or integrated management approaches. To date, the advice for a good management is based on the use of resistant varieties, chemical treatment of seeds, rotation of crops, and the solar treatment of nursery beds to reduce the inoculums of pathogens in the soil. Several fungicides have been reported as being effective against the mentioned soil-borne fungi, but none is specifically registered to control or prevent the disease and/or symptoms in fennel crops. Moreover, intensive use of fungicides in agriculture has raised public concern over the environment and human health [6].

For these reasons, several researchers have focused on biological control as a promising alternative approach to controlling soil-borne diseases in sustainable and organic agriculture. Indeed, many microbial antagonists have been demonstrated to be able to control a large number of pathogens such as fungi and bacteria. For instance, *Trichoderma* spp. as biocontrol agents (BCA) are the most important and are more used against different fungal pathogens such as *Sclerotinia sclerotiorum* and *Phytophthora nicotianae* [7] and *Chalara thielavioides* [8]; *Pseudomonas* spp. is used against *Verticillium dahliae* [9] and *Bacillus* spp. against several phytopathogens [10]. Moreover, recently, plant growth promoting rhizobacteria (PGPR) have been demonstrated to promote plant growth, support nutrition and suppress different plant diseases [11], especially those that are gram-positive and endospore-forming due to being resistant to heat, drying, radiation and toxic chemicals [12,13]. In particular, *Streptomyces* spp. are the most spread gram-positive filamentous bacteria that are ubiquitous in the soil as free-living organisms and symbionts of plants and animals [14]. They are known for producing a wide variety of active biological compounds and are used in agriculture as plant growth promoters, and to be effective as a BCA against a large number of plant pathogens [15,16].

Although a few studies have been carried out to assess the biocontrol efficacy and mechanism of *Streptomyces* spp. [17,18], to date there has been no research related to the control of soil-borne pathogens on fennel crops.

Therefore, the main aims of the present study were: (i) to identify the *Streptomyces* strain isolated and used in this study using molecular tools; (ii) to assess its antagonistic and biostimulant activities; (iii) to assess its putative toxicity/pathogenicity on cucumber cotyledons; (iv) to determine its biocontrol efficacy in vitro and in vivo conditions on fennel seedlings against five fungal soil-borne pathogens, *Athelia rolfsii*, *Fusarium oxysporum*, *Plectosphaerella ramiseptata*, *Sclerotinia sclerotiorum* and *Verticillium dahliae*.

2. Results

2.1. Morphological and Molecular Identification

The amplicon of the CARA17 strain obtained during PCR analysis produced a fragment of 669 bp. The sequence of this amplicon analyzed in GenBank (Acc. n. ON548415) resulted in having a similarity of 100% with *Streptomyces albidoflavus* group.

The ITS sequence of the *Plectosphaerella* strain analyzed in GenBank by the BLAST tool resulted in having a similarity of 100% with the reference strain of *Plectosphaerella ramiseptata* (CBS 131861; Acc. n. JQ246953).

2.2. Inhibitory Activity against Fungal Pathogens in Dual Cultures

According to Shapiro–Wilk tests, the data followed normal distributions for all CARA17 exposition times (14, 21 and 28 days), with W values of 0.77, 0.79 and 0.78 ($p < 0.05$), respectively. The Levene tests showed that the homogeneity of variance was also significant for all three CARA17 exposition times (F = 3.65/3.64/4.19; $p < 0.05$).

One-way ANOVA demonstrated that significant differences in inhibitory activity (IA) by the CARA17 strain against fungal pathogens were observed. The IAs played by the CARA17 strain in dual cultures with fungal pathogens is reported in Table 1. It is possible to observe that the percentage values of IA decreased from 14 at 28 days after inoculation with CARA17 on a medium. In general, the *Streptomyces* strain was able to inhibit the mycelial growth of all fungal pathogens tested with variable percentage values of IA. *Sclerotina sclerotiorum* was the most sensible to antagonistic activity played by the CARA17 strain, as the IAs recorded were 100% at 14, 21 and 28 days after its inoculation. *Athelia rolfsii* resulted in being the least inhibited by the CARA17 strain at all inoculation times. The remaining fungal soil-borne pathogens, *V. dahliae*, *F. oxysporum* and *P. ramiseptata*, resulted in being well controlled with percentage values of IA, such as 71.30% (*V. dahliae*), 82.73% (*P. ramiseptata*) and 84.71% (*F. oxysporum*) after 28 days of inoculation (Table 1; Figure 1).

Table 1. Inhibition activity percentage (IA) by CARA17 strain against mycelial growth of soil-borne pathogens in vitro (dual cultures).

Fungal Soilborne Isolates	Inibition Activity (IA) %					
	14 Days [a]		21 Days		28 Days	
	Mean	Min–Max [c]	Mean	Min–Max	Mean	Min–Max
Athelia rolfsii	15.08 D [b]	14.92–15.25	4.31 D	4.12–4.50	0.00 D	0.00–0.00
Verticillium dahliae	87.42 C	86.31–86.54	77.15 BC	76.96–77.34	71.30 C	71.28–71.32
Plectosphaerella ramiseptata	96.35 B	96.33–96.74	84.82 B	84.58–85.07	82.73 B	82.27–83.19
Fusarium oxysporum	88.26 D	88.63–87.89	84.91 B	82.32–87.50	84.71 B	80.82–88.60
Sclerotinia sclerotiorum	100.00 A	100.00–100.00	100.00 A	100.00–100.00	100.00 A	100.00–100.00

[a] Exposition of fungal strains vs. CARA17 grown on medium at different colony ages; [b] Data followed by different capital letters within the column are significantly different (Fisher's tests; $p < 0.01$); [c] Minimum and maximum values detected (five observations).

Figure 1. Phytotoxicity/pathogenicity assays carried out on young leaf cotyledones of cucumber (**a**), and inhibition activity played by CARA17 strain against *Athelia rolfsii* (**b**), *Fusarium oxysporum* (**c**), *Plectosphaerella ramiseptata* (**d**), *Sclerotinia sclerotiorum* (**e**), and *Verticillium dahliae* (**f**) after 28 days of exposition of fungal pathogens vs. CARA17 strain grown on medium.

2.3. Assessment of Toxicity/Pathogenicity of CARA17 on Cucumis sativus L. cotyledons In Vitro

To ascertain if the CARA17 strain was phytotoxic for fennel seedlings, no symptoms were observed after 3, 6 and 12 days from inoculations when aliquots of its propagules were placed on young cotyledons of cucumber (Figure 1).

2.4. Assessment of Antagonistic Effectiveness of CARA17 Strain on Foeniculum vulgare L. Seedlings against Fungal Soil-Borne Pathogens In Vivo

The results obtained from in vivo experiments carried out on fennel seedlings are reported in Table 2. The symptoms observed on the root consisted of browning, rot, and growth reduction, while on the leaves they consisted of general growth reduction and yellowing of the epigeal portion (Figure 2). The DS values collected by observations carried out on the root of seedlings treated by dipping in CARA17 inoculum suspension were 0.0 for *A. rolfsii* and *S. sclerotiorum*, from 0.6 to 0.9 for *P. ramiseptata*, *V. dahliae* and *F. oxysporum*.

Disease severity values observed on roots increased for experiments where the CARA17 strain was added to a pot after the transplanting of fennel seedlings. In particular, DS values varied from 0.6 to 1.9 against *S. sclerotiorum*, *V. dahliae*, *F. oxysporum* and *P. ramiseptata*. No DS value was produced by *A. rolfsii*. The DS values were higher when no CARA17 inoculum was used against the fungal soil-borne pathogens. The most aggressive fungal pathogen was *A. rolfsii* with a DS value of 4.8, and the less aggressive *S. sclerotiorum* with a DS value of 2.1. No DS values were recorded from fennel seedlings used as control not treated. The DS values recorded from leaves showed the same trend observed on the root, although they resulted in being lower than in those collected from roots (Table 2). The percentages of re-isolation of fungal soil-borne pathogens were significantly higher when the plants were not treated with both kinds CARA17 inoculations (100.00% for *A. rolfsii*; 80.00% for *V. dahliae*), while they were lower when the seedlings were treated with CARA17 by dipping inoculation (13.3% for all fungal pathogens except for *V. dahliae*), and were major with CARA17 inoculated by pouring into pots after transplanting (26.6% for *A. rolfsii*, and 80.00% for *V. dahliae*) (Table 2).

Figure 2. Effectiveness of CARA17 strain as antagonist for controlling disease severity by dipping treatment and after transplanting of fennel seedlings. Fennel seedlings not treated (**a**), treated with CARA17 strain by dipping (**b**) and after transplanting (**c**). Fennel seedlings treated by only fungal pathogen (1); fungal pathogen with CARA17 strain by dipping (2) and after transplanting (3). Inoculation with: *Athelia rolfsii* (**d1–d3**); *Fusarium oxysporum* (**e1–e3**); *Plectosphaerella ramiseptata* (**f1–f3**); *Sclerotinia sclerotiorum* (**g1–g3**); *Verticillium dahliae* (**h1–h3**).

Table 2. Effectiveness of CARA17 strain as antagonist for controlling disease severity (DS) against fungal soil-borne pathogens on fennel seedlings.

	Treatment with CARA17 Strain	Fungal Pathogen	Description Symptoms Recodered					
			Root	DS	Leaves	DS	Re-Isolation from Root (%)
TEST 1	Dipping	*Athelia rolfsii*	No disease symptoms	0.0	No disease symptoms	0.0	13.3
	Dipping	*Fusarium oxysporum*	Light root browning	0.9	Light growth reduction and yellowing	0.6	13.3
	Dipping	*Plectosphaerella ramiseptata*	Light root growth reduction	0.6	Light apical yellowing	0.4	13.3
	Dipping	*Sclerotinia sclerotiorum*	No disease symptoms	0.0	No disease symptoms	0.0	13.3
	Dipping	*Verticillium dahliae*	Light root growth reduction	0.8	Light growth reduction	0.7	26.6
	Dipping	No fungal pathogen	No disease symptoms	0.0	No disease symptoms	0.0	-
TEST 2	After transplanting	*Athelia rolfsii*	No disease symptoms	0.0	No disease symptoms	0.0	26.6
	After transplanting	*Fusarium oxysporum*	Root growth reduction	1.9	Growth reduction and yellowing	1.2	40.0
	After transplanting	*Plectosphaerella ramiseptata*	Root browning	1.9	Growth reduction and yellowing	1.5	86.6
	After transplanting	*Sclerotinia sclerotiorum*	Light root growth reduction	0.6	No disease symptoms	0.0	33.3
	After transplanting	*Verticillium dahliae*	Root growth reduction	1.8	Growth reduction	1.8	80.0
	After transplanting	No fungal pathogen	No disease symptoms	0.0	No disease symptoms	0.0	-
	No treatment	*Athelia rolfsii*	Root browning and rot	4.8	Growth reduction and yellowing	2.7	100.0
	No treatment	*Fusarium oxysporum*	Root browning	2.7	Growth reduction and yellowing	1.8	86.6
	No treatment	*Plectosphaerella ramiseptata*	Root browning	2.9	Growth reduction and yellowing	1.7	93.3
	No treatment	*Sclerotinia sclerotiorum*	Root browning and rot	2.1	Growth reduction	1.4	93.3
	No treatment	*Verticillium dahliae*	Root growth reduction	2.3	Growth reduction	1.5	80.0
	Control	No fungal pathogen	No disease symptoms	0.0	No disease symptoms	0.0	-

2.5. Plant Growth Promotion by Streptomyces Strain In Vivo

Table 3 also reports the data related to the promotion of plant growth by the evaluation of fresh and dry biomasses (related to the entire plants). According to Shapiro–Wilk tests, the data from fresh and dry biomasses followed normal distributions (fresh biomass: $W = 0.90$, $p < 0.01$; dry biomass: $W = 0.88$; $p < 0.01$). The Levene tests showed that the homogeneity of variances, from the in vivo antagonism experiments (TEST1 and TEST2) carried out on fennel seedlings, were significant (fresh biomass: $F = 3.74$, $p < 0.01$; dry biomass: $F = 3.75$, $p < 0.01$). One-way ANOVA highlighted the inoculation by the CARA17 strain significantly promoting the growth of fennel seedlings also in the presence of fungal soil-borne pathogens.

Table 3. Effectiveness of CARA17 strain on promotion of plant growth by biomass production of fennel seedlings.

	Treatment with CARA17 Strain	Fungal Pathogen	Growth Promotion by Biomass * (gr)			
			Fresh		Dry	
			Mean **	SD [§]	Mean	SD
TEST 1	Dipping	*A. rolfsii*	93.11 E [†]	5.13	11.69 D	2.64
	Dipping	*F. oxysporum*	85.07 D	3.97	6.90 B	0.83
	Dipping	*P. ramiseptata*	85.57 D	4.14	7.15 BC	1.71
	Dipping	*S. sclerotiorum*	87.27 D	7.04	8.23 C	0.92
	Dipping	*V. dahliae*	85.15 D	3.33	6.86 B	0.99
	Dipping	No fungal pathogen	165.03 H	15.60	14.43 E	2.47
TEST 2	After transplanting	*A. rolfsii*	69.95 C	10.02	8.81 C	2.13
	After transplanting	*F. oxysporum*	60.05 C	2.37	6.25 B	0.57
	After transplanting	*P. ramiseptata*	45.20 A	7.00	4.93 AB	0.61
	After transplanting	*S. sclerotiorum*	87.27 D	5.29	7.35 BC	0.99
	After transplanting	*V. dahliae*	69.97 C	11.14	6.34 B	2.16
	After transplanting	No fungal pathogen	122.86 G	15.76	12.74 DE	1.74
	No treatment	*A. rolfsii*	38.13 A	5.69	4.15 A	0.34
	No treatment	*F. oxysporum*	47.69 AB	3.69	5.09 AB	0.24
	No treatment	*P. ramiseptata*	48.33 AB	4.56	6.03 B	0.61
	No treatment	*S. sclerotiorum*	44.08 A	4.68	4.89 AB	0.70
	No treatment	*V. dahliae*	45.51 A	6.20	5.00 AB	0.29
	Control	No fungal pathogen	101.61 F	7.00	11.66 D	1.61

* Biomass includes epigeal and hypogeal plant tissues; ** mean values of 15 replicates; [§] Standard Deviation; [†] Data followed by different capital letters within the column are significantly different (Fisher's tests; $p < 0.01$).

The lowest biomasses produced (fresh, 38.13 g; dry, 4.15 g) were recorded from seedlings treated only with *A. rolfsii* with respect to other fungal soil-borne pathogens used alone without any antagonistic treatment (Figure 2). Indeed, the fresh/dried biomasses varied from 44.08/4.89 to 48.33/6.03 g, when the seedlings were treated with the other fungal soil-borne pathogens. In general, the biomasses resulted in being higher when the fennel seedlings were subjected to dipping in CARA17 inoculum rather than to being poured into pots after transplanting. In particular, the trials where the CARA17 strain was used as the dipping inoculation of seedlings inoculated with *A. rolfsii* as fungal soil-borne pathogen, allowed obtaining the highest biomasses (fresh, 93.11 g; dry, 11.69 g). The biomass values (fresh/dry) from other trials treated with the dipping of CARA17 inoculum and other fungal pathogens varied from 85.07/6.90 g (*F. oxysporum*) to 87.27/8.23 g (*S. sclerotiorum*).

The treatment of the CARA17 strain by dipping without fungal pathogens allowed the fennel seedlings to reach significantly higher biomass values (fresh/dry, 165.03/14.43 g) than during treatment after transplanting (fresh/dry, 122.86/12.74 g), and no antagonistic treatment (fresh/dry, 101.61/11.66 g) (Table 3).

3. Discussion

Since in agriculture, low environmental impact control means that phytopathogenic agents are increasingly in demand to reduce the negative effects of chemicals such as pesticide residues in plant products for human use and to preserve the natural ecosystems, the present work reports preliminary results obtained from biological assays carried out in in vivo and in vitro conditions. It is known that biological control by microorganisms such as *Streptomyces* spp. confirms that they are a promising tool for the management of various microbial diseases causing severe losses of agricultural yields. For this reason, the development of biological control (BCAs) and plant growth promoting agents are needed [19]. To date, a very large number of *Streptomyces* strains as antagonists have been used to promote plant growth and control soil-borne phytopathogens [20]. In the present work, a *Streptomyces* strain named CARA17 isolated from some healthy roots of grapevine affected by grapevine trunk diseases was subjected to identification and then used to ascertain its capacities to control five soil-borne phytopathogens, known to be ubiquitous and polyphagous. With cultural and molecular tools, the strain CARA17 was found to belong to the *Streptomyces albidoflavus* group, which was used to ascertain if it was able to control *A. rolfsii*, *F. oxysporum*, *P. ramiseptata*, *S. sclerotiourum* and *V. dahliae* as severe fungal phytopathogens. Antagonism assays carried out in dual cultures in in vitro conditions showed that the *Streptomyces* CARA17 strain was capable of inhibiting the mycelial growth of soil-borne pathogens used as targets with varying degrees of inhibition. As it was observed that the CARA17 strain could inhibit the mycelial growth after some days, in order to emphasize its antagonistic capability, it was placed in Petri dishes 14, 21 and 28 days before the placing of mycelia plugs of phytopathogens. Indeed, it was possible to assess that the CARA17 strain showed a high ability to control the fungal mycelium starting from 14 days after its placement in a Petri dish. It is probable that the CARA17 strain releases bioactive antimicrobic compounds in an artificial medium some days after its placement according to Kaur et al. [21], who found that *Streptomyces* spp. are able to produce a large number of metabolic compounds. In particular, these researchers assessed the antifungal activity of *Streptomyces* spp. against *Fusarium moniliforme* which is responsible for Fusarium wilt in tomato plants in both in vitro and in vivo experiments. Moreover, other investigations reported the effectiveness of *Streptomyces* strains for managing rice blast disease caused by the fungus *Magnaporthe oryzae* [22] in greenhouse conditions, and considering these actinobacteria excellent candidates as biocontrol agents. According to the latest research, our findings allowed us to observe that the CARA17 strain was able to inhibit *S. sclerotiorum* all three times (14, 21 and 28 days of exposition) while against the other fungal pathogens the antagonistic effectiveness was weakly reduced but still significant, and decreased weakly from 14 days to 28 days of exposition. In dual culture, only *A. rolfsii* was already poorly controlled at 14 days of exposition, and not inhibited at 21 and 28 days.

In in vivo conditions, the CARA17 strain was also demonstrated to be efficacious at controlling all fungal phytopathogens used in the present work, although the best effectiveness consisted of the control of *Scleotinia sclerotiorum* for both trials—the dual culture and the greenhouse experiment. Lower but significant biocontrol action acted on by the CARA17 strain was detected against *F. oxysporum*, *P. ramiseptata* and *V. dahliae*. Similar results have been discussed by Colombo et al. [15] when they used two *Streptomyces* strains to control *Fusarium graminearum* as an agent of FRR (Fusarium root rot), FFR (Fusarium foot rot) and FHB (Fusarium head blight) in greenhouse and open field conditions. While no protection was assessed against *A. rolfsii*, a considerable effect of growth promotion in fennel seedlings was observed. The biological control played by the CARA17 strain on fennel seedlings artificially inoculated with fungal soil-borne pathogens was more clear in terms of the protection of seedlings by dipping inoculation before transplanting than by pouring the inoculation into pots after transplanting. The CARA17 strain was probably more able to protect the root of fennel seedlings by dipping inoculation, because the antifungal compounds present in the inoculum solution were immediately adsorbed by roots enhancing the hyperparasitism mechanism and inducing systemic resistance [23,24].

Further, the greenhouse experiments determined significant effects of *Streptomyces albidoflavus* CARA17 strain on the growth parameters of the fennel seedlings, such as the hypogeal and epigeal portions and the fresh and dry weights of the biomasses. Therefore, the data obtained from the in vivo experiments suggested that the CARA17 strain might be used as a biological control agent (BCA) and plant growth promoting agent (PGPA) when used as a propagules inoculum containing spores or mycelium. It is known that *Streptomyces* spp. are included in the PGPR microbial community (Plant Growth Promoting Rhizobacteria), and they are actively or passively involved in plant growth promotion [25]. Therefore, in this work, by the fresh and dry biomass weights obtained, it was possible to assess the ability of the CARA17 strain: (a) to promote the fennel seedlings growth, because it may act as a biofertilizer; (b) to facilitate the tolerance to biotic (phytopathogens used in vivo conditions) and abiotic stresses (no organic and mineral fertilizers), because the disease severity indices collected from the fennel seedlings were significantly reduced due to the presence of the CARA17 strain.

Moreover, this study allowed us, for the first time, to associate *Plectosphaerella ramiseptata*, as a soil-borne pathogen, with fennel plants, and to ascertain its ability to cause significant damages to seedlings consisting of root browning, leaf yellowing and plant growth reduction. Similar symptoms and the disease severity caused by *P. ramiseptata* have also been reported in tomato, pepper, basil and parsley [26,27].

The preliminary results discussed here encourage further studies to assess whether the CARA17 strain is able to produce putative antifungal compounds, to extract secondary metabolites as putative resistance inducers, and to verify if biocontrol and growth promotion actions can be improved and increased by formulation with other PGPR microorganisms.

4. Materials and Methods

4.1. Isolation and Identification of Actinomycetes Strains

During a survey carried out from 2015 to 2019 on the root diseases of grapevine plants affected by decline and apoplexy, caused by fungal agents of grapevine root (Black foot) such as *Dactylonectria* spp., *Ilyonectria* spp. and trunk diseases (GTD) such as *Phaeomoniella chlamydospora* and *Phaeoacremonium* spp. [28], other unknown microorganisms were isolated including not-sporulated fungi, bacteria and *Actinomycetes*. Among the latter microbial agents, a consistent number (37 isolates, corresponding to 1.6% of microorganisms isolated) of presumptive *Streptomyces* strains were observed and recorded. All *Actinomycetes* cultures were subjected to purification techniques by spreading over Petri dishes of agar and water (AW). After an overnight incubation at 28 ± 3 °C, single germinating spores or small pieces of hyphae were transferred to Petri dishes with fresh potato-dextrose-agar (PDA; 39 g/L, Oxoid) for molecular identification. For this purpose, a representative strain of *Actinomycetes* showing a putative antimicrobial activity, called CARA17, was used for molecular characterization. Genomic DNA of the CARA17 strain was extracted from a 15-day-old culture growing on PDA at 28 ± 3 °C in the dark, according to Carlucci et al. [29].

For preliminary molecular identifications, the primer pairs used were 16SAct1F (5′ CGC GGC CTA TCA GCT TGT TG 3′) and 16SAct1R (5′ CCG TAC TCC CCA GGC GGG G 3′) of 16S rDNA for the amplification of the 16S ribosomal RNA (R RNA) region [30]. The amplification was made according to the following PCR protocol: $1\times$ PCR buffer, 2.5 mM MgCl2, 200 µM of each nucleotide, 2.5 pmol of each primer, 0.25 U Taq polymerase, 0.5 µL DMSO, and a 30–50 ng DNA template taken to a total volume of 25 µL. The Taq polymerase, nucleotides and buffers were supplied by Eurofins Genomics (Milan, Italy). The amplification conditions were: initial denaturation for 5 min at 94 °C, followed by 35 cycles of denaturation for 1 min at 95 °C, annealing for 30 s at 58 °C, elongation for 1 min at 72 °C and the final extension step for 10 min at 72 °C. Five microliters of amplicon were analyzed by electrophoresis at 100 V for 30 min in 1.5 % (*w/v*) agarose gels in $1\times$ TAE buffer (40 mM Tris, 40 mM acetate, 2 mM EDTA, pH 8.0). The gels were stained with ethidium bromide and were visualized in a Gel Doc EZ System under UV light (Biorad, Hercules, CA, USA). The PCR products were purified before DNA sequencing, using Nucleo Spin Extract II purification kits (Macherey-Nagel, Germany), according to the manufacturer's instructions. Both strands of the PCR products were sequenced by Eurofins Genomics (Ebersberg, Germany). The nucleotide sequences obtained were manually edited using BioEdit v.7.0.9 (http://www.mbio.ncsu.edu/ BioEdit accessed on 8 Febrary 2021). The consensus sequence was compared with those available in the GenBank database, using the Basic Local Alignment Search Tool (BLAST, http://www.ncbi.nlm.nih.gov/ accessed on 24 March 2021) to confirm the preliminary morphological identification and to ascertain the sequence similarity searches.

4.2. Fungal Soil-Borne Phytopathogens

To assess the antagonistic activity of the *Streptomyces* strain CARA17 in vitro and in vivo conditions, fungal soil-borne pathogens such as *Athelia rolfsii*, *Fusarium oxysporum*, *Sclerotinia sclerotiorum* and *Verticillium dahliae* strains were used, which were all isolated from fennel plants during previous surveys carried out in the northern Apulia region (2017–2020). The taxonomic identity of all fungal species here used was assessed earlier with molecular tools and DNA extraction according to Carlucci et al. [29] (data not shown).

From the root and collar of fennel plants, a conspicuous amount of isolates morphologically attributed to *Plectosphaerella* genus were collected; a representative strain was included in the in vitro and in vivo trials. For this purpose, the collection of *Plectosphaerella* isolates was subjected to DNA extraction, as mentioned above, and to molecular screening by MSP-PCR using the M13 minisatellite primer (5′-GAGGGTGGCGGTTCT-3′) [31]. MSP-PCR profiles were generated according to Santos and Phillips [32]. The DNA banding patterns were analyzed using the Bionumerics v.5.1 software (Applied Maths, A Biomerieux company, Sint-Martens-Latem, Belgium), with calculations of Pearson's correlation coefficients and the unweighted pair group method with arithmetic means. The reproducibility levels were calculated by comparing the banding profiles obtained for the M13 primer. For this purpose, 10% of the strains were chosen from any cluster at random, and their profiles were analyzed again. The MSP dendrogram generated one clade, from which one isolate was chosen as a representative and was molecularly characterized according to Carlucci et al. [33] (data not shown).

All fungal strains used here are maintained in the laboratory of the Plant Pathology and Diagnosis of Department of Sciences Agriculture, Food, Natural resources and Engineering (DAFNE) at the University of Foggia, Italy.

4.3. Inhibitory Activity against Fungal Pathogens in Dual Cultures

An agar-mycelium disc (5 mm diameter) of CARA17, taken from the edge of a 21-day-old colony grown on PDA, was put in a PDA Petri dish at 15 mm from the center and was left to grow at $25 \pm 3\ °C$ in darkness. After 14, 21 and 28 days incubation, an agar-mycelium disc (5 mm diameter) from each fungal pathogen was put at 15 mm from the center in front of the agar disc with the CARA17 strain. Five replicates were performed for each fungal strain, and only the plates inoculated with the pathogen and the sterile agar disk were used as a control. The dual cultures were kept at $21 \pm 3\ °C$ for 15 days in darkness. The inhibitory activity (IA) was calculated as the percentage of mycelium growth inhibition compared to the control by the formula $[(R1-R2)/R1] \times 100$, where R1 was the radius measurement from the center of the colony of fungal pathogen towards the edge of the control Petri plate (without CARA17), and R2 was the radius from the center towards the edge of the fungal colony in the direction of the antagonist CARA17, respectively, according to Kunova et al. [34].

The percentage data of inhibition activity were arcsine root-square transformed in Excel 2007 by the formula DEGREES(ASIN(SQRT(X))), where X is the percentage value. One-way ANOVA analysis was performed using Statistica, version 6 (StatSoft, Hamburg,

Germany) to assess the significant differences of inhibition activity values. Fisher's test was used as a post-hoc test ($p = 0.01$).

4.4. Assessment of Toxicity/Pathogenicity of CARA17 Strain on Cucumis sativus L. Cotyledons In Vitro

The *Streptomyces* strain CARA17 was assayed for its putative ability to cause damage to horticultural crops by the inoculation of a suspension of its propagules (solution at 0.2% of Tween 20 in sterile water) on young cucumber cotyledons. The cotyledons were taken from seeds germinated on peat loam sterilized at 121 °C for 30 min three times at intervals of 24 h. They were gently disinfected by immersion in 70 % ethanol for 5 min, rinsed in sterile distilled water and dried on sterile paper. Four disinfected cotyledons were gently placed in Petri dishes containing water-agar (0.3%) and were inoculated with a drop of 20 µL of *Streptomyces* strain CARA17 propagules suspension at 1×10^6 cfu/mL. As a control assay, the cotyledons were inoculated with sterile Tween 20 solution (0.2%). This assay was replicated six times and kept at 21 ± 3 °C in darkness, and the cotyledons were inspected after 3, 6, 9 and 12 days to ascertain putative toxicity and/or pathogenicity symptoms.

4.5. Assessment of Antagonistic Effectiveness of CARA17 Strain on Foeniculum vulgare L. Seedlings against Fungal soil-borne pathogens In Vivo

The inoculum solution with the CARA17 strain propagules was prepared by collecting spores and small fragments of mycelium scraped from surface of 21-day-old *Streptomyces* colonies grown on PDA medium at 28 ± 3 °C in darkness until reaching a concentration of 1×10^7 cfu/mL in sterile Tween 20 solution (0.2%).

Inoculation preparation of fungal soil-borne pathogens. The inoculum solution with each fungal soil-borne pathogen was prepared as described above for the CARA17 strain, scraping from the surface of 21-day-old colonies grown on PDA medium at 21 ± 3 °C in darkness until reaching a concentration of 1×10^7 cfu/mL in sterile Tween 20 solution (0.2%).

The experimental design was performed as two independent batches at the end of August and consisted of two different inoculation kinds, where the horticultural host target was represented by 30-day-old seedlings of *Foeniculum vulgare* var. DONATELLO F1 (HM.CLAUSE Vegetable Seeds).

The first experiment (TEST1) consisted of preliminarily dipping the fennel seedlings in the inoculum solution of the CARA17 strain for 30 min, before transplanting them in a pot containing 1.5 kg of soil and peat (3:1), (sterilized early twice at 121 °C for 30 min and kept for 20 days in a controlled chamber at 25 ± 3 °C, 70% relative humidity, and under natural light), and wetting them with 500 mL of irrigation water. Subsequently, 50 mL of inoculum solution of each fungal soil-borne pathogen was poured into the soil of each pot around the collar of the fennel seedlings.

The second experiment (TEST2) consisted of inoculation with 50 mL of inoculum solution of each fungal soil-borne pathogen after seedlings transplantation into wet soil with 500 mL of irrigation water. After 48 h, 50 mL of CARA17 inoculum solution was poured into the soil around the collar of the fennel seedlings.

As control trials, for both experiments (TEST1 and TEST2), pots containing fennel seedlings treated with the CARA17 strain, treated with each fungal pathogen, and not treated with either the CARA17 strain or the fungal soil-borne pathogens, were prepared. Each trial was replicated fifteen times. The pots with fennel seedlings were placed in a greenhouse with temperature and humidity not conditioned. During the growth of the seedlings, no fertilizers, pesticides or fungicides were used. They were only subjected to irrigation practice when necessary, using the same water volumes for each pot. After 100 days, the fennel plants were gently removed from the pots, the roots and collars were carefully washed, and the presence/absence of browning symptoms observed on the root and collar were evaluated and described using an empiric scale from 0 to 5, where 0 = no symptoms observed; 1 = 1–20%; 2 = 21–40%; 3 = 41–60%; 4 = 61–80%; and

5 = 81–100%. The disease severities (DS) on the roots and collar were determined according to the following formula:

$$DS = \frac{\sum(Number\ of\ observ \times values\ of\ scores)}{Total\ number\ of\ cases}. \tag{1}$$

All fungi underwent re-isolation from the root, collar and stem of the inoculated plants to fulfil Koch's postulates.

4.6. Plant Growth Promotion by Streptomyces Strain In Vivo

To assess putative plant growth promotion by the CARA17 strain on fennel plants, during both the experiments described above (TEST1 and TEST2), all plants were cut at the basis of the collar, and all tissues of epigeal (collar, stem and leaves) and hypogeal portions (root), after a careful washing, were separately weighed and put into a stove at 105 °C until reaching a constant dry weight. For all trials, the average weight of 15 fennel plants (epigeal and hypogeal portions) were calculated. To determine whether these data followed normal distributions, the Shapiro–Wilk test (W test) was used. The homogeneity of variance of the datasets was assessed on the basis of fresh and dried weights of fennel seedlings using the Levene test. One-way ANOVA was performed using Statistica v. 6 (StatSoft, Hamburg, Germany) to determine the significant differences in fresh and dried weights recorded during both kinds of antagonist inoculation (by dipping and by pouring after transplanting). Fisher's test was used for the comparison of the treatment means, at $p < 0.01$.

Author Contributions: Conceptualization, A.C. and M.L.R.; methodology, A.C., M.L.R., D.C. and F.L.; software, A.C. and M.L.R.; validation, A.C. and M.L.R.; formal analysis, A.C., M.L.R., D.C. and F.L.; investigation, A.C., M.L.R. and D.C.; resources, A.C.; data curation, A.C. and M.L.R.; writing—original draft preparation, A.C.; writing—review and editing, A.C. and M.L.R.; supervision, A.C. and M.L.R.; project administration, A.C.; funding acquisition, A.C. All authors have read and agreed to the published version of the manuscript.

Funding: This research received no external funding.

Institutional Review Board Statement: Not applicable.

Informed Consent Statement: Not applicable.

Data Availability Statement: Not applicable.

Conflicts of Interest: The authors declare no conflict of interest.

References

1. ISTAT (Istituto Nazionale di Statistica). Ortive: Finocchio in Pien'aria. 2021. Available online: http://dati.istat. (accessed on 17 May 2022).
2. Shaker, G.A.; Alhamadany, H.S. Isolation and identification of fungi which infect fennel *Foeniculum vulgare* Mill. and its impact as antifungal agent. Bull. *Iraq nat. Hist. Mus.* **2015**, *13*, 31–38.
3. Choi, I.Y.; Kim, J.H.; Kim, B.S.; Park, M.J.; Shin, H.D. First report of Sclerotinia stem rot of fennel caused by *Sclerotinia sclerotiorum* in Korea. *Plant Dis.* **2016**, *100*, 223. [CrossRef]
4. Khare, M.N.; Tiwari, S.P.; Sharma, Y.K. Disease problems in fennel (*Foeniculum vulgare* Mill) and fenugreek (*Trigonella foenum graceum* L.) cultivation and their management for production of quality pathogen free seeds. *Int. J. Seed Spices* **2014**, *4*, 11–17.
5. Ghoneem, K.M.; Saber, W.A.; Elwakil, M.A. Alkaline seed bed an innovative technique for manifesting *Verticillium dahliae* on fennel seeds. *Plant Pathol. J.* **2009**, *8*, 22–26. [CrossRef]
6. Dias, M.C. Phytotoxicity: An overview of the physiological responses of plants exposed to fungicides. *J. Bot.* **2012**, *2012*, 135479. [CrossRef]
7. Chilosi, G.; Aleandri, M.A.; Luccioli, E.; Stazi, S.R.; Marabottini, R.; Morales-Rodríguez, C.; Vettraino, A.N.; Vannini, A. Suppression of soil-borne plant pathogens in growing media amended with espresso spent coffee grounds as a carrier of *Trichoderma* spp. *Sci. Hortic.* **2020**, *259*, 108666. [CrossRef]
8. Kowalska, B.; Smolińska, U.; Szczech, M.; Winciorek, J. Application of organic waste material overgrown with *Trichoderma atroviride* as a control strategy for *Sclerotinia sclerotiorum* and *Chalara thielavioides* in soil. *J. Plant Prot. Res.* **2017**, *57*, 205–211. [CrossRef]

9.   Oktay, E.; Kemal, B. Biological control of Verticillium wilt on cotton by the use of fluorescent *Pseudomonas* spp. under field conditions. *Biol. Control.* **2010**, *53*, 39–45.

10. Khan, M.; Salman, M.; Jan, S.A.; Shinwari, Z.K. Biological control of fungal phytopathogens: A comprehensive review based on *Bacillus* species. *MOJ Biol. Med.* **2021**, *6*, 90–92. [CrossRef]

11. Basu, A.; Prasad, P.; Das, S.N.; Kalam, S.; Sayyed, R.Z.; Reddy, M.S.; El Enshasy, H. Plant Growth Promoting Rhizobacteria (PGPR) as Green Bioinoculants: Recent Developments, Constraints, and Prospects. *Sustainability* **2021**, *13*, 1140. [CrossRef]

12. Comasriu, J.; Vivesrego, J. Cytometric monitoring of growth, sporogenesis and spore cell sorting in *Paenibacillus polymyxa* (formerly *Bacillus polymyxa*). *J. Appl. Microbiol.* **2002**, *92*, 475–481. [CrossRef] [PubMed]

13. Emmert, E.A.; Handelsman, J. Biocontrol of plant disease: A (gram-) positive perspective. *FEMS Microbiol. Lett.* **1999**, *171*, 1–9. [CrossRef] [PubMed]

14. Seipke, R.F.; Kaltenpoth, M.; Hutchings, M.I. *Streptomyces* as symbionts: An emerging and widespread theme? *FEMS Microbiol. Rev.* **2012**, *36*, 862–876. [CrossRef] [PubMed]

15. Colombo, E.M.; Kunova, A.; Pizzatti, C.; Saracchi, M.; Cortesi, P.; Pasquali, M. Selection of an endophytic *Streptomyces* sp. strain DEF09 from wheat roots as a biocontrol agent against Fusarium graminearum. *Front. Microbiol.* **2019**, *10*, 2356. [CrossRef]

16. Liu, D.; Yan, R.; Fu, Y.; Wang, X.; Zhang, J.; Xiang, W. Antifungal, plant growth-promoting, and genomic properties of an endophytic actinobacterium *Streptomyces* sp. NEAU-S7GS2. *Front. Microbiol.* **2019**, *10*, 2077. [CrossRef] [PubMed]

17. Boukaew, S.; Prasertsan, P. Suppression of rice sheath blight disease using a heat stable culture filtrate from *Streptomyces philanthi* RM-1-138. *Crop Prot.* **2014**, *61*, 1–10. [CrossRef]

18. Kunova, A.; Cortesi, P.; Saracchi, M.; Migdal, G.; Pasquali, M. Draft genome sequences of two *Streptomyces albidoflavus* strains DEF1AK and DEF147AK with plant growth promoting and biocontrol potential. *Ann. Microbiol.* **2021**, *71*, 2–8. [CrossRef]

19. Hartman, A.; Schmid, M.; van Tuinen, D.; Berg, G. Plant driven selection of microbes. *Plant Soil* **2009**, *321*, 235–257. [CrossRef]

20. Faheem, M.; Raza, W.; Zhong, W.; Nan, Z.; Shen, Q.; Xu, Y. Evaluation of the biocontrol potential of *Streptomyces goshikiensis* YCXU against *Fusarium oxysporum* f. sp. *niveum*. *Biol. Control* **2015**, *81*, 101–110. [CrossRef]

21. Kaur, T.; Rani, R.; Manhas, R.K. Biocontrol and plant growth promoting potential of phylogenetically new *Streptomyces* sp. MR14 of rhizospheric origin. *AMB Express* **2019**, *9*, 125. [CrossRef]

22. Law, J.W.F.; Ser, H.L.; Khan, T.M.; Chuah, L.H.; Pusparajah, P.; Chan, K.G.; Goh, B.H.; Lee, L.H. The Potential of Streptomyces as Biocontrol Agents against the Rice Blast Fungus, *Magnaporthe oryzae* (*Pyricularia oryzae*). *Front. Microbiol.* **2017**, *8*, 3. [CrossRef] [PubMed]

23. De-Oliveira, M.F.; Da Silva, M.G.; Van Der Sand, S.T. Anti-phytopathogen potential of endophytic actinobacteria isolated from tomato plants (*Lycopersicon esculentum*) in southern Brazil, and characterization of *Streptomyces* sp. R18, a potential biocontrol agent. *Res. Microbiol.* **2010**, *161*, 565–572. [CrossRef] [PubMed]

24. Passari, A.K.; Mishra, V.K.; Gupta, V.K.; Yadav, M.K.; Saikia, R.; Singh, B.P. In vitro and in vivo plant-growth-promoting activities and DNA fingerprinting of antagonistic endophytic actinomycetes associates with medicinal plants. *PLoS ONE* **2015**, *10*, 0139468. [CrossRef]

25. Etesami, H.; Maheshwar, D.K. Use of plant growth promoting rhizobacteria (PGPRs) with multiple plant growth promoting traits in stress agriculture: Action mechanisms and future prospects. *Ecotoxicol. Environ. Safe.* **2018**, *156*, 225–246. [CrossRef] [PubMed]

26. Raimondo, M.L.; Carlucci, A. Characterization and pathogenicity assessment of Plectosphaerella species associated with stunting disease on tomato and pepper crops in Italy. *Plant Pathol.* **2018**, *67*, 626–641. [CrossRef]

27. Raimondo, M.L.; Carlucci, A. Characterization and pathogenicity of *Plectosphaerella* spp. collected from basil and parsley in Italy. *Phytopathol. Mediterr.* **2018**, *57*, 284–295.

28. Carlucci, A.; Lops, F.; Mostert, L.; Halleen, F.; Raimondo, M.L. Occurrence fungi causing black foot on young grapevines and nursery rootstock plants in Italy. *Phytopathol. Mediterr.* **2017**, *56*, 10–39.

29. Carlucci, A.; Raimondo, M.L.; Cibelli, F.; Phillips, A.J.L.; Lops, F. *Pleurostomophora richardsiae*, *Neofusicoccum parvum* and *Phaeoacremonium aleophilum* associated with a decline of olives in southern Italy. *Phytopathol. Mediterr.* **2013**, *52*, 517–527.

30. Shariffah-Muzaimah, S.A.; Idris, A.S.; Madihah, A.Z.; Dzolkhifli, O.; Kamaruzzaman, S.; Maizatul-Suriza, M. Characterization of *Streptomyces* spp. isolated from the rhizosphere of oil palm and evaluation of their ability to suppress basal stem rot disease in oil palm seedlings when applied as powder formulations in a glasshouse trial. *World J. Microb. Biot.* **2018**, *34*, 15. [CrossRef]

31. Meyer, W.; Mitchell, T.G.; Freedman, E.Z.; Vilgalys, R. Hybridization probes for conventional DNA fingerprinting used as single primers in the polymerase chain reaction to distinguish strains of *Cryptococcus neoformans*. *J. Clin. Microbiol.* **1993**, *31*, 2274–2280. [CrossRef]

32. Santos, J.M.; Phillips, A.J.L. Resolving the complex of *Diaporthe* (*Phomopsis*) species occurring on *Foeniculum vulgare* in Portugal. *Fungal Divers.* **2009**, *34*, 111–125.

33. Carlucci, A.; Raimondo, M.L.; Santos, J.; Phillips, A.J.L. *Plectosphaerella* species associated with root and collar rots of horticultural crops in southern Italy. *Persoonia* **2012**, *28*, 34–48. [CrossRef] [PubMed]

34. Kunova, A.; Bonaldi, M.; Saracchi, M.; Pizzatti, C.; Chen, X.; Cortesi, P. Selection of *Streptomyces* against soil borne fungal pathogens by a standardized dual culture assay and evaluation of their effects on seed germination and plant growth. *BMC Microbiol.* **2016**, *16*, 272. [CrossRef] [PubMed]

plants

Article

Rhizospheric *Actinomycetes* Revealed Antifungal and Plant-Growth-Promoting Activities under Controlled Environment

Hazem S. Elshafie and Ippolito Camele *

School of Agricultural, Forestry, Food and Environmental Sciences, University of Basilicata, Viale dell'Ateneo Lucano 10, 85100 Potenza, Italy; hazem.elshafie@unibas.it
* Correspondence: ippolito.camele@unibas.it; Tel.: +39-0971205544; Fax: +39-0971205503

Abstract: *Actinomycetes* has large habitats and can be isolated from terrestrial soil, rhizospheres of plant roots, and marine sediments. *Actinomycetes* produce several bioactive secondary metabolites with antibacterial, antifungal, and antiviral properties. In this study, some *Actinomycetes* strains were isolated from the rhizosphere zone of four different plant species: rosemary, acacia, strawberry, and olive. The antagonistic activity of all isolates was screened in vitro against *Escherichia coli* and *Bacillus megaterium*. Isolates with the strongest bioactivity potential were selected and molecularly identified as *Streptomyces* sp., *Streptomyces atratus*, and *Arthrobacter humicola*. The growth-promoting activity of the selected *Actinomycetes* isolates was in vivo evaluated on tomato plants and for disease control against *Sclerotinia sclerotiorum*. The results demonstrated that all bacterized plants with the studied *Actinomycetes* isolates were able to promote the tomato seedlings' growth, showing high values of ecophysiological parameters. In particular, the bacterized seedlings with *Streptomyces* sp. and *A. humicola* showed low disease incidence of *S. sclerotiorum* infection (0.3% and 0.2%, respectively), whereas those bacterized with *S. atratus* showed a moderate disease incidence (7.6%) compared with the positive control (36.8%). In addition, the ability of the studied *Actinomycetes* to produce extracellular hydrolytic enzymes was verified. The results showed that *A. humicola* was able to produce chitinase, glucanase, and protease, whereas *Streptomyces* sp. and *S. atratus* produced amylase and pectinase at high and moderate levels, respectively. This study highlights the value of the studied isolates in providing bioactive metabolites and extracellular hydrolytic enzymes, indicating their potential application as fungal-biocontrol agents.

Keywords: biocontrol; phytopathogens; bioactive substances; microbial biostimulants; antagonistic activity; *Actinobacteria*

Citation: Elshafie, H.S.; Camele, I. Rhizospheric *Actinomycetes* Revealed Antifungal and Plant-Growth-Promoting Activities under Controlled Environment. *Plants* **2022**, *11*, 1872. https://doi.org/10.3390/plants11141872

Academic Editors: Carlos Agustí-Brisach and Eugenio Llorens

Received: 30 June 2022
Accepted: 13 July 2022
Published: 18 July 2022

Publisher's Note: MDPI stays neutral with regard to jurisdictional claims in published maps and institutional affiliations.

1. Introduction

Recently, new agrochemical drugs have been registered in agriculture field, but they can have different negative effects on plants, the environment, and humans. Furthermore, several phytopathogenic microorganisms have become resistant to some agrochemicals, which requires the development of new antimicrobial agents to avoid this serious phenomenon [1,2]. Currently, many scientists all over the world are trying to discover new natural drugs of plant or microbial origin [3–7]. Many plant and microorganisms produce different bioactive secondary metabolites that can potentially be used in the agro-pharmaceutical industry as efficient alternatives for several chemical pesticides [3,8–10].

The soil is a rich matrix of living microorganisms and is a valuable resource of biological control agents [11–13]. The rhizosphere, which is made up of aggregates containing accumulated organic matter, is a repository of microbial activity in the soil. The rhizosphere has great importance because it can support large populations of active microorganisms [14]. Furthermore, soil microorganisms provide an excellent source for important bioactive products [15]. There is growing interest in using bacteria for medicinal and agricultural purposes due to their ability to produce a wide range of biologically active substances with antibiotic, fungicidal, herbicidal, hydrolytic enzymatic, antitumor, antivirals, and immune-suppressant activities [16–18]. Recently, pathogen resistance has necessitated the discovery of new antimicrobial agents effective against bacteria and fungi. There is strong interest in screening new microorganisms from different habitats for antimicrobial activity in order to discover new and promising antibiotics in the treatment against multi-drug resistant pathogens (MDRPs).

Actinomycetes, a type of unicellular Gram-positive bacteria, are widely distributed in nature from different habitats and are well-known and important producers of several bioactive secondary metabolites, antibiotics, and growth-promoting factors [19]. *Actinomycetes* are very similar to fungi, though they form hyphae much smaller than fungi [19,20]. The phylum *Actinobacteria* is considered one of the important groups of *Actinomycetes* [21,22]. Girão et al. [23] reported that many thousands of bioactive substances have been identified from *Actinobacteria*, especially those from terrestrial sources. The produced bioactive metabolites from *Actinomycetes*, especially those from terrestrial sources, represent about the 45% of known microbial bioactive metabolites [23,24]. In addition, Girão et al. [23] studied the antimicrobial activity of the organic extracts from some *Actinobacteria* isolated from *Laminaria ochroleucahe* and concluded that several isolates were able to inhibit the growth of *Candida albicans* and *Staphylococcus aureus*. *Streptomyces*, among the *Actinobacteria*, is considered an important genus able to produce the majority of the identified bioactive compounds, as reported by Berdy [25].

The isolation and biochemical characterization of *Actinomycetes* may allow finding new bioactive substances for pharmaceutical and agricultural purposes. The main objectives of the current study were to: (i) isolate and identify new strains of *Actinomycetes* from different soil habitats; (ii) evaluate the in vitro antagonistic effect of the tested isolates against some common phytopathogens; and (iii) evaluate the in vivo growth-promoting effect of the most bioactive isolates and their antifungal activity against *Sclerotinia sclerotiorum* on tomato seedlings.

2. Results

2.1. Isolation and Preliminary Screening

The isolation from the soil samples allowed obtaining ten pure *Actinomycetes* isolates (Table 1). All isolates were preliminarily evaluated for their antagonistic activity against the two target microorganisms (*Escherichia coli* and *Bacillus megaterium*). The isolates AC1 and RS3 showed the highest biological activity against both tested microorganisms, whereas FG1 showed moderate activity against both tested microorganisms (Table 1). The OL2 isolate showed the highest activity against *E. coli* and the most promising activity against *B. megaterium* (Table 1). Based on the obtained results, the isolates AC1, RS3, and OL2 were selected for molecular identification and further biological assays.

2.2. Molecular Identification

The amplification with the primers Y1/Y2 produced amplicons with molecular weight of about 434 bp. No amplicons were observed in the negative control. The amplified DNA were sequenced (BMR Genomics, Padova, Italy), and the obtained sequences were compared with those available in GenBank nucleotide archive using Basic Local Alignment Search Tool software (BLAST) (RKV, MD, USA). The results of sequences analysis of AC1, RS3, and OL2 showed high similarity percentages to the sequences of *Streptomyces* sp., *Streptomyces atratus*, and *Arthrobacter humicola*, respectively, present in GenBank with the following accession numbers: ON241810, ON241816, and ON241806, respectively.

Table 1. Antagonistic activity of the *Actinomycetes* isolates.

Isolates	Antagonistic Activity	
	E. coli	*B. megaterium*
AC1 *	+++	+++
AC2	-	-
AC3	+	+
RS1	+	-
RS2	-	-
RS3 *	+++	+++
FG1	+	++
FG2	+	-
OL1	+	-
OL2 *	+++	++

Note: +++, very high activity; ++, high activity; +, moderate activity; -, no activity. AC1, AC2, and AC3 were isolated from acacia rhizosphere; RS1, RS2, and RS3 were isolated from rosemary rhizosphere; FG1 and FG2 were isolated from strawberry rhizosphere; OL1 and OL2 were isolated from olive rhizosphere. *, isolates that showed the highest antagonistic effect.

2.3. Extracellular Hydrolytic Enzymes

The results showed that all studied isolates were able to produce some extracellular hydrolytic enzymes (Table 2). In particular, the highest significant hydrolytic activity of chitinase (chitin azure), glucanase, and protease was observed in the case of *A. humicola*, where the diameter of the hydrolysis zones was 31.5, 36.0, and 21.5 mm, respectively. On the other hand, *Streptomyces* sp. and *S. atratus* showed the highest significant activity of amylase with a diameter of hydrolysis area of 37.5 and 42.0 mm, respectively, whereas the same two isolates showed moderate activity for pectinase with a diameter of hydrolysis area of 14.0 and 10.5 mm, respectively. However, *S. atratus* and *A. humicola* did not show either glucanase or pectinase activity, respectively. None of the three tested isolates showed hydrolytic activity for chitinase (chitin from crab shells) and polygalacturanase.

Table 2. Extracellular hydrolytic enzymes produced by the tested *Actinomycetes* isolates.

Enzyme	Substrates	Staining	Diameter of Hydrolysis Area (mm)		
			AC1 *Streptomyces* sp.	RS3 *S. atratus*	OL2 *A. humicola*
Chitinase	Chitin azure (1%)	Congo red (0.03%)	23.0 ± 2.3 b	0.0 ± 0.0 c	31.5 ± 1.7 a
	Chitin crab shells (1%)	Congo red (0.03%)	0.0 ± 0.0	0.0 ± 0.0	0.0 ± 0.0
Amylase	Soluble starch (1%)	Lugol solution [a]	37.5 ± 2.9 a	42.0 ± 1.2 a	28.0 ± 3.5 b
Glucanase	Lichenan (0.2%)	Congo red (0.03%)	22.0 ± 2.3 b	0.0 ± 0.0 c	36.0 ± 1.2 a
Pectinase	Pectin (0.5%)	CTAB [b] (2%)	14.0 ± 1.2 a	10.5 ± 1.7 a	0.0 ± 0.00 b
Protease	Skim milk (1%)	-	14.5 ± 2.9 b	12.5 ± 2.9 b	21.5 ± 1.7 a
Polygalacturanase	Polygalacturonic acid (1%)	Ruthenium red (0.1%)	0.0 ± 0.0	0.0 ± 0.0	0.0 ± 0.0

[a] Lugol solution was prepared as follows: 0.35 g iodide + 0.66 g potassium iodide KI in 100 mL dis. H_2O; [b] CTAB: hexadecyltrimethylammonium bromide; values followed by different letters in each row for each tested enzyme were significantly different according to *Tukey's* B multiple comparison test post hoc test at $p < 0.05$.

2.4. In Vivo Growth Promoting and Disease Control

2.4.1. Eco-Physiological Parameters

The results revealed that all studied *Actinomycetes* isolates were able to stimulate the growth of bacterized tomato seedlings, which showed higher values of eco-physiological parameters in comparison with the negative control (non-bacterized plants), as represented in Table 3. In particular, seedlings inoculated with *Streptomyces* sp. and *A. humicola* showed the highest significant values ($p < 0.05$) of number of leaves, shoot length, shoot fresh weight, and shoot dry weight. The eco-physiological parameters of bacterized tomato seedlings artificially infected with *S. sclerotiorum* are reported in Table 4. In particular, seedlings inoculated with *Streptomyces* sp. and *A. humicola* demonstrated high values ($p < 0.05$) of number of leaves, twigs, shoot fresh weight, and shoot dry weight. However, *S. atratus* showed a moderate growth-promoting effect on tomato seedlings, especially in terms of the number of twigs, shoot length, and total shoot dry weight.

Table 3. Effect of *Actinomycetes* isolates on eco-physiological parameters of tomatoes (health control).

Actinomycetes Isolates	Eco-Physiological Parameters				
	TN (*n*)	SL (cm)	LN (*n*)	SFW (g)	SDW (g)
Cont. -ve	8 ± 1.4 a	36.05 ± 3.2 ab	116 ± 6.9 b	150.02 ± 4.1 b	15.33 ± 1.4 b
AC1: *Streptomyces* sp.	8 ± 0.9 a	39.01 ± 4.1 a	195 ± 11.8 a	204.00 ± 13.4 a	33.02 ± 4.5 a
RS3: *Streptomyces atratus*	6 ± 1.2 a	38.25 ± 7.1 a	123 ± 13.6 b	119.33 ± 8.0 c	15.78 ± 1.9 b
OL2: *Arthrobacter humicola*	7 ± 1.0 a	46.00 ± 3.2 a	151 ± 7.2 a	184.01 ± 7.9 a	24.76 ± 2.7 a

Note: TN: twig number; SL: shoot length; LN: leaf number: SFW and SDW: fresh and dry weight of shoot systems, respectively. Values followed by different letters in each vertical column for each measured parameter were significantly different according to *Tukey's* B multiple comparison test post hoc test at $p < 0.05$.

Table 4. Effect of *Actinomycetes* isolates on eco-physiological parameters of tomatoes (artificially infected with *S. sclerotiorum*).

Actinomycetes Isolates	Eco-Physiological Parameters				
	TW (*n*)	SL (cm)	LN (*n*)	SFW (g)	SDW (g)
Cont. -ve	5 ± 0.2 b	42.32 ± 0.3 a	161 ± 1.0 bc	142.33 ± 1.0 c	24.04 ± 0.2 ab
AC1: *Streptomyces* sp.	8 ± 0.1 a	54.31 ± 0.4 a	333 ± 2.3 a	240.12 ± 2.5 a	30.67 ± 0.3 a
RS3: *Streptomyces atratus*	5 ± 0.1 b	44.34 ± 0.8 a	210 ± 1.8 b	214.67 ± 1.9 ab	23.00 ± 1.0 ab
OL2: *Arthrobacter humicola*	8 ± 0.1 a	51.76 ± 0.2 a	477 ± 3.7 a	304.65 ± 0.8 a	29.67 ± 0.6 a

Note: TN: twig number; SL: shoot length; LN: leaf number: SFW and SDW: fresh and dry weight of shoot systems, respectively. Values followed by different letters in each vertical column for each measured parameter were significantly different according to *Tukey's* B multiple comparison test post hoc test at $p < 0.05$.

2.4.2. Disease Control

The bacterized plants with *Streptomyces* sp. and *A. humicola* did not show any symptoms on their leaves and roots after the infection with *S. sclerotiorum*. The disease indexes of the plants bacterized with *Streptomyces* sp. and *A. humicola* were 0.3% and 0.2% (Figure 1), whereas the control effects were 99.2% and 99.5%, respectively (Figure 2). The seedlings bacterized with *S. atratus* showed a moderate disease index of 7.6% (Figure 1) and a control effect of 79.5% (Figure 2). Regarding the positive control (plants inoculated only with *S. sclerotiorum*), the results showed the development of leaf yellowing and chlorosis at 20 DAI, where the leaf chlorotic zone of infected tomato plants became necrotic. Moreover, complete leaf wilting and root necrosis was also observed at 35 DAI. In particular, a significantly higher symptomatic leaves percentage ($p < 0.05$) was observed in the positive control, where the disease index was 36.8% compared with the negative control and bacterized plants with *Actinomycetes* isolates (Figure 1). *S. sclerotiorum* was always re-isolated from the inoculated plants.

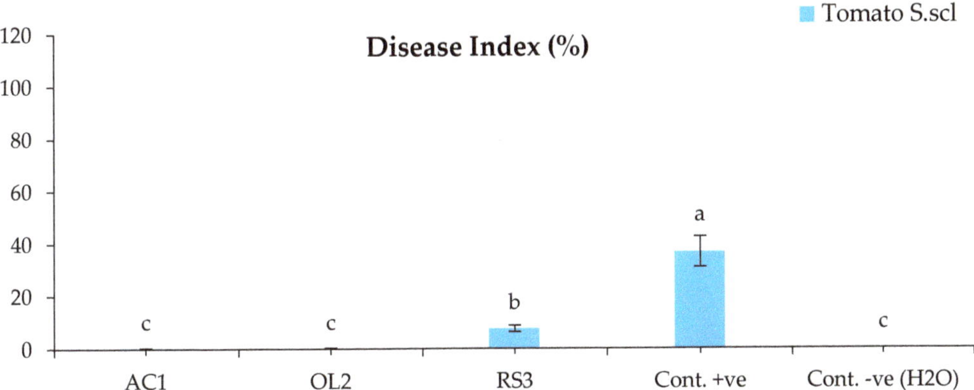

Figure 1. Disease index of tomato inoculated with *S. sclerotiorum*. Bars with different letters indicate mean values significantly different at $p < 0.05$ according to *Tukey's* B test. Data are expressed as mean of 3 replicates ± SDs. DI (%) = [Σ (Scale × No. of SL)/(HS × TL)] × 100 (Equation (1)). AC1, OL2, and RS3 are *Streptomyces* sp., *Arthrobacter humicola*, and *Streptomyces atratus*, respectively.

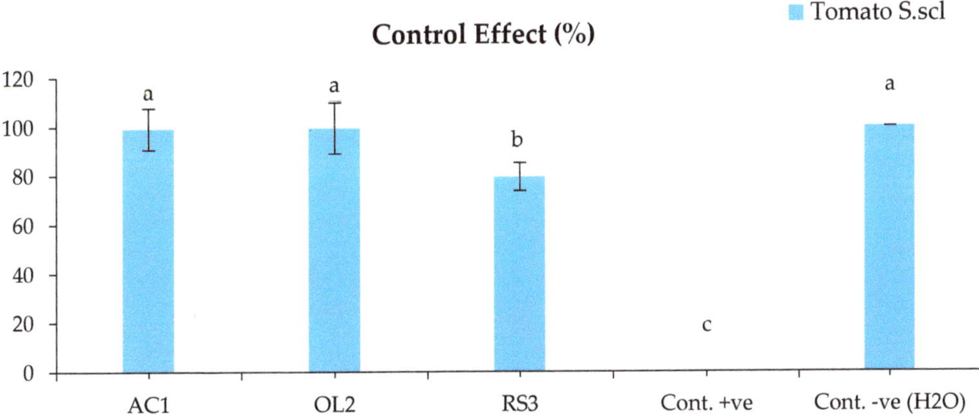

Figure 2. Control effect of the tomato inoculated with *S. sclerotiorum*. Bars with different letters indicate mean values significantly different at $p < 0.05$ according to Tukey's test. Data are expressed as mean of 3 replicates ± SDs. CE (%) = 100 × (DI-P-DI-B)/DI-P (Equation (2)).

3. Discussion

The obtained results proved that the studied *Actinomycetes* isolates were able to promote the growth of tomato plants by improving the eco-physiological characteristics and also are promising for the biocontrol of *S. sclerotiorum* on tomato seedlings. In particular, the biological activity and growth-promoting effect of the studied *Actinomycetes* strains may be due to their ability to produce some bioactive metabolites, such as growth hormones, which enhance the tomato seedlings' growth [26–29]. The application of microbial plant stimulants is considered an important strategy for sustainable agriculture systems for enhancing plant growth and increasing production, especially under abiotic stress [30].

Sousa and Olivares [31] concluded that plant-growth-promoting Streptomyces (PGPS) was able to biostimulate plant growth by direct and indirect pathways such as phytohormones production, phosphate solubilization, and alleviation of various abiotic stresses. In particular, endophytic *Actinobacteria* can biostimulate the secretion of plant growth hormones such as indole acetic acid (IAA), as reported by Manulis et al. [32] and Dochhil et al. [33].

The treatments with *Streptomyces* sp. and *Arthrobacter humicola* showed high reduction in disease symptoms on tomato seedlings against the tested pathogenic fungi. Furthermore, the bacterization treatments induced a significant disease protection of tomato seedlings compared with non-bacterized plants against fungal infection with *S. sclerotiorum*. These results are in agreement with those of several researchers who reported that many soil-borne *Actinomycetes* are able to reduce the growth of some pathogenic fungi such as *Colletotrichum gloeosporioides*, *C. capsici*, and *Fusarium solani* f. sp. *pisi* [34–37].

The production of hydrolytic enzymes can also play an important role in controlling phytopathogenic fungi [38,39]. The cell wall lytic enzymes glucanase and protease can contribute to the degradation of fungal cell wall (skeletal) components through embedment in its protein matrix [40,41]. In addition, Ordentlich et al. [40] reported that chitinase and other lytic enzymes produced by *Serratia marcescens* were able to control the pathogenic fungus *Sclerotium rolfsii*, causing the release of β-glucanase, which can increase the chitinolytic activity in hyphal degradation [42].

Chaudhary et al. [43] studied the antagonistic activity of some *Actinomycetes* strains isolated from different niche habitats of Sheopur (India) and observed that some strains were highly active against *Bacillus cereus*, *Enterococcus faecalis*, *Shigella dysenteriae*, *Streptococcus pyogenes*, *Staphylococcus saprophyticus*, *S. epidermidis*, methicillin-resistant *Staphylococcus*, and *S. xylosus*. The same authors also reported that all studied isolates were able to inhibit the extracellular growth of tested microorganisms, whereas they were not able to inhibit intracellular growth of mycelium. The latter phenomena may be due to the production of some bioactive secondary metabolites that may not reach to the intracellular cells of the tested bacteria and hence were not able to denature their cell walls [43].

In a recent study conducted by Odumosu et al. [42], it was reported that some species of *Streptomyces* sp. showed promising antibacterial activity against some food and human pathogens such as *S. aureus*, *E. coli*, *Klebsiella pneumonia*, and *Salmonella typhi*. The same authors also chemically analyzed the secondary metabolites produced by the studied species using GC-MS and verified their antibiotic properties may be used in novel antimicrobials. The antifungal activity of *Streptomyces* strains may also be due to their ability to produce some bioactive secondary metabolites such as isoikarugamycin, a novel polycyclic tetramic acid macrolactam produced by *Streptomyces zhaozhouensis* active against *C. albicans*, as reported by Lacret et al. [44].

4. Materials and Methods

4.1. Soil Sampling and Isolation

For isolation of *Actinomycetes*, 200 g subsamples were collected from the rhizosphere zone of four different plant species: rosemary (3 samples), acacia (3 samples), strawberry (2 samples), and olive (2 samples) from Potenza (Basilicata region, southern Italy). Each soil sample was air-dried on the benches for one week and sieved through a 250 μm pore sieve (Glenammer, Scotland, UK). The samples were further held in a hot-air oven at 121 °C for 1 h to prevent the growth of other microorganisms. The isolation was carried out following the membrane filter technique using Difco™ *Actinomycetes* Isolation Agar (Sparks, MD, USA) [45] with some minor modifications. The cultivated plates were incubated for 4 days at 28 °C until the *Actinomycetes* become visible. The prepared nutrient media was supplemented with 100 μg/mL cycloheximide to suppress eventual growth of fungi. All obtained isolates were cultured in triplicates and further purified for obtaining the pure cultures, which were conserved on slant agar nutrient glycerol (ANG) tubes at 4 °C for further biological assays. The obtained isolates were initially examined based on their microscopic morphological features with a light microscope. For exact identification, the obtained isolates were further analyzed by the molecular method.

4.2. Antagonistic Activity

The studied isolates were verified for their biological activity against *E. coli* and *B. megaterium* using the cross-streak method as reported by Odumosu et al. [42]. Briefly, a single, small mass from a fresh culture (24 h) of each studied isolate was streaked in the center of a Petri dish containing King'B (KB) nutrient media [46] and then incubated at 37 °C for 48 h. Successively, the plates were inoculated with the tested microorganisms by a single streak at a perpendicular close to the initial inoculum of each studied Actinomycetes isolate. All plates were incubated at 37 °C and the antagonistic activity was evaluated after 24 h. The bacterial antagonistic activity was recorded as follows: very high activity (+++); high activity (++); moderate activity (+); no activity (-). The most bioactive isolates were selected for molecular identification and further in vitro and in vivo biological assays.

4.3. Molecular Identification

The bacterial isolates that demonstrated potentially antagonistic effects against the tested target microorganisms were previously morphologically identified under a light microscope (60×) and then by the molecular method based on the analysis of genomic DNA (gDNA) sequences. The gDNA of each studied isolate was extracted using a Qiagen Genomic DNA Kit (Qiagen, Heidelberg, Germany). The extracted gDNA was amplified using the universal primers for bacteria Y1/Y2 (Table 5). The PCR reaction was carried out in a final volume of 25 μL containing: 200 ng DNA, 0.2 μL of 1 U *Taq* DNA polymerase, 2.5 μL *Taq* buffer (20 mM $MgCl_2$), 5 μL of each primer (2.5 μM), 5 μL of dNTPs (4 mM) and ultrapure dH_2O for a final volume of 25 μL. Both the concentration and purity of the total DNA extracted from each sample were measured using a Nano-drop (Thermo Fisher Scientific, Waltham, MA USA). Each DNA sample was subjected to PCR amplification following the cycling profile: 94 °C for 5 min (initial denaturation), followed by 34 cycles of 94 °C for 30 c (denaturation), 57 °C for 30 s (annealing), and 72 °C for 1 min (extension), with a final extension step of 5 min at 72 °C. The amplified DNA, stained by Bromophenol blue (3 μL/10 μL), was applied for agarose gel electrophoresis (1.2%) stained by SYBR green dye (4 μL/100 gel). The obtained amplicons were directly sequenced and compared with those available in the GenBank nucleotide archive using BLAST software [47].

4.4. Extracellular Hydrolytic Enzymes

The enzymatic activity of the studied *Actinomycetes* isolates was screened by carrying out an assay of extracellular hydrolytic enzymes on KB media supplemented with the below specific substrates for each enzyme: chitin azure (1%) or chitin from crab shells (1%) for chitinase [48]; skim milk (1%) for protease [48]; and lichenan (0.2%) for glucanase [49]. In addition, soluble starch (1%), pectin (0.5%), and polygalacturonic acid (1%) were used for amylase, pectinase, and polygalacturanase, respectively [50,51]. All plates were incubated at 30 °C for 96 h and then flooded with specific staining solutions as follows: Congo red (0.03%) for chitinase and glucanase; lugol solution for amylase; CTAB: hexadecyltrimethylammonium bromide (2%) for pectinase and ruthenium red (0.1%) for polygalacturanase. The enzymatic activity was taken as evidence of the appearance of hydrolysis clear zones around the colonies, and their diameters were measured in millimeters.

4.5. In Vivo Growth Promoting Effect and Disease Control

An in vivo pot experiment was carried out in a greenhouse (School of Agricultural, Forestry, Food and Environmental Sciences-SAFE, University of Basilicata, Potenza, Italy) to evaluate the growth-promoting effect (GPE) of the tested *Actinomycetes* isolates on tomato plants, and the disease control (DC) of the most bioactive isolates was studied against *S. sclerotiorum*.

The pot experiment was carried out in a glass greenhouse at 25 °C for a 15-h photoperiod. Each pot was 20 cm high and 25 cm wide, and previously sterilized with 1.2% sodium hypochlorite for 5 min, rinsed twice with distilled water, and filled with a growing medium mixture (compost/peat moss, 1:1). Seeds of *Solanum lycopersicum* L. cv. *cerasiforme* were surface sterilized by ethanol (70%) and sowed in a cell tray. The temperature and relative humidity in the greenhouse remained stable at 25 ± 2 °C and 70–80%, respectively, for the duration of the experiment.

For the *Actinomycetes* treatment, an initial nutrient culture of peptone yeast calcium agar (PY-Ca) was prepared for the tested isolates and incubated for 5 days at $28 \pm 2\,°C$. A suspension of each studied isolate was prepared by inoculating 10^6 CFU/mL from the original culture into minimal mineral (MM) media prepared as follows: (g/L) 10.5 K_2HPO_4, 4.5 KH_2PO_4, 1.0 $(NH_4)2SO_4$, 0.5 $Na_3C_6H_5O_7 \times 2H_2O$, 0.2 $MgSO_4$ and 5.0 dextrose. The pH value was adjusted at 7.0. The suspensions were then incubated for 7 days at $28 \pm 2\,°C$. The broth cultures were poured into the rhizosphere zone of tomato seedlings (100 mL/pot) 15 days after germination (DAG).

For the fungal artificial infection, Ø 5 mm agar discs from a pure fresh culture (96 h) of *S. sclerotiorum* were inoculated in a sterilized flask filled with potato dextrose broth (PDB) and incubated for 7 days at $22 \pm 2\,°C$. After that, 50 mL of the incubated broth was inoculated in the rhizosphere zone of the seedlings 10 days after the *Actinomycetes* treatment. Ten seedlings were used as the negative health control. The whole experiment was repeated twice with five replicates per treatment. The experimental pots were distributed in a randomized block design in the greenhouse and watered once a day.

For the eco-physiological parameters, plant growth was monitored at the end of the experiment, about 40 DAG, by measuring stem length (SL) in centimeters, number of leaves (NL), number of twigs (NT), the total fresh weights of shoots (TFwS) in grams, and total dry weight of shoots (TDwS) in grams. Regarding the evaluation of the disease incidence, tomato plants were monitored daily, fifteen days after the infection (DAI), to observe the eventual appearance of disease symptoms. The disease incidence was assessed using the following scale (0 = no symptoms observed; 1 = 1 to 20% of leaf chlorosis; 2 = 21 to 50% of leaf chlorosis; 3 = 51 to 80% of leaf chlorosis; 4 ≥ 80% of leaf chlorosis), as reported by Elshafie et al. [4]. The infection percentage (IP %) was measured using Equation (1), whereas the disease index (DI %) and the control effect (CE %) were calculated using Equations (2) and (3), respectively, as described by Lee et al. [52].

$$\text{IP \%} = (\text{SL/TL}) \times 100 \tag{1}$$

$$\text{DI \%} = \left[\sum (\text{Scale} \times \text{No. of SL})/(\text{HS} \times \text{TL})\right] \times 100 \tag{2}$$

$$\text{CE \%} = 100 \times (\text{DI.P} - \text{DI.B})/\text{DI.P} \tag{3}$$

where SL is symptomatic leaves; TL is total number of leaves; HS is highest scale; DI-P is disease index of infection; DI-B is disease index of control.

4.6. Statistical Analysis

The obtained results were subjected to one-way ANOVA for the statistical analysis. The significance level was checked by applying the Tukey's B post hoc multiple comparison test with a probability of $p < 0.05$ using Statistical Package for the Social Sciences (SPSS) version 13.0, 2004 (Chicago, IL, USA).

Table 5. Primers used in this study.

Primers	Sequences	Target	Amplified Fragment (kb)	Gene	Reference
Y1	5′-TGGCTCAGAACGAACGCTGGCGGC-3′	Bacteria	0.43	16S rDNA	Darrasse et al. [53]
Y2	5′-CCCACTGCTGCCTCCCGTAGGAGT-3′				

5. Conclusions

The obtained results of the current research confirmed the promising biological activity of *Actinomycetes*, particularly of the genus *Streptomyces*. This study also underlined the usefulness of the new isolated strains for producing some important bioactive metabolites and extracellular hydrolytic enzymes; hence, they can be effectively used as biocontrol agents against *S. sclerotiorum*. Furthermore, the studied isolates also demonstrated an important plant-growth-promoting effect, which may be due to the production of phytohormones. Further studies remain necessary to identify and biochemically characterize the produced bioactive metabolites from the *Actinomycetes* isolates and evaluate their biological effects against other serious phytopathogens.

Author Contributions: Conceptualization, H.S.E. and I.C.; methodology, H.S.E.; formal analysis, H.S.E.; investigation, H.S.E. and I.C.; data curation, H.S.E. and I.C.; writing—original draft preparation, H.S.E.; writing—review and editing, H.S.E. and I.C. All authors have read and agreed to the published version of the manuscript.

Funding: This research received no external funding.

Informed Consent Statement: Not applicable.

Data Availability Statement: Not applicable.

Conflicts of Interest: The authors declare no conflict of interest.

References

1. Urban, C.; Segal-Maurer, S.; Rahal, J.J. Considerations in control and treatment of nosocomial infections due to multidrug resistant *Acinetobacter baumannii*. *Clin. Infect. Dis.* **2003**, *36*, 1268–1274. [CrossRef] [PubMed]
2. Paterson, D.L.; Ko, W.C.; Von Gottberg, A.; Mohapatra, S.; Casellas, J.M.; Goossens, H.; Mulazimoglu, L.; Trenholme, G.; Klugman, K.P.; Bonomo, R.A.; et al. International prospective study of Klebsiella pneumoniae bacteremia: Implications of extended-spectrum beta-lactamase production in nosocomial Infections. *Ann. Inter. Med.* **2004**, *140*, 26–32. [CrossRef] [PubMed]
3. Gerwick, B.C.; Sparks, T.C. Natural products for pest control: An analysis of their role, value and future. *Pest. Manag. Sci.* **2014**, *70*, 1169–1185. [CrossRef] [PubMed]
4. Elshafie, H.S.; Sakr, S.; Bufo, S.A.; Camele, I. An attempt of biocontrol the tomato-wilt disease caused by Verticillium dahliae using *Burkholderia gladioli* pv. *agaricicola* and its bioactive secondary metabolites. *Int. J. Plant Biol.* **2017**, *8*, 57–60.
5. Della Pepa, T.; Elshafie, H.S.; Capasso, R.; De Feo, V.; Camele, I.; Nazzaro, F.; Scognamiglio, M.R.; Caputo, L. Antimicrobial and phytotoxic activity of *Origanum heracleoticum* and *O. majorana* essential oils growing in Cilento (Southern Italy). *Molecules* **2019**, *24*, 2576. [CrossRef]
6. Gruľová, D.; Caputo, L.; Elshafie, H.S.; Baranová, B.; De Martino, L.; Sedlák, V.; Camele, I.; De Feo, V. Thymol chemotype *Origanum vulgare* L. essential oil as a potential selective bio-based herbicide on monocot plant species. *Molecules* **2020**, *25*, 595. [CrossRef]
7. Elshafie, H.S.; Camele, I.; Sofo, A.; Mazzone, G.; Caivano, M.; Masi, S.; Caniani, D. Mycoremediation effect of Trichoderma harzianum strain T22 combined with ozonation in diesel-contaminated sand. *Chemosphere* **2020**, *252*, 126597. [CrossRef]
8. Elshafie, H.S.; Viggiani, L.; Mostafa, M.S.; El-Hashash, M.A.; Bufo, S.A.; Camele, I. Biological activity and chemical identification of ornithine lipid produced by *Burkholderia gladioli* pv. *agaricicola* ICMP 11096 using LC-MS and NMR analyses. *J. Biol. Res.* **2017**, *90*, 96–103.
9. Vitti, A.; Elshafie, H.S.; Logozzo, G.; Marzario, S.; Scopa, A.; Camele, I.; Nuzzaci, M. Physico-chemical Characterization and biological activities of a digestate and a more stabilized digestate-derived compost from agro-waste. *Plants* **2021**, *10*, 386. [CrossRef]
10. Elshafie, H.S.; Caputo, L.; De Martino, L.; Sakr, S.H.; De Feo, V.; Camele, I. Study of bio-pharmaceutical and antimicrobial properties of pomegranate (*Punica granatum* L.) Leathery Exocarp Extract. *Plants* **2021**, *10*, 153. [CrossRef]
11. Camele, I.; Elshafie, H.S.; Caputo, L.; Sakr, S.H.; De Feo, V. *Bacillus mojavensis*: Biofilm formation and biochemical investigation of its bioactive metabolites. *J. Biol. Res.* **2019**, *92*, 39–45. [CrossRef]
12. Sofo, A.; Elshafie, H.S.; Camele, I. Structural and functional organization of the root system: A comparative study on five plant species. *Plants* **2020**, *9*, 1338. [CrossRef] [PubMed]
13. D'Ippolito, I.; Mang, S.M.; Elshafie, H.S.; Camele, I.; Scillitani, G.; Mastrodonato, M.; Sofo, A.; Mininni, A.N.; Xylogiannis, E. Morpho-anatomical and microbiological analysis of kiwifruit roots with KVDS symptoms. *Acta Hortic.* **2022**, *1332*, 131–136. [CrossRef]
14. El-Sayed, W.S.; Akhkha, A.; El-Naggar, M.Y.; Elbadry, M. In vitro antagonistic activity, plant growth promoting traits and phylogenetic affiliation of rhizobacteria associated with wild plants grown in arid soil. *Front. Microbiol.* **2014**, *5*, 651. [CrossRef] [PubMed]
15. Rani, A.; Saini, K.C.; Bast, F.; Mehariya, S.; Bhatia, S.K.; Lavecchia, R.; Zuorro, A. Microorganisms: A Potential Source of Bioactive Molecules for Antioxidant Applications. *Molecules* **2021**, *26*, 1142. [CrossRef] [PubMed]
16. Jeffrey, L.S.H.; Norzaimawati, A.N.; Rosnah, H. Prescreening of bioactivities from actinomycetes isolated from forest peat soil in Sarawak. *J. Trp. Agric. Fd. Sci.* **2011**, *39*, 245–254.
17. Manivasagan, P.; Kang, K.H.; Sivakumar, K.; Li-Chan, E.C.; Oh, H.M.; Kim, S.K. Marine Actinobacteria: An important source of bioactive natural products. *Environ. Toxicol. Pharmacol.* **2014**, *38*, 172–188. [CrossRef]

18. Vikineswary, S.; Nadaraj, P.; Wong, W.H.; Balabaskaran, S. Actinomycetes from a tropical mangrove ecosystem—Antifungal activity of selected strains. *Asia Pac. J. Mol. Biol. Biotechnol.* **1997**, *5*, 81–86.

19. Pepper, I.L.; Gentry, T.J. Chapter 4—"Earth Environments". In *Environmental Microbiology*, 3rd ed.; Academic Press: Cambridge, MA, USA, 2015; pp. 59–88.

20. Holt, J.G.; Krieg, N.R.; Sneath, P.H.A.; Staley, J.T.; Williams, S.T. *Bergey's Manual of Determinative Bacteriology*, 9th ed.; Williams and Wilkins: Baltimore, MD, USA, 1994.

21. Manivasagan, P.; Venkatesan, J.; Sivakumar, K.; Kim, S.K. Pharmaceutically active secondary metabolites of marine *Actinobacteria*. *Microbiol. Res.* **2014**, *169*, 262–278. [CrossRef]

22. Marcolefas, E.; Leung, T.; Okshevsky, M.; McKay, G.; Hignett, E.; Hamel, J.; Aguirre, G.; Blenner-Hassett, O.; Boyle, B.; Lévesque, R.C.; et al. Culture-Dependent bioprospecting of bacterial isolates from the canadian high arctic displaying antibacterial activity. *Front. Microbiol.* **2019**, *10*, 1836. [CrossRef]

23. Girão, M.; Ribeiro, I.; Ribeiro, T.; Azevedo, I.C.; Pereira, F.; Urbatzka, R.; Leão, P.N.; Carvalho, M.F. Actinobacteria Isolated from *Laminaria ochroleuca*: A source of new bioactive compounds. *Front. Microbiol.* **2019**, *10*, 683. [CrossRef] [PubMed]

24. Solecka, J.; Zajko, J.; Postek, M.; Rajnisz, A. Biologically active secondary metabolites from Actinomycetes. *Cent. Eur. J. Biol.* **2012**, *7*, 373–390. [CrossRef]

25. Berdy, J. Bioactive microbial metabolites. *J. Antibiot.* **2005**, *58*, 1. [CrossRef] [PubMed]

26. Vurukonda, S.S.; Giovanardi, D.; Stefani, E. Plant growth promoting and biocontrol activity of *Streptomyces* spp. as endophytes. *Int. J. Mol. Sci.* **2018**, *19*, 952. [CrossRef] [PubMed]

27. Rajan, B.M.; Kannabiran, K. Extraction and identification of antibacterial secondary metabolites from marine *Streptomyces* sp. VITBRK2. *Int. J. Mol. Cell Med.* **2014**, *3*, 130–137.

28. Khan, A.; Halo, B.A.; Elyassi, A.; Ali, S.; Al-Hosni, K.; Hussain, J.; Al-Harrasi, A.; Lee, I. Indole acetic acid and ACC deaminase from endophytic bacteria improves the growth of *Solanum lycopersicum*. *Electron. J. Biotechnol.* **2016**, *21*, 58–64. [CrossRef]

29. Moon, Y.S.; Ali, S. Isolation and identification of multi-trait plant growth–promoting rhizobacteria from coastal sand dune plant species of Pohang beach. *Folia Microbiol.* **2022**, *67*, 523–533. [CrossRef]

30. Ali, S.; Moon, Y.S.; Hamayun, M.; Khan, M.A.; Bibi, K.; Lee, I. Pragmatic role of microbial plant biostimulants in abiotic stress relief in crop plants. *J. Plant Interact.* **2022**, *17*, 705–718. [CrossRef]

31. Sousa, J.A.A.; Olivares, F.L. Plant growth promotion by *Streptomycetes*: Ecophysiology, mechanisms and applications. *Chem. Biol. Technol. Agric.* **2016**, *3*, 24. [CrossRef]

32. Manulis, S.; Epstein, E.; Shafrir, H.; Lichter, A.; Barash, I. Biosynthesis of indole-3-acetic acid via the indole-3-acetamide pathway in *Streptomyces* spp. *Microbiology* **1994**, *140*, 1045–1050. [CrossRef]

33. Dochhil, H.; Dkhar, M.S.; Barman, D. Seed germination enhancing activity of endophytic Streptomyces isolated from indigenous ethno-medicinal plant *Centella asiatica*. *Int. J. Pharm. Biol. Sci.* **2013**, *4*, 256–262.

34. Sacramento, D.R.; Coelho, R.R.; Wigg, M.D.; Linhares, L.F.; Santos, M.G.; Semedo, L.T.; Silca, A.J. Antimicrobial and antiviral activities of an actinomycetes (*Streptomyces* sp.) isolated from a Brazilian tropical forest soil. *World J. Microbiol. Biotechnol.* **2004**, *20*, 225–229. [CrossRef]

35. Prapagdee, B.; Kuekulvong, C.; Mongkolsuk, S. Antifungal potential of extracellular metabolites produced by *Streptomyces hygroscopicus* against phytopathogenic fungi. *Int. J. Biol. Sci.* **2008**, *4*, 330–337. [CrossRef] [PubMed]

36. Suwan, N.; Boonying, W.; Nalumpang, S. Antifungal activity of soil actinomycetes to control chilli anthracnose caused by *Colletotrichum gloeosporioides*. *J. Agric. Technol.* **2012**, *8*, 725–737.

37. Soltanzadeh, M.; Nejad, M.S.; Bonjar, G.H. Application of soil-borne actinomycetes for biological control against fusarium wilt of chickpea (*Cicer arietinum*) caused by *Fusarium solani* f. sp. *Pisi. J. Phytopathol.* **2016**, *164*, 967–978. [CrossRef]

38. Elshafie, H.S.; Camele, I.; Racioppi, R.; Scrano, L.; Iacobellis, N.S.; Bufo, S.A. In vitro antifungal activity of *Burkholderia gladioli* pv. *agaricicola* against some phytopathogenic fungi. *Int. J. Mol. Sci.* **2012**, *13*, 16291–16302.

39. Djebaili, R.; Pellegrini, M.; Bernardi, M.; Smati, M.; Kitouni, M.; Del Gallo, M. Biocontrol activity of actinomycetes strains against fungal and bacterial pathogens of *Solanum lycopersicum* L. and *Daucus carota* L.: In vitro and in planta antagonistic activity. *Biol. Life Sci. Forum* **2021**, *4*, 27.

40. Ordentlich, A.; Elad, Y.; Chet, I. The role of chitinase of *Serratia marcescens* in biocontrol *Sclerotium rolfsii*. *Phytopathology* **1988**, *78*, 84–87.

41. Saligkarias, I.D.; Gravanis, F.T.; Harry, A.S. Biological control of *Botrytis cinerea* on tomato plants by the use of epiphytic yeasts *Candida guilliermondii* strains 101 and US 7 and *Candida oleophila* strain I-182: II. A study on mode of action. *Biol. Control* **2002**, *25*, 151–161. [CrossRef]

42. Odumosu, B.T.; Buraimoh, O.M.; Okeke, C.J.; Ogah, J.O.; Michel, F.C., Jr. Antimicrobial activities of the *Streptomyces ceolicolor* strain AOBKF977550 isolated from a tropical estuary. *J. Taibah Uni. Sci.* **2017**, *11*, 836–841. [CrossRef]

43. Chaudhary, H.S.; Yadav, J.; Shrivastava, A.R.; Singh, S.; Singh, A.K.; Gopalan, N. Antibacterial activity of actinomycetes isolated from different soil samples of Sheopur (A city of central India). *J Adv. Pharm. Technol. Res.* **2013**, *4*, 118–123. [CrossRef]

44. Lacret, R.; Oves-Costales, D.; Gómez, C.; Díaz, C.; de la Cruz, M.; Pérez-Victoria, I.; Vicente, F.; Genilloud, O.; Reyes, F. New ikarugamycin derivatives with antifungal and antibacterial properties from *Streptomyces zhaozhouensis*. *Mar. Drugs* **2014**, *13*, 128–140. [CrossRef] [PubMed]

45. Rahman, M.A.; Islam, M.Z.; Islam, M.A. Antibacterial activities of *Actinomycete* isolates collected from soils of Rajshahi, Bangladesh. *Biotechnol. Res. Int.* **2011**, *2011*, 857925. [CrossRef] [PubMed]
46. King, E.O.; Ward, M.K.; Raney, D.E. Two simple media for the demonstration of pyocyanin and fluorescin. *J. Lab. Clin. Med.* **1954**, *44*, 301–307. [PubMed]
47. Altschul, S.F.; Madden, T.L.; Schaffer, A.A.; Zhang, J.; Zhang, Z.; Miller, W.; Lipman, D.J. Gapped BLAST and PSIBLAST: A new generation of protein database search programs. *Nucleic Acids Res.* **1997**, *25*, 3389–3402. [CrossRef] [PubMed]
48. Tahtamouni, M.E.W.; Hameed, K.M.; Saadoun, I.M. Biological control of *Sclerotinia sclerotiorum* using indigenous chitolytic actinomycetes in Jordan. *J. Plant Pathol.* **2006**, *22*, 107–114. [CrossRef]
49. Teather, R.M.; Wood, P.J. Use of Congo red-polysacchatide interaction in enumeration and characterization of cellulolytic bacteria from the bovine rumen. *Appl. Environ. Microbiol.* **1998**, *43*, 777–780. [CrossRef]
50. Bhardwaj, V.; Garg, N. Exploitation of Micro-Organisms for Isolation and Screening of Pectinase from Environment. In Proceedings of the 8th International Conference, Making Innovation Work for Society: Linking, Leveraging and Learning, Kuala Lumpur, Malaysia, 1–3 November 2010.
51. Evmert, M.K.; Gabriele, B.G.; Piechulla, B. Volatiles of bacterial antagonists inhibit mycelial growth of the plant pathogen *Rhizoctonia solani*. *Arch. Microbiol.* **2007**, *187*, 351–360.
52. Lee, K.J.; Kamala-Kannan, S.; Sub, H.S.; Seong, C.K.; Lee, G.W. Biological control of Phytophthora blight in red pepper (*Capsicum annuum* L.) using *Bacillus subtilis*. *World J. Microb. Biotecnol.* **2008**, *24*, 1139–1145. [CrossRef]
53. Darrasse, A.; Priou, S.; Kotoujansky, A.; Bertheau, Y. PCR and Restriction Fragment Length Polymorphism of a pel Gene as a Tool to Identify Erwinia carotovora in Relation to Potato Diseases. *Appl. Environ. Microbiol.* **1994**, *60*, 1437–1443. [CrossRef]

Article

Cyanobacteria-Mediated Immune Responses in Pepper Plants against *Fusarium* Wilt

Amer Morsy Abdelaziz [1,*], Mohamed S. Attia [1,*], Marwa S. Salem [2], Dina A. Refaay [3], Wardah A. Alhoqail [4,*] and Hoda H. Senousy [5]

1   Botany & Microbiology Department, Faculty of Science, Al-Azhar University, Cairo 11884, Egypt
2   Botany & Microbiology Department, Faculty of Science, Al-Azhar University (Girls Branch), Cairo 11884, Egypt
3   Botany Department, Faculty of Science, Mansoura University, Mansoura 35516, Egypt
4   Department of Biology, College of Education, Majmaah University, Majmaah 11952, Saudi Arabia
5   Botany and Microbiology Department, Faculty of Science, Cairo University, Giza 12613, Egypt
*   Correspondence: amermorsy@azhar.edu.eg (A.M.A.); drmohamedsalah92@azhar.edu.eg (M.S.A.); w.alhoqail@mu.edu.sa (W.A.A.); Tel.: +20-010-0857-8963 (A.M.A.)

Abstract: Research in plant pathology has increasingly focused on developing environmentally friendly, effective strategies for controlling plant diseases. Cyanobacteria, including *Desmonostoc muscorum*, *Anabaena oryzae*, and *Arthrospira platensis*, were applied to *Capsicum annuum* L. to induce immunity against *Fusarium* wilt. Soil irrigation and foliar shoots (FS) application were used in this investigation. The disease symptoms, disease index, osmotic contents, total phenol, Malondialdehyde (MDA), hydrogen peroxide (H_2O_2), antioxidant enzymes (activity and isozymes), endogenous hormone content, and response to stimulation of defense resistance in infected plants were assessed. Results demonstrated that using all cyanobacterial aqueous extracts significantly reduced the risk of infection with *Fusarium oxysporum*. One of the most effective ways to combat the disease was through foliar spraying with *Arthrospira platensis*, *Desmonostoc muscorum*, and *Anabaena oryzae* (which provided 95, 90, and 69% protection percent, respectively). All metabolic resistance indices increased significantly following the application of the cyanobacterial aqueous extracts. Growth, metabolic characteristics, and phenols increased due to the application of cyanobacteria. Polyphenol oxidase (PPO) and peroxidase (POD) expressions improved in response to cyanobacteria application. Furthermore, treatment by cyanobacteria enhanced salicylic acid (SA) and Indole-3-Acetic Acid (IAA) in the infected plants while decreasing Abscisic acid (ABA). The infected pepper plant recovered from *Fusarium* wilt because cyanobacterial extract contained many biologically active compounds. The application of cyanobacteria through foliar spraying seems to be an effective approach to relieve the toxic influences of *F. oxysporum* on infected pepper plants as green and alternative therapeutic nutrients of chemical fungicides.

Keywords: pepper; *Fusarium*; cyanobacteria; antioxidant enzymes; plants immunity; salicylic acid

Citation: Abdelaziz, A.M.; Attia, M.S.; Salem, M.S.; Refaay, D.A.; Alhoqail, W.A.; Senousy, H.H. Cyanobacteria-Mediated Immune Responses in Pepper Plants against *Fusarium* Wilt. *Plants* **2022**, *11*, 2049. https://doi.org/10.3390/plants11152049

Academic Editors: Carlos Agustí-Brisach and Eugenio Llorens

Received: 23 June 2022
Accepted: 19 July 2022
Published: 5 August 2022

Publisher's Note: MDPI stays neutral with regard to jurisdictional claims in published maps and institutional affiliations.

1. Introduction

The food issues worldwide result from the increased human population, besides plant pathogens resulting in whole or partial harm to crop yields [1]. Most roots of the *Solanaceae* family and other plants have suffered from soil fungal plant pathogens that cause harmful effects on morphological, physiological, molecular, and yield properties. Pepper (*Capsicum annuum* L.) is a hardy plant cultivated extensively worldwide. The annual production of Egyptian pepper is 623,221 tons, with a total cultivation area of 41,047 hectares [2]. Pepper crops worldwide, including in Egypt, are being destroyed by soil-borne diseases such as *F. oxysporum*, which causes significant losses in quantity and quality [3]. Despite the effectiveness of synthetic fungicides in eliminating *Fusarium* and minimizing the harmful effects, the ecological troubles and the increased fungal resistance to these chemical fungicides are evident. We should not fail to note that the excessive use of pesticides has led to more serious problems than the disease itself, as it has negatively affected humans, animals, the environment, and healthy microbial communities in soil and plants [4–6]. Therefore, biological control of *Fusarium* wilt through different non-pathogenic microorganisms such as cyanobacteria, fungi, yeast, and bacteria are good techniques [7,8]. Plant resistance means preventing or limiting the progression of its damage, whether biotic or abiotic [9,10]. Biological agents can induce systemic pepper plant resistance. Inducers of resistance affect anatomical structures, morphology, or the making of certain chemical composites that obstruct the pathogen or minimize the severity of stress [11,12]. Structure and chemical weapons may be present in the plant regardless of whether a pathogen is attacking it. These weapons may also originate from an attack on the plant by a pathogen or stress [13]. The destruction of *F. oxysporum* in-vitro and in vivo is similar to the activity of mancozeb chemical fungicides through inhibition of fungal growth and sporulation [14].

Algae, including cyanobacteria, act as bio-protectants and bio-stimulants for crop enhancement [15,16] by destructing the structure or function of the membrane of plant pathogens, devastating pathogenic enzymes, and the defeat of protein synthesis [17]. *Desmonostoc muscorum* is an effective bio-fungicide to control some plant pathogens such as *Alternaria porri* in-vitro [18]. It also can inhibit the radial *Fusarium* mycelial growth [19]. On the other hand, *Arthrospira platensis* has antifungal activity against *F. oxysporum* through polyphenols production in in-vitro [20] and in-vivo against *Moringa Fusarium* wilt [21]. *Arthrospira platensis* extract contains phenolics resulting in antifungal activity [22]. *Anabaena*, when applied to seeds, resulted in the protection of root diseases from fungal pathogens such as *Fusarium* [23] through different mechanisms such as phosphate solubilization and indole-3-acetic acid (IAA), ammonia, hydrogen cyanide (HCN), and enzyme production [24]. Today, the risk of fungal plant disease is one of the most urgent issues [25,26]. In light of our increased understanding of environmental issues, we must look for feasible and easy-to-use solutions to *Fusarium*'s harmful effects on plants. One of the most famous pathogens of fungal diseases, *F. oxysporum*, hurts crops, especially vegetables [8,27]. Cyanobacteria could be a viable alternative to synthetic fungicides in the fight against phytopathogenic fungi because they produce bioactive metabolites with high antifungal efficiency, particularly polyphenols and flavonoids [28,29]. The use of growth-stimulating cyanobacteria is a common strategy for researchers to enhance and improve the defense capacity and physiological immunity of plants as well as the bioavailability of minerals in the soil [30].

Anabaena sp. and *Oscillatoria nigro-viridis* recorded the highest phenolics and flavonoid levels and the most elevated antifungal activity among the investigated cyanobacterial strains in recent research [31]. Accordingly, we examined phenolics and flavonoid levels in the three cyanobacterial strains used in this study to determine which cyanobacterial strains are best for reducing the risk of pepper plant infection with *Fusarium oxysporum*. However, this study's major purpose was to explore cyanobacteria's activity to minimize the harmful effect of pepper *Fusarium* wilt by enhancing pepper plant immunity.

2. Materials and Methods

2.1. Source of Pathogen

The pathogen was received from Al-Azhar University's Regional Center for Mycology and Biotechnology (RCMB). The pathogen was grown on PDA media and incubated at $28 \pm 2\,°C$ for 5 days before being preserved at $4\,°C$. According to [2], the pathogenic fungus inoculum was ready.

2.2. Growth Conditions of Cyanobacteria

This study uses three isolates related to cyanobacteria: *Desmonostoc muscorum* HSSASE1 KT277784, *Anabaena oryzae* HSSASE6 KT277789, and *Arthrospira platensis* HSSASE5 KT277788. Cyanobacterial samples were obtained from the botany and microbiology department, science faculty, Cairo University, Egypt. These strains were previously isolated from Egyptian soil, showing a significant antifungal activity and recording elevated levels of phenolics and flavonoids in vitro. The axenic cultures of the tested isolates were identified and deposited in GenBank under accession numbers according to [32]. Cyanobacterial strains were cultured in BG11 medium [33], but *Arthrospira platensis* was cultivated in Zarrouk medium [34]. A shaker incubator was used to grow all the cyanobacterial strains, which have been maintained in highly controlled growth conditions. Cyanobacterial cultures were incubated under constant illumination of (150 ± 10 µmol photons m^{-2} s^{-1}) at 27 ± 2 °C, pH = 7 for *Desmonostoc muscorum*, *Anabaena oryzae*, and at 34 ± 2 °C, pH = 8.7 for *Arthrospira platensis* and a continuous 5% CO_2 airflow was provided via an air pump. After 14 days, total biomass was harvested at the end of the growth stationary phase by centrifugation at $4200 \times g$ for 10 min, and then pellets were rinsed with water and lyophilized.

2.3. Preparation of Cyanobacteria Extracts

After 14 days of cultivation, freeze-dried cyanobacteria biomass was exposed to aqueous extraction [35]. After washing 100 mg of the cyanobacterial dry biomass in sterile distilled water, the amount was dissolved in 12.5 mL phosphate buffer (0.1 M pH 6.0) for 10 min before sonication (5 s pulses of 8 W over 30 s, on ice). The phosphate buffer solution did not affect extract nutrient composition because it was employed to control and maintain system pH. In addition, the extraction tubes were kept at 4 °C for 24 h. Aqueous extracts were obtained by centrifugation at $8000 \times g$ for 10 min and then freeze-dried.

2.4. Total Phenolic Content (TPC) of Tested Cyanobacteria

Folin–Ciocalteu technique [36] has been employed. For assessment of the total phenolic content of each cyanobacterial strain extract, the Gallic acid (0–500 mg L^{-1}) was used to construct a standard calibration curve. The absorbance of the sample was tested to a blank at 760 nm. Gallic acid equivalent (GAE)/g extract was used to measure total phenolic content (TPC).

2.5. Total Flavonoids Content (T.F.C.s) of Tested Cyanobacteria

Total flavonoid contents were measured for the tested cyanobacterial strains using a colorimetric method [37]. Five hundred microliters of the cyanobacterial extract were dissolved in 2 mL methanol and then mixed with 3 mL of water, 100 µL of potassium acetate (1 M), and 100 µL of aluminum chloride. After that, the samples were left in the dark for 30 min. The mixture's absorbance was measured at 415 nm. The total flavonoid content was calculated using a standard curve as (mg/concentration of Quercetin Equivalent (QE) obtained from calibration curve) mg Q.C.E./g of extract.

2.6. Experimental Design

The Agricultural Research Center in Giza, Egypt, provided three-week-old pepper seedlings. The consolidated seedlings were transplanted into 40×40 cm plastic pots, each treatment containing 6 seedlings. The pots in the green plastic house had a 1:3 mixture of sand and clay, totaling 7 kg. The pots were kept in the greenhouse at a temperature of 22 °C during the daylight hours and 18 °C at the nighttime, with a relative humidity of 70–85%. Except for the healthy control pot, the pathogenic fungus *F. oxysporum* (10^7 spores mL) was introduced into the soil after planting. For five days, the plants were irrigated routinely. Then, before and after flowering, a one-handed pressure sprayer was used to spray microalgae suspensions to the leaves of the plants three times (20 mL per plant once every week) at a concentration of 1 g L^{-1}. Roots were soaked in 1 g of dried algae extract per kg of soil. All plants were irrigated every 72 h for the period of the experiment. The pots were arranged in three duplicates in a completely randomized design: T1-healthy control (sowing pepper seedlings in sterilized soil), T2-infected control (sowing the pepper seedlings in sterilized soil inoculated with *F. oxysporum*), T3-infected plants treated with *Desmonostoc muscorum* through the soil, T4-infected plants treated with *Anabaena oryzae* through the soil, T5-infected plants treated with *Arthrospira platensis* through the soil, T6-infected plants treated with *Desmonostoc muscorum* through the foliar spray, T7-infected plants treated with *Anabaena oryzae* through the foliar spray and T8-infected plants treated with *Arthrospira platensis* through foliar application. For plant resistance evaluation, morphological and biochemical signals from plant samples were analyzed 45 days after sowing, and the disease was assayed.

2.7. Disease Symptoms and Disease Index

The disease symptoms were observed 45 days after sowing. The disease index and plant protection were assessed using a score consisting of five classes, as described in [38] with minor adjustments. (1) minor yellowing of lower leaves, (2) moderate yellow plant, (3) wilted plant with browning of vascular bands, and (4) severely stunted and damaged plants. It's worth noting that the percent disease index (PDI) was determined using a five-grade scale and the formula below. PDI = (1n1+ 2n2 + 3n3 + 4n4)100/4nt, where n1–n4 represents the number of plants in each class and nt represents the total number of plants examined. In addition, the following formula was used to obtain % Protection (P percent): P percent = A − B/A 100 percent, where A is the PDI in infected control plants and B is the PDI in infected plants treated with cyanobacteria.

2.8. Resistance Indicators in Pepper Plant

2.8.1. Morphological Resistance Indicators

Shoot length, root length, and the number of leaves were recorded.

2.8.2. Photosynthetic Pigment Determination

To determine the presence of chlorophyll a, chlorophyll b, and carotenoids in fresh pepper newly leaves (one leaf for each replicate), the photosynthetic pigments were tested according to [39,40]. Throughout this technique, photosynthetic pigments were extracted from fresh leaves (0.5 g) using 50 mL of acetone (80%), then the green color was determined spectrophotometrically at 665, 649, and 470 nm after the extract was filtered.

2.8.3. Estimation of Osmolytes Content

Soluble Sugar Determination

The dried shoot's soluble sugar content was estimated using the method [41,42]. The dried shoots (0.5 g) from each treatment were diluted with 5 mL of 30% trichloroacetic acid (TCA) and 2.5 mL of 2% phenol and filtered through filter paper, then 1 mL of the filtrate was treated with 2 mL of anthrone reagent (2 g anthrone/L of 95% H_2SO_4). 620 nm was used to determine the produced blue-green color.

Soluble Protein Estimation

The dry shoot's soluble protein content was estimated using the method described in [43]. 5 mL of 2% phenol and 10 mL of deionized water were used to extract the dried pepper shoots. One mL of this extract was combined with 5 mL of alkaline reagent (50 mL of 2% Na_2CO_3 prepared in 0.1N NaOH and 1 mL of 0.5% $CuSO_4$ prepared in 1% potassium sodium tartrate) and 0.5 mL of Folin's reagent (diluted by 1:3 *v/v*). After 30 min, a color change could be seen at 750 nm.

Proline Content Determination

The proline content of the dried shoot was determined using the technique [44]. The dried shoots (0.5 g) were digested by 10 mL (3%) of sulfosalicylic acid in this technique. In a boiling water bath, 2 mL of the filtrate was mixed with 2 mL of ninhydrin acid and 2 mL of glacial acetic acid for an hour. Then the mixture was placed in an ice bath to stop the reaction. 4 mL of toluene was added to the mix, then the absorbance at 520 nm was determined.

Total Phenol

A previous technique [26,45] was used to determine the total phenol content. One g of dried pepper shoots were extracted in 5–10 mL of 80% ethanol for at least 24 h. After dehydrating the alcohol, the remaining residue was extracted thrice using 5–10 mL of 80% ethanol each time. The purified extract was then filled to a capacity of 50 mL with 80% ethanol, and then 0.5 mL of the extract was mixed well with 0.5 mL of Folin's reagent and shaken for 3 min. After that, 3 mL of purified water and 1 mL of saturated sodium carbonate solution were added and thoroughly mixed. The blue color was detected at 725 nm after 1 h.

2.8.4. Estimation of Malondialdehyde (MDA) and Hydrogen Peroxide (H_2O_2) Contents

The method [41] was used to determine the amount of MDA in fresh pepper leaves. New leaf samples (0.5 g) were extracted with 5% TCA and centrifuged for 10 min at $4000\times g$. 2 mL of the extract was combined with 2 mL of 0.6% thiobarbituric acid (TBA) solution and placed in a water bath for 10 min. After cooling, the generated color's absorbance was measured at 532, 600, and 450 nm. The following equation was used to calculate the MDA content:

$$6.45 \times (A_{532} - A_{600}) - 0.56 \times A_{450}.$$

Fresh pepper leaves were tested for H_2O_2 content [46]. Fresh pepper leaves (0.5 g) were added to 4 mL cold acetone, then 3 mL of the extract was mixed with 1 mL of 0.1% titanium dioxide in 20 percent (*v:v*) H_2SO_4, then the mixture was centrifuged at $6000\times g$ for 15 min. The yellow color generated at 415 nm was detected.

2.8.5. Antioxidant Enzymes Activities Assay

POD and PPO enzyme activity was evaluated in this study to obtain a clear indication of defense-related enzymes. The peroxidase activity (POD) and polyphenol oxidase (PPO) enzymes were assayed according to [38], respectively.

2.8.6. Isozyme Electrophoresis

Peroxidase (POD) isozyme electrophoresis was analyzed according to the technique [26], while polyphenol oxidase (PPO) isozyme was recorded using the method [47]. The Gel Doc VILBER LOURMAT approach was used to evaluate and investigate gels. While the gels were saturated, the banding shape was videotaped, and the band's number was declared in each gel lane and computed and correlated with each other. The Helena Densitometer Model Junior 24 was used to determine quantitative band quantity and strength changes.

2.8.7. Endogenous Hormones (IAA, SA, and ABA) Contents

Plant hormones are natural organic compounds that affect growth and metabolism, affecting all external manifestations and chemical reactions. These act as chemical signals to activate or inhibit plant growth. As mentioned in [47], the Indole acetic acid (IAA), Salicylic acid (SA), and Abscisic acid (ABA) contents were measured in the terminal buds of both treated and control plants.

2.9. Statistical Analysis

Analyses were conducted using one-way variance (ANOVA). The statistically signifi-cant differences between treatments with a *p*-value of 0.05 or lower, the LSD test was used with CoStat (CoHort, Monterey, CA, USA) [48].

3. Results

3.1. Total Phenolics and Flavonoids of the Applied Cyanobacteria

The total phenolics and flavonoids of each cyanobacteria species were evaluated at the end of the stationary phase of growth. It can be shown in (Figure 1) that the total phenolics and flavonoids vary with species. The highest total phenolic content was determined in

A. platensis at 35.39 ± 0.19 mg GAE/g, followed by *A. oryzae* at 32.71 ± 19 mg GAE/g. In terms of flavonoids, *A. oryzae* had the highest content (5.58 ± 0.29 mg QCE/g), followed by *A. platensis* (5.34 ± 0.33 mg QCE/g). On the other hand, *D. muscorum* recorded the lowest phenolics and flavonoid content (30.47 ± 0.31 mg GAE/g and 4.73 ± 0.16 mg QCE/g, respectively).

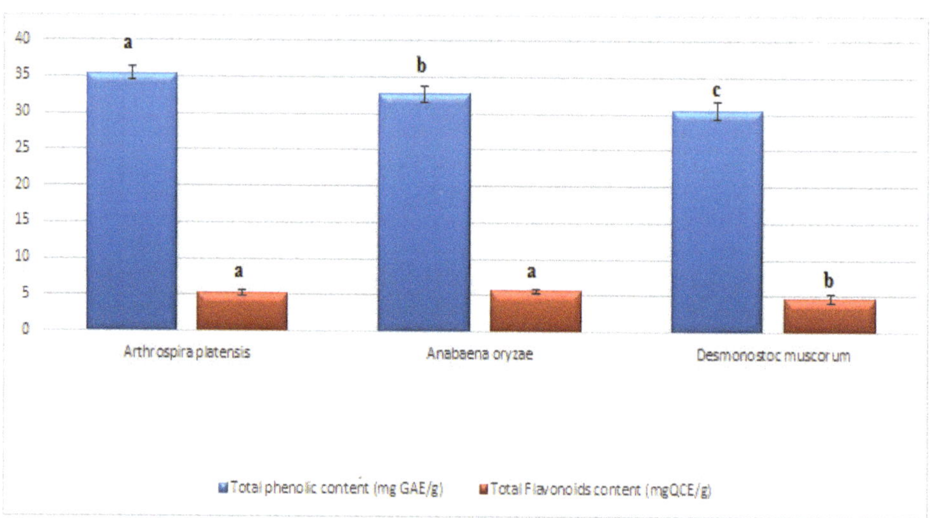

Figure 1. Variation in the total phenolics and flavonoid content of different applied cyanobacteria in the soil application and foliar spraying. Data represent mean ± SD, n = 3. (a–c Letters revered to significant in statically analysis).

3.2. Effect of Cyanobacteria on Disease Index of Infected Pepper Plants

Data presented in Table 1 reported that *F. oxysporum* shows a highly destructive effect on pepper plants that caused typical wilt symptoms with DI 83.33%. Applying cyanobacterial filtrate in the soil and foliar spraying highly protected plants against *Fusarium* wilt and lowered the *Fusarium* wilt disease symptoms caused by *F. oxysporum* compared with infected pepper plants. Data showed that spraying cyanobacteria was more effective than irrigation in soil, while foliar spraying with *A. platensis* showed high protection against *Fusarium* infection (95%), followed by *D. muscorum* (90%), then *A. oryzae* (69.9%).

Table 1. Effect of cyanobacteria on disease index of the infected pepper plant with *Fusarium* wilt.

Treatments	Method of Application	Disease Symptoms Classes					DI (Disease Index) (%)	Protection (%)
		0	1	2	3	4		
D. muscorum		2	1	2	0	1	37.5	54.99
A. oryzae	Through soil	1	3	1	1	0	33.33	60.03
A. platensis		2	1	3	0	0	29.16	65
D. muscorum		5	0	1	0	0	8.3	90
A. oryzae	Through Foliar	3	0	3	0	0	25	69.9
A. platensis		5	1	0	0	0	4.16	95
Control infected		0	0	1	2	3	83.33	0

3.3. Resistance Indicators in Pepper Plant

3.3.1. Morphological Indicators

The results in (Figure 2) indicated that *Fusarium*-wilt hurt all vegetative pepper growth compared to healthy control. *F. oxysporum* minimized shoot length by 46.89%, root length by 48.45%, and the number of leaves by 45.07%, respectively. Regarding the effect of cyanobacteria, it was observed that infected plants treated with *D. muscorum, A. oryzae,* and *A. platensis* through different modes, i.e., soil and foliar spraying showed promising recovery. It was determined that *A. platensis* was the most effective cyanobacteria for recovering plant height by 70.64% and 69.53%, root lengths by 92.94% and 92.94%, and the number of leaves by 76.92% and 58.97% through the soil and foliar application, respectively.

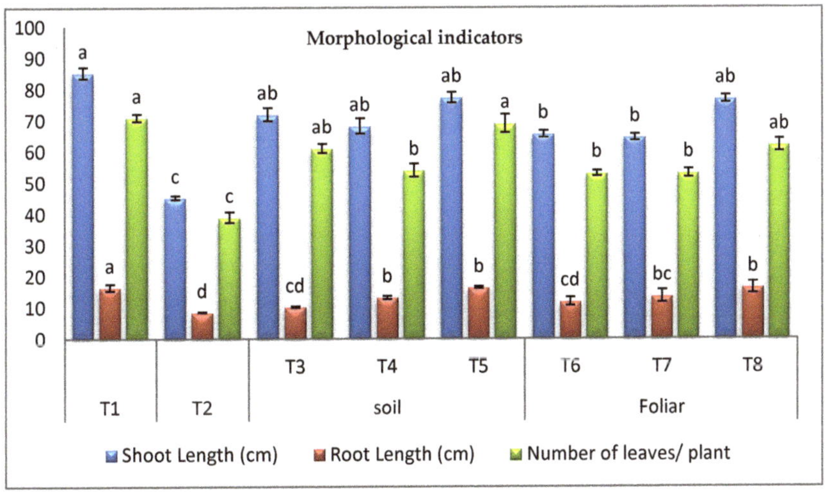

Figure 2. Effect of cyanobacteria on morphological traits of the infected pepper plant with *Fusarium* wilt (T1-healthy control, T2-infected control, T3-infected plants treated with *D. muscorum* soil, T4-infected plants treated with *A. oryzae* soil, T5-infected plants treated with *A. platensis* soil, T6-infected plants treated with *D. muscorum* foliar spray, T7-infected plants treated with *A. oryzae* foliar spray and T8-infected plants treated with *A. platensis* foliar application.) (Data represent mean ± SD, n = 3), (a–d Letters revered to significant in statically analysis).

3.3.2. Photosynthetic Pigments

Results in (Figure 3) showed that chlorophyll a (Chl a) and b (Chl b) were highly inhibited in *Fusarium* infected pepper plants by 39.41% and 59.46%, respectively. On the other hand, the present study showed that the level of carotenoids in infected pepper plants increased compared to the healthy control. However, the application of cyanobacteria through the soil and foliar application to the infected plants significantly increased the level of carotenoids over the infected control plants. It was found that foliar and soil application of *A. platensis* was the most effective way to enhance infected plants' levels of Chl a, b (84.81% and 191.01%). On the other hand, *D. muscorum* foliar spraying was the most efficient method for increasing the carotenoids in infected plants (123.00%).

3.3.3. Osmolytes (Soluble Sugar, Soluble Protein, and Proline) Contents

Results in (Table 2) clearly showed *F. oxysporum* infection caused a major reduction in contents of soluble sugars by 38.50% and soluble protein by 57.13% over the healthy control. Still, it caused proline increment by 11.84% over the healthy control. The infected pepper plants treated with cyanobacteria ameliorated total carbohydrates, protein, and proline contents. Cyanobacteria treatment through different modes of application showed a high response. It increased the Accumulation of osmolytes (soluble sugar, soluble protein, and proline) contents compared to the infected control plants.

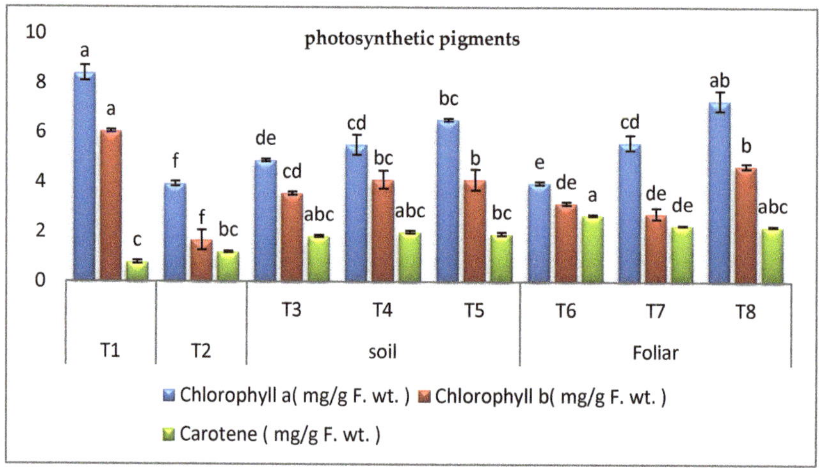

Figure 3. Effect of cyanobacteria on photosynthetic pigments of the infected pepper plant with *Fusarium* wilt (T1-healthy control, T2-infected control, T3-infected plants treated with *D. muscorum* soil, T4-infected plants treated with *A. oryzae* soil, T5-infected plants treated with *A. platensis* soil, T6-infected plants treated with *D. muscorum* foliar spray, T7-infected plants treated with *A. oryzae* foliar spray and T8-infected plants treated with *A. platensis* foliar application.) (Data represent mean ± SD, n = 3), (a–f Letters revered to significant in statically analysis).

Table 2. Effect of cyanobacteria on osmolytes (soluble sugar, soluble protein, and proline) mg g^{-1} DW) contents of the infected pepper plant with *Fusarium* wilt. Data presented as means ± SD (n = 3). Data followed by letters are significantly different in the LSD test at $p \leq 0.05$.

Treatments	Method of Application	Total Carbohydrate	Total Protein	Free Proline
Healthy control		30 ± 2.4 [a]	34.2 ± 0.2 [a]	0.76 ± 0.002 [d]
Infected control		18.45 ± 1.62 [c]	14.66 ± 1.6 [f]	0.85 ± 0.001 [e]
D. muscorum		19.55 ± 0.82 [c]	29.08 ± 0.4 [b]	0.913 ± 0.001 [a]
A. oryzae	Through soil	27.7 ± 0.9 [a]	22.26 ± 1 [e]	0.86± 0.001 [b]
A. platensis		24.02 ± 2.62 [b]	29.56 ±0.3 [b]	0.91 ± 0.002 [a]
D. muscorum		19.74 ± 1.4 [c]	26.46 ± 0.9 [c]	0.862± 0.001 [c]
A. oryzae	Through Foliar	20.7 ± 1.7 [c]	24.8 ± 0.7 [d]	0.863 ± 0.001 [c]
A. platensis		21.38 ± 1.26 [bc]	28.56 ± 0.7 [b]	0.864 ± 0.001 [bc]
LSD at 0.05		2.967	1.507	0.0029

3.3.4. Stress Biomarkers

The results in (Figure 4) showed that pepper *Fusarium* wilt disease resulted in a rise in phenolics over the healthy control plant. On the other hand, it was observed that infected plants treated with cyanobacteria exhibited a significant increase in phenolics over the infected control plants. The data (Figure 4) indicated that the highest increase in phenolics level was recorded by the infected plants treated with *A. platensis* in the soil treatment. It was observed that cyanobacteria supplementation reduced the generation of H_2O_2 and lipid peroxidation (MDA) significantly over the infected control plants (Figure 4). Accumulation of H_2O_2 increased in infected control plants, causing an increase in lipid peroxidation (MDA) over the healthy control plants. Supplementation of infected plants with cyanobacteria reduced the generation of H_2O_2 over the infected control plants leading to a declined lipid peroxidation (MDA) (Figure 4). The data revealed that the most effective cyanobacterial treatment was foliar spraying with *A. platensis*.

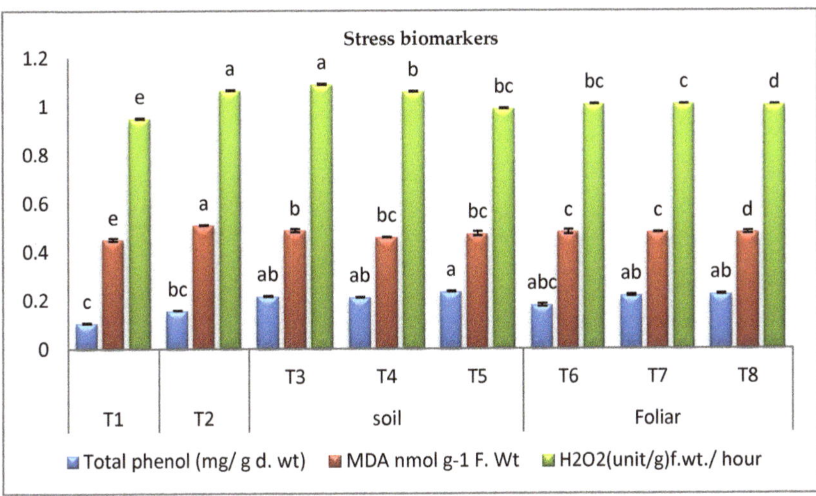

Figure 4. Effect of cyanobacteria on total phenol, Malondialdehyde (MDA), and H_2O_2 contents of infected pepper plants (T1-healthy control, T2-infected control, T3-infected plants treated with *D. muscorum* soil, T4-infected plants treated with *A. oryzae* soil, T5-infected plants treated with *A. platensis* soil, T6-infected plants treated with *D. muscorum* foliar spray, T7-infected plants treated with *A. oryzae* foliar spray and T8-infected plants treated with *A. platensis* foliar application.) (Data represent mean $\pm$ SD, n = 3), (a–e Letters revered to significant in statically analysis).

3.3.5. Oxidative Enzymes Activity

Fusarium wilt disease increased the activities of POD and PPO enzymes in the infected control plants over the healthy control plants; however, applying cyanobacteria through different modes to the infected plants significantly enhanced the activities of PPO and POD enzymes over the infected control plants. The data recorded in (Figure 5) revealed that the maximal increase in the activities of PPO and POD enzymes was observed in the infected pepper plants treated with *A. platensis* and *A. oryzae* through the foliar spraying, respectively. According to our findings, the antioxidant enzymes in infected plants treated with cyanobacteria in various modes, such as soil mode and foliar spraying, significantly increased, but the foliar spraying was more effective.

3.3.6. Antioxidant Isozymes

The antioxidant isozymes (POD and PPO) could be detected using native PAGE. The results (Figures 6 and 7) revealed the expression of 4 PPO and 9 POD isozymes in the leaves of pepper plants. For PPO isozymes, infected pepper plants treated with *A. platensis* through the soil application showed the strongest PPO expression as it produced 5 bands, including 4 high-intensity bands at Rf (0.487, 0.658, 0.748, and 0.895) and one medium intensity band at Rf (0.950), followed by *A. platensis* through the foliar spraying and *A. oryzae* through the soil application as they showed the same number and density of 4 bands. Referring to POD isozymes, it was observed that treatment of *D. muscorum* (foliar) application showed the strongest POD expression as it produced 11 bands, including four high-intensity bands at Rf (0.274, 0.444, 0.586, and 0.812), five moderated intensity bands at Rf (0.160, 0.244, 0.654, 0.707, and 0.880), and two weak intensity bands at Rf (0.142 and 0.944). In addition, applying cyanobacteria through foliar spraying showed the same number and density of 10 bands for all used cyanobacteria.

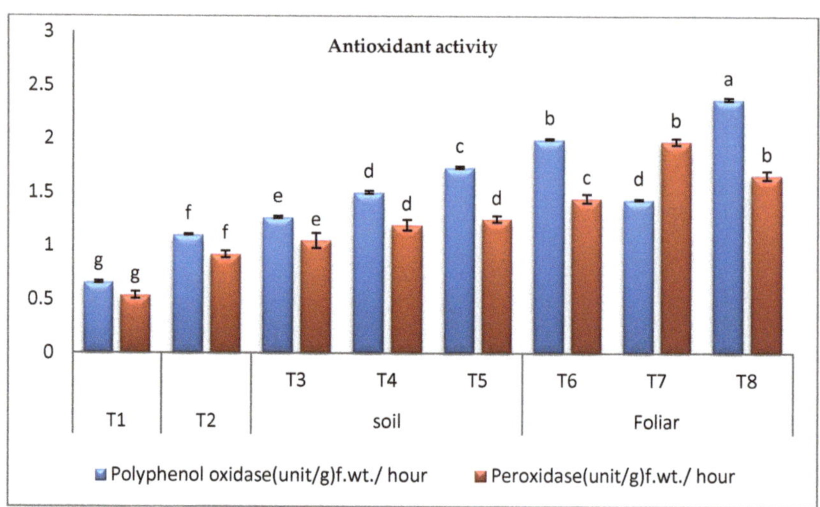

Figure 5. Effect of cyanobacteria on the activity of oxidative enzymes in the infected pepper plants (T1-healthy control, T2-infected control, T3-infected plants treated with *D. muscorum* soil, T4-infected plants treated with *A. oryzae* soil, T5-infected plants treated with *A. platensis* soil, T6-infected plants treated with *D. muscorum* foliar spray, T7-infected plants treated with *A. oryzae* foliar spray and T8-infected plants treated with *A. platensis* foliar application.) (Data represent mean ± SD, n = 3), (a–g Letters revered to significant in statically analysis).

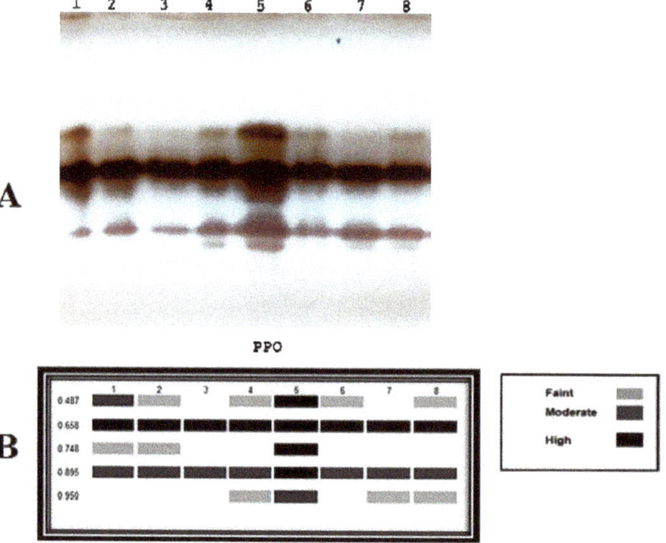

Figure 6. The effect of *F. oxysporum* and the application of cyanobacteria on (**A**) Polyphenol oxidase isozyme. (**B**) Ideogram analysis of PPO isozyme of the infected pepper plants. 1 = Healthy control; 2 = Infected control; 3 = Infected plant + *D. muscorum* (soil); 4 = Infected plant + *A. oryzae* (soil); 5 = Infected plant + *A. platensis* (soil); 6 = Infected plant + *D. muscorum* (foliar); 7 = Infected plant + *A. oryzae* (foliar); 8 = Infected plant + *A. platensis* (foliar).

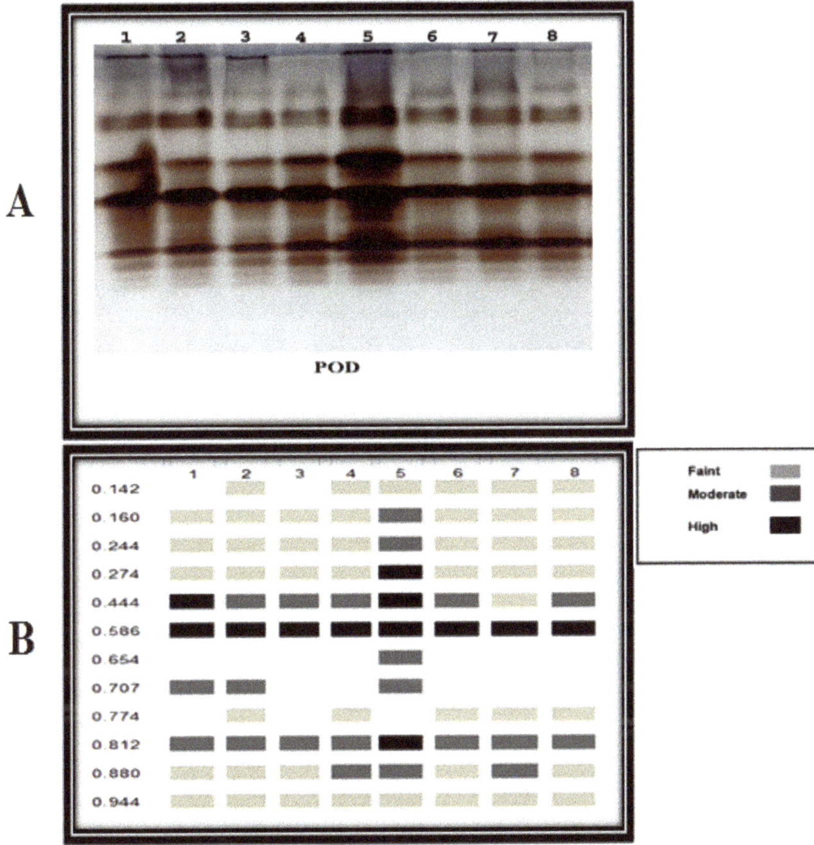

Figure 7. The effect of *F. oxysporum* and the application of cyanobacteria on (**A**) Peroxidase isozyme. (**B**) Ideogram analysis of POD isozyme of the infected pepper plants. 1 = Healthy control; 2 = Infected control; 3 = Infected plant + *D. muscorum* (soil); 4 = Infected plant + *A. oryzae* (soil); 5 = Infected plant + *A. platensis* (soil); 6 = Infected plant + *D. muscorum* (foliar); 7 = Infected plant + *A. oryzae* (foliar); 8 = Infected plant + *A. platensis* (foliar).

3.3.7. Endogenous Hormones

The results in (Figure 8) highlighted that *Fusarium* infection decreased IAA but increased SA and ABA contents in the infected control plants over the healthy control plants. However, applying cyanobacteria through different modes to the infected plants significantly increased the levels of IAA and SA hormones but decreased ABA concentrations over the infected control plants. The data (Figure 8) revealed that the maximal increase in the IAA contents was observed in the infected plants treated with *D. muscorum* and *A. oryzae* through the soil treatment mode. In addition, the highest growth in the SA contents was recorded by the infected plants treated with *A. platensis* through the foliar spraying. On the other hand, infected plants treated with *A. platensis* through foliar spraying showed the least reduction in ABA concentration in our research.

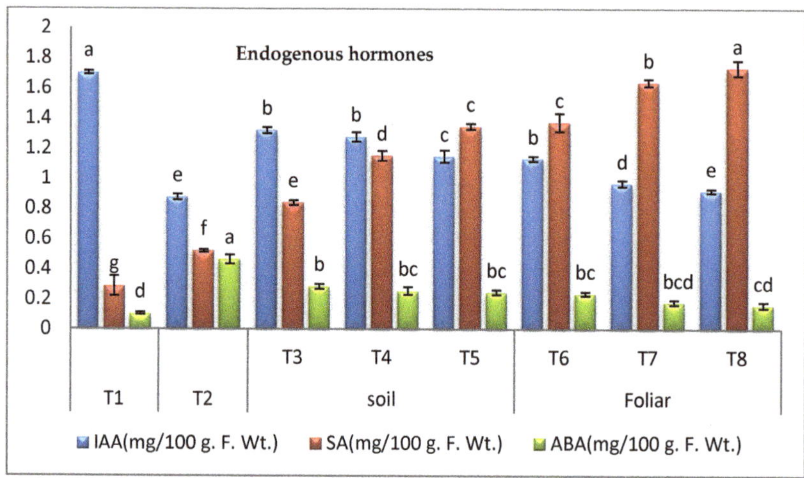

Figure 8. Effect of cyanobacteria on endogenous hormones (IAA, SA, and ABA) of infected pepper plants (T1-healthy control, T2-infected control, T3-infected plants treated with *D. muscorum* soil, T4-infected plants treated with *A. oryzae* soil, T5-infected plants treated with *A. platensis* soil, T6-infected plants treated with *D. muscorum* foliar spray, T7-infected plants treated with *A. oryzae* foliar spray and T8-infected plants treated with *A. platensis* foliar application.) (Data represent mean ± SD, n = 3), (a–g Letters revered to significant in statically analysis).

4. Discussion

Disease severity was the first guide to govern systemic resistance in treated plants by cyanobacteria. Data presented in this study reported that *F. oxysporum* shows a highly destructive effect on pepper plants that caused typical wilt symptoms with DI 83.33%, similar to earlier studies on the same pathogenic fungus [40]. Using cyanobacterial extract to treat *Fusarium* wilt-infected pepper plants greatly reduced the symptoms of the disease, which is the primary criterion for assessing resistance in the pepper plant. Data showed that foliar spraying with cyanobacteria extracts was more effective than putting them in soil; foliar spraying with *A. platensis* showed high protection against *Fusarium* infection (95%), followed by *D. muscorum* (90%), then *A. oryzae* (69.9%), these results are consistent with [49], which stated that foliar spraying was more effective than soil application, where the nutrients in the foliar spray are absorbed up directly by the leaves of the plant.

Moreover, the results in this study revealed that the highest total phenolic and flavonoid contents were determined in *Arthrospira platensis*, which provided the pepper plant with the highest percentage of protection through foliar spraying. This result may recommend choosing cyanobacterial strains with elevated levels of phenolics and flavonoids to stimulate immune responses in pepper plants against *Fusarium* wilt. The results explained that spraying with cyanobacteria caused an improvement in pepper plant resistance against biotic stresses, including fungal pathogens [49]. Furthermore, antifungal activity of *A. platensis* [50], *D. muscorum* [51], and *A. oryzae* [18] against fungal pathogens were also reported. Reducing symptoms and severity of infection is one of the most important goals of using fungicides, whether chemical or biological [52]. Undoubtedly, applying natural or biological factors such as cyanobacteria is more environmentally friendly. Interestingly, the use of cyanobacteria understudy led to a reduction in symptoms, and a reduction in the severity of infection was reflected positively on the health of plant growth. These results are consistent with several scientific studies [25,53] that confirm that cyanobacteria contain many antioxidants, proteins, vitamins, hormones, and antimicrobials.

The results indicated that *F. oxysporum* hurt all vegetative pepper growth compared to healthy control. *F. oxysporum* minimized shoot length by 46.89%, root length by 48.45%, and several leaves by 45.07%, respectively. Therefore, the infection of the pepper plant with

F. oxysporum caused a significant inhibition of all growth parameters, where our findings align with those of a huge number of other researchers [3,54]. Regarding the effect of cyanobacteria, it was observed that infected plants treated with *D. muscorum*, *A. oryzae*, and *A. platensis* in soil and foliar spraying exhibited promising recovery. It was determined that *A. platensis* was the most effective cyanobacteria for recovering plant height by 70.64% and 69.53%, root lengths by 92.94%, 92.94%, and the number of leaves by 76.92%, and 58.97% through the soil and foliar application, respectively. These results agree with a previous study [21], which reported that the application of *A. platensis* improved plant growth through polysaccharides production. The use of cyanobacteria to enhance crop growth has been proposed as a potential executive performance in crop enhancement. [55]. These results align with [35], who found that treating plants with cyanobacteria greatly improved their vegetative growth. The increase in vegetative growth and crop yield with cyanobacteria could mainly be due to the release of plant nutrients like N, P, and K and the excretion of plant growth regulators (auxin, gibberellins), amino acids, and vitamins [56].

After applying cyanobacteria, photosynthetic pigments had become a clear positive indicator of sufficient treatments. By analyzing data from this investigation, it was obvious that chlorophylls a and b (Chl a and b) were severely inhibited in *Fusarium* infected pepper plants by 58.22% and 59.46%, respectively. This reduction in chlorophyll was induced by the production of reactive oxygen species (ROS) after the attack with *Fusarium*, which destroyed chlorophyll contents, preventing the plants from capturing sunlight and reducing photosynthesis [38,57]. Chlorophyll disruption, reduced chlorophyll synthesis, and thylakoid membrane strength are also diminished [58]. On the other hand, the present study showed that the level of carotenoids in pepper plants increased by 70% in response to *Fusarium* infection. These findings are consistent with those of other studies [3,6,39,59,60], which found that the content of carotenoids in plants increased dramatically in response to *Fusarium* infection. In terms of cyanobacteria's beneficial impacts, it was discovered that diseased plants improved after applying *D. muscorum*, *A. oryzae*, and *A. platensis* through various modalities. It was found that foliar and soil application of *A. platensis* was the most effective way to enhance infected plants' Chl a, b, and carotenoids levels. Increased chlorophyll contents in infected plants treated with cyanobacterial strains could be resulted from the higher amount of atmospheric nitrogen assimilation by cyanobacteria then transported to pepper plant tissues [59].

Moreover, cyanobacteria supply decreased ethylene production and chlorophyll, stimulated the synthesis of carotenoids which defend chlorophyll from oxidation, and increased chlorophyll content [60]. These results are in agreement with the results reported in this study. Photosynthetic protection may have been supplied by improved synthesis of carotenoids due to enhancing ROS scavenging [61].

In the current study, the results clearly showed *F. oxysporum* infection caused a major reduction in contents of soluble sugars by 38.50% and soluble protein by 57.13% over the healthy control. Still, it caused proline increment by 11.84% over the healthy control. These results agree with the previous studies [3,62]. The infected pepper plants treated with cyanobacteria showed amelioration in the contents of total carbohydrates, protein, and proline. These results may be explained by the potency of cyanobacteria to secrete complex heteropolymers, polysaccharides, and lipopolysaccharides to induce defense-related gene expression [63,64]. Cyanobacteria-treated plants were capable of fighting against *Fusarium* infection by accumulating more proline, which protects proteins from oxidation [65]. The osmolytes (soluble sugar, proline, and soluble protein) levels in the cyanobacteria-treated plants were significantly higher than those in the infected control plants. Also, oil organic carbon, nutrient absorption, and nitrogen fixation were all improved due to the cyanobacteria treatment [66]. As a result, our findings reveal that osmolyte content significantly increased when infected plants were treated with cyanobacteria, as previously explained [49,62]. This increase in soluble proteins can be explained by activating plant defense systems when pathogens are challenged.

According to the findings of this research, pepper *Fusarium* wilt disease resulted in a rise in phenolics over the healthy control plant. On the other hand, it was observed that infected plants treated with cyanobacteria exhibited a marked increase in phenolics over the infected control plants. The present data indicated that the highest growth in phenolic level was recorded by the infected plants treated with *A. platensis* in the soil treatment. The increased accumulation of phenolics by treatment with cyanobacteria resulted in stress tolerance of the pepper plant against *Fusarium* infection [66,67]. It was observed that cyanobacteria supplementation reduced the generation of H_2O_2 and lipid peroxidation (MDA) significantly over the infected control plants. Accumulation of H_2O_2 increased in infected control plants, causing an increase in lipid peroxidation (MDA) over the healthy control plants. Supplementation of infected plants with cyanobacteria reduced the generation of H_2O_2 over the infected control plants leading to a declined lipid peroxidation (MDA). The data revealed that the most effective cyanobacterial treatment was foliar spraying with *A. platensis*. The findings here are comparable to a previous study conducted by [26], which reported that the application of biological stimulators under stress conditions decreased MDA.

The activity of POD and PPO was assessed to identify enzymes involved in protecting the infected plant. *Fusarium* wilt disease increased the activities of POD and PPO enzymes in the infected control plants over the healthy control plants; however, the application of cyanobacteria through different modes to the infected plants significantly enhanced the PPO and POD enzyme activity over the infected control plants. The recorded data revealed that the maximal increase in the activities of PPO and POD enzymes was detected in the infected pepper plants treated with *A. platensis* and *A. oryzae* through the foliar spraying, respectively. Protective enzymes such as POD and PPO are the most significant in biotic stress response [68]. These enzymes are involved in the early stages of plant resistance to various stressors and the synthesis of phenolic compounds. According to our findings, the antioxidant enzymes in infected plants treated with cyanobacteria in multiple modes, such as soil mode and foliar spraying, significantly increased. The plant displayed distinct strategies for coping with stress by increasing the activity level of some antioxidant enzymes in the cell to maintain a low concentration of reactive oxygen species [26,69].

By detecting the antioxidant isozymes (POD and PPO) by native PAGE, the results showed that a significant part of the plant's response to various stresses is the isozyme substance, which also serves as an important metabolic regulator. Isozyme is a clear indication of the occurrence of resistance, as it plays an important role in mitigating or limiting free radicals that result from oxidative explosions. The findings of this research fully agree with the other scientific reports [70,71]. In the presence of a *Fusarium* wilt disease, cyanobacteria can induce gene expression in infected plants similarly to antioxidant enzymes. It can produce chemicals activating plant immunity, such as phenols and natural hormones [63].

In this study, results highlighted that *Fusarium* infection decreased IAA but increased SA and ABA contents in the infected control plants over the healthy control plants. However, applying cyanobacteria through different modes to the infected plants significantly increased the levels of IAA and SA hormones but decreased ABA concentrations over the infected control plants. Plant hormones are natural organic compounds that affect growth and metabolism, thereby affecting all external manifestations and chemical reactions. These act as chemical signals to activate or inhibit plant growth. When plants are exposed to infections, a considerable modulation in the biosynthesis of hormones occurs, affecting various growth processes [38]. Phytohormones produced by the cyanobacteria have an important function in controlling plant fungal diseases, activating several genes responsible for systemic resistance in plants [49]. The modulation of plant hormone contents is essential in plant defense reactions against *Fusarium* wilt [72]. The SA plays an important role as a plant hormone, as it stimulates growth and works to activate the chemical and synthetic resistance of the plants against any pathogen, increases the absorption of nutrients, and increases the process of photosynthesis [73,74]. Recently, SA has been used externally or internally as inducers against various plant pathogens [75,76]. SA has been described as a key molecule in the signal transduction pathways of the biological stress response by using cyanobacteria to activate and increase SA, IAA as these hormones act as antimicrobial substances [76,77]. The significance of ABA in plant disease resistance is unclear when compared to that of the plant hormones jasmonic acid and salicylic acid, both of which play key roles in disease resistance [78]. Similarly to our findings, previous research has recorded the accumulation of ABA during the infection of sugar beets by fungi [79].

5. Conclusions

It is concluded that *Fusarium* wilt generated oxidative destruction and developed into reduced growth and dropped physiological performance. Applying cyanobacteria to *Fusarium* infected pepper plants through soil application or foliar spraying activated the immunity of infected pepper plants. The infected plants treated with cyanobacteria showed enhanced photosynthetic pigments and accumulation of osmoprotectants, phenols, and antioxidant systems that also act as a scavenging tool to remove the excess ROS under *Fusarium* infection. Therefore, it could be applied in agricultural fields through soil or foliar application. As far as we know, this is the first evidence to report that cyanobacteria metabolites influence the isozymes of pepper plants attacked with *Fusarium*, and supplementary molecular findings can provide information on the impact of cyanobacteria on the metabolism of the plant under biotic stress. This study supports the positive application of cyanobacteria in protecting pepper plants under fungal infection; however, further studies are required to unravel actual mechanisms.

Author Contributions: Conceptualization, M.S.A., A.M.A. and H.H.S. Methodology, A.M.A., W.A.A., H.H.S., D.A.R. and M.S.S. Software, D.A.R., A.M.A., M.S.S. and M.S.A.; Formal analysis, H.H.S., M.S.A., A.M.A., W.A.A. and M.S.S. Investigation, D.A.R., A.M.A. and M.S.A.; Resources, A.M.A., M.S.A. and H.H.S.; Data Curation, A.M.A. and M.S.A.; Writing—original draft preparation, M.S.A., A.M.A., W.A.A. and H.H.S.; Writing—Review and Editing, A.M.A., D.A.R., H.H.S., M.S.S. and W.A.A.; Supervision, M.S.A. All authors have read and agreed to the published version of the manuscript.

Funding: The author would like to thank the Deanship of Scientific Research at Majmaah University for supporting this work under Project Number No. R-2022-231.

Institutional Review Board Statement: Not applicable.

Informed Consent Statement: Not applicable.

Data Availability Statement: The data presented in this study are available in the article.

Acknowledgments: The authors are grateful to Abd Almoneam Sharaf. Mohamed M. Ali and Amr S. Mohamed for their help during the study, critical reading of the manuscript, and discussions. The authors are also grateful to the Botany & Microbiology Department Faculty of Science, Al-Azhar University 11884 Nasr City, Cairo, Egypt & Botany, and Microbiology Department, Faculty of Science, Cairo University for the technical support offered during this work.

Conflicts of Interest: The authors declare no conflict of interest.

References

1. Tijjani, A.; Khairulmazmi, A. Global Food Demand and the Roles of Microbial Communities in Sustainable Crop Protection and Food Security: An Overview. *Role Microb. Communities Sustain.* **2021**, *29*, 81–107.
2. El-Feky, N.; Essa, T.; Elzaawely, A.A.; El-Zahaby, H.M. Antagonistic activity of some bioagents against root rot diseases of pepper (*Capsicum annum* L.). *Environ. Biodivers. Soil Secur.* **2019**, *3*, 103–104. [CrossRef]
3. Abdelaziz, A.M.; Dacrory, S.; Hashem, A.H.; Attia, M.S.; Hasanin, M.; Fouda, H.M.; Kamel, S.; ElSaied, H. Protective role of zinc oxide nanoparticles based hydrogel against wilt disease of pepper plant. *Biocatal. Agric. Biotechnol.* **2021**, *35*, 102083. [CrossRef]
4. Rani, L.; Thapa, K.; Kanojia, N.; Sharma, N.; Singh, S.; Grewal, A.S.; Srivastav, A.L.; Kaushal, J. An extensive review on the consequences of chemical pesticides on human health and environment. *J. Clean. Prod.* **2020**, *283*, 124657. [CrossRef]
5. Bhandari, G. An Overview of Agrochemicals and Their Effects on Environment in Nepal. *Appl. Ecol. Environ. Sci.* **2014**, *2*, 66–73. [CrossRef]
6. Alhakim, A.A.; Hashem, A.; Abdelaziz, A.M.; Attia, M.S. Impact of plant growth promoting fungi on biochemical defense performance of Tomato under Fusarial infection. *Egypt. J. Chem.* **2022**. [CrossRef]

7. Aldinary, A.M.; Abdelaziz, A.M.; Farrag, A.A.; Attia, M.S. Biocontrol of tomato Fusarium wilt disease by a new Moringa endophytic Aspergillus isolates. *Mater. Today Proc.* **2021**. [CrossRef]

8. Attia, M.S.; El-Wakil, D.A.; Hashem, A.H.; Abdelaziz, A.M. Antagonistic Effect of Plant Growth-Promoting Fungi Against Fusarium Wilt Disease in Tomato: In vitro and In vivo Study. *Appl. Biochem. Biotechnol.* **2022**, 1–19. Available online: https://link.springer.com/content/pdf/10.1007/s12010-022-03975-9.pdf (accessed on 3 March 2022). [CrossRef]

9. Rausher, M.D. Co-evolution and plant resistance to natural enemies. *Nature* **2001**, *411*, 857–864. [CrossRef]

10. Attia, M.S.; El-Sayyad, G.S.; Elkodous, M.A.; Khalil, W.F.; Nofel, M.M.; Abdelaziz, A.M.; Farghali, A.A.; El-Batal, A.I.; El Rouby, W.M. Chitosan and EDTA conjugated graphene oxide antinematodes in Eggplant: Toward improving plant immune response. *Int. J. Biol. Macromol.* **2021**, *179*, 333–344. [CrossRef] [PubMed]

11. Witzell, J.; Martín, J.A. Phenolic metabolites in the resistance of northern forest trees to pathogens—Past experiences and future prospects. *Can. J. For. Res.* **2008**, *38*, 2711–2727. [CrossRef]

12. Vargas-Hernandez, M.; Macias-Bobadilla, I.; Guevara-Gonzalez, R.; Rico-Garcia, E.; Ocampo-Velazquez, R.; Avila-Juarez, L.; Torres-Pacheco, I. Nanoparticles as Potential Antivirals in Agriculture. *Agriculture* **2020**, *10*, 444. [CrossRef]

13. Ab Rahman, S.F.S.; Singh, E.; Pieterse, C.M.; Schenk, P.M. Emerging microbial biocontrol strategies for plant pathogens. *Plant Sci.* **2018**, *267*, 102–111. [CrossRef] [PubMed]

14. Kim, J.; Kim, J.-D. Inhibitory effect of algal extracts on mycelial growth of the tomato-wilt pathogen, *Fusarium oxysporum* f. sp. *lycopersici*. *Mycobiology* **2008**, *36*, 242–248. [CrossRef] [PubMed]

15. Lee, S.-M.; Ryu, C.-M. Algae as New Kids in the Beneficial Plant Microbiome. *Front. Plant Sci.* **2021**, *12*, 599742. [CrossRef]

16. Refaay, D.A.; El-Marzoki, E.M.; Abdel-Hamid, M.I.; Haroun, S.A. Effect of foliar application with *Chlorella vulgaris*, *Tetradesmus dimorphus*, and *Arthrospira platensis* as biostimulants for common bean. *J. Appl. Phycol.* **2021**, *33*, 3807–3815. [CrossRef]

17. Swain, D.L.; Singh, D.; Horton, D.E.; Mankin, J.S.; Ballard, T.C.; Diffenbaugh, N.S. Remote Linkages to Anomalous Winter Atmospheric Ridging Over the Northeastern Pacific. *J. Geophys. Res. Atmos.* **2017**, *12*, 194–209. [CrossRef]

18. Abdel-Hafez, S.I.I.; Abo-Elyousr, K.A.M.; Abdel-Rahim, I.R. Fungicidal activity of extracellular products of cyanobacteria against *Alternaria porri*. *Eur. J. Phycol.* **2015**, *50*, 239–245. [CrossRef]

19. Morales-Jiménez, M.; Gouveia, L.; Yáñez-Fernández, J.; Castro-Muñoz, R.; Barragán-Huerta, B.E. Production, Preparation and Characterization of Microalgae-Based Biopolymer as a Potential Bioactive Film. *Coatings* **2020**, *10*, 120. [CrossRef]

20. Hlima, H.B.; Bohli, T.; Kraiem, M.; Ouederni, A.; Mellouli, L.; Michaud, P.; Abdelkafi, S.; Smaoui, S. Combined effect of *Spirulina platensis* and *Punica granatum* peel extacts: Phytochemical content and antiphytophatogenic activity. *Appl. Sci.* **2019**, *9*, 5475. [CrossRef]

21. Imara, D.A.; Zaky, W.H.; Ghebrial, E.W. Performance of Soil Type, Cyanobacterium *Spirulina platensis* and *Biofertilizers* on Controlling Damping-off, Root Rot and Wilt Diseases of Moringa (*Moringa oleifera* Lam.) in Egypt. *Egypt. J. Phytopathol.* **2021**, *49*, 10–28. [CrossRef]

22. Souza, M.M.D.; Prietto, L.; Ribeiro, A.C.; Souza, T.D.D.; Badiale-Furlong, E. Assessment of the antifungal activity of *Spirulina platensis* phenolic extract against *Aspergillus flavus*. *Ciência e Agrotecnologia* **2011**, *35*, 1050–1058. [CrossRef]

23. El-Mougy, N.S.; Abdel-Kader, M.M. Effect of Commercial Cyanobacteria Products on the Growth and Antagonistic Ability of Some Bioagents under Laboratory Conditions. *J. Pathog.* **2013**, *2013*, 838329. [CrossRef] [PubMed]

24. Afify, A.; Ashour, A. Use of Cyanobacteria for Controlling Flax Seedling Blight. *J. Agric. Chem. Biotechnol.* **2018**, *9*, 259–261. [CrossRef]

25. Prasanna, R.; Chaudhary, V.; Gupta, V.; Babu, S.; Kumar, A.; Singh, R.; Shivay, Y.S.; Nain, L. Cyanobacteria mediated plant growth promotion and bioprotection against Fusarium wilt in tomato. *Eur. J. Plant Pathol.* **2013**, *136*, 337–353. [CrossRef]

26. Abdelaziz, A.M.; El-Wakil, D.A.; Attia, M.S.; Ali, O.M.; AbdElgawad, H.; Hashem, A.H. Inhibition of Aspergillus flavus Growth and Aflatoxin Production in *Zea mays* L. Using Endophytic Aspergillus fumigatus. *J. Fungi* **2022**, *8*, 482. [CrossRef] [PubMed]

27. Dutta, A.; Mandal, A.; Kundu, A.; Malik, M.; Chaudhary, A.; Khan, M.R.; Shanmugam, V.; Rao, U.; Saha, S.; Patanjali, N.; et al. Deciphering the Behavioral Response of *Meloidogyne incognita* and *Fusarium oxysporum* toward Mustard Essential Oil. *Front. Plant Sci.* **2021**, *12*, 1791. [CrossRef]

28. Harun, R.; Yip, J.W.; Thiruvenkadam, S.; Ghani, W.A.; Cherrington, T.; Danquah, M.K. Algal biomass conversion to bioethanol–a step-by-step assessment. *Biotechnol. J.* **2014**, *9*, 73–86. [CrossRef]

29. Singh, D.P.; Prabha, R.; Verma, S.; Meena, K.K.; Yandigeri, M. Antioxidant properties and polyphenolic content in terrestrial cyanobacteria. *3 Biotech* **2017**, *7*, 134. [CrossRef]

30. Prasanna, R.; Bidyarani, N.; Babu, S.; Hossain, F.; Shivay, Y.S.; Nain, L. Cyanobacterial inoculation elicits plant defense response and enhanced Zn mobilization in maize hybrids. *Cogent Food Agric.* **2015**, *1*, 998507. [CrossRef]

31. Senousy, H.H.; El-Sheekh, M.M.; Saber, A.A.; Khairy, H.M.; Said, H.A.; Alhoqail, W.A.; Abu-Elsaoud, A.M. Biochemical Analyses of Ten Cyanobacterial and Microalgal Strains Isolated from Egyptian Habitats, and Screening for Their Potential against Some Selected Phytopathogenic Fungal Strains. *Agronomy* **2022**, *12*, 1340. [CrossRef]

32. Senousy, H.H.; Ellatif, S.A.; Ali, S. Assessment of the antioxidant and anticancer potential of different isolated strains of cyanobacteria and microalgae from soil and agriculture drain water. *Environ. Sci. Pollut. Res.* **2020**, *27*, 18463–18474. [CrossRef]

33. Rippka, R.; Deruelles, J.; Waterbury, J.B.; Herdman, M.; Stanier, R.Y. Generic Assignments, Strain Histories and Properties of Pure Cultures of Cyanobacteria. *Microbiology* **1979**, *111*, 1–61. [CrossRef]

34. Zarrouk, C. Contribution a l'etude d'une Cyanophycee. Influence de Divers Facteurs Physiques et Chimiques sur la Croissance et la Photosynthese de *Spirulina mixima*. Ph.D. Thesis, University of Paris, Paris, France, 1966.

35. Bello, A.S.; Ben-Hamadou, R.; Hamdi, H.; Saadaoui, I.; Ahmed, T. Application of Cyanobacteria (*Roholtiella* sp.) Liquid Extract for the Alleviation of Salt Stress in Bell Pepper (*Capsicum annuum* L.) Plants Grown in a Soilless System. *Plants* **2021**, *11*, 104. [CrossRef]

36. Jain, S.; Jain, A.; Jain, S.; Malviya, N.; Jain, V.; Kumar, D. Estimation of total phenolic, tannins, and flavonoid contents and antioxidant activity of *Cedrus deodara* heart wood extracts. *Egypt. Pharm. J.* **2015**, *14*, 10. [CrossRef]

37. Chang, C.-C.; Yang, M.H.; Wen, H.M.; Chern, J. Estimation of total flavonoid content in propolis by two complementary colorimetric methods. *J. Food Drug Anal.* **2002**, *10*, 3.

38. Farrag, A.; Attia, M.S.; Younis, A.; Abd Elaziz, A.M.A. Potential impacts of elicitors to improve tomato plant disease resistance. *Azhar. Bull. Sci.* **2017**, *9*, 311–321.

39. Abdelaziz, A.M.; Salem, S.S.; Khalil, A.M.A.; El-Wakil, D.A.; Fouda, H.M.; Hashem, A.H. Potential of biosynthesized zinc oxide nanoparticles to control Fusarium wilt disease in eggplant (*Solanum melongena*) and promote plant growth. *BioMetals* **2022**, *35*, 601–616. [CrossRef]

40. The Chlorophylls. 1966. Available online: https://www.elsevier.com/books/the-chlorophylls/vernon/978-1-4832-3289-8 (accessed on 12 March 2022).

41. Badawy, A.; Alotaibi, M.; Abdelaziz, A.; Osman, M.; Khalil, A.; Saleh, A.; Mohammed, A.; Hashem, A. Enhancement of Seawater Stress Tolerance in Barley by the Endophytic Fungus *Aspergillus ochraceus*. *Metabolites* **2021**, *11*, 428. [CrossRef]

42. Umbreit, W.W.; Burris, R.H.; Stauffer, J.F. Manometric Techniques: A Manual Describing Methods Applicable to the Study of Tissue Metabolism. Available online: https://www.car.chula.ac.th/display7.php?bib=b1247738 (accessed on 12 March 2022).

43. Lowry, O.H.; Rosebrough, N.J.; Farr, A.L.; Randall, R.J. Protein measurement with the Folin phenol reagent. *J. Biol. Chem.* **1951**, *193*, 265–275. [CrossRef]

44. Irigoyen, J.J.; Einerich, D.W.; Sanchez-Diaz, M. Water stress induced changes in concentrations of proline and total soluble sugars in nodulated alfalfa (*Medicago sativa*) plants. *Physiol. Plant.* **1992**, *84*, 55–60. [CrossRef]

45. Dai, G.; Andary, C.; Cosson-Mondolot, L.; Boubals, D. Polyphenols and Resistance of Grapevines to Downy Mildew. *Acta Hortic.* **1993**, *381*, 763–766. [CrossRef]

46. Mukherjee, S.P.; Choudhuri, M.A. Implications of water stress-induced changes in the levels of endogenous ascorbic acid and hydrogen peroxide in *Vigna* seedlings. *Physiol. Plant.* **1983**, *58*, 166–170. [CrossRef]

47. Knegt, E.; Bruinsma, J. A rapid, sensitive and accurate determination of indolyl-3-acetic acid. *Phytochemistry* **1973**, *12*, 753–756. [CrossRef]

48. Snedecor, G.W.; Cochran, W.G. *Statistical Methods*, 7th ed.; Iowa State University USA: Ames, IA, USA, 1980; pp. 80–86.

49. Righini, H.; Roberti, R. Algae and Cyanobacteria as Biocontrol Agents of Fungal Plant Pathogens. In *Plant Microbe Interface*; Springer: Cham, Switzerland, 2019; pp. 219–238.

50. Usharani, G.; Srinivasan, G.; Sivasakthi, S.; Saranraj, P. Antimicrobial activity of *Spirulina platensis* solvent extracts against pathogenic bacteria and fungi. *Adv. Biol. Res.* **2015**, *9*, 292–298.

51. Kulik, M.M. The potential for using cyanobacteria (blue-green algae) and algae in the biological control of plant pathogenic bacteria and fungi. *Eur. J. Plant Pathol.* **1995**, *101*, 585–599. [CrossRef]

52. Bai, G.; Shaner, G. Management and resistance in wheat and barley to Fusarium head blight. *Annu. Rev. Phytopathol.* **2004**, *42*, 135–161. [CrossRef]

53. Gunupuru, L.R.; Patel, J.S.; Sumarah, M.W.; Renaud, J.B.; Mantin, E.G.; Prithiviraj, B. A plant biostimulant made from the marine brown algae *Ascophyllum nodosum* and chitosan reduce Fusarium head blight and mycotoxin contamination in wheat. *PLoS ONE* **2019**, *14*, e0220562. [CrossRef]

54. Ahmed, A.F.; Attia, M.S.; Faramawy, F.; Salaheldin, M.M. Impact of PGPM in Pepper Immune-Response to Fusarium Wilt Disease. *Saudi J. Pathol. Microbiol. (SJPM)* **2018**, *3*, 5–15.

55. Kaul, S.; Choudhary, M.; Gupta, S.; Dhar, M.K. Engineering Host Microbiome for Crop Improvement and Sustainable Agriculture. *Front. Microbiol.* **2021**, *12*, 635917. [CrossRef]

56. Tan, C.-Y.; Dodd, I.C.; Chen, J.E.; Phang, S.-M.; Chin, C.F.; Yow, Y.-Y.; Ratnayeke, S. Regulation of algal and cyanobacterial auxin production, physiology, and application in agriculture: An overview. *J. Appl. Phycol.* **2021**, *33*, 2995–3023. [CrossRef]

57. Wang, L.Y.; Liu, J.L.; Wang, W.X.; Sun, Y. Exogenous melatonin improves growth and photosynthetic capacity of cucumber under salinity-induced stress. *Photosynthetica* **2016**, *54*, 19–27. [CrossRef]

58. He, Y.; Shi, Y.; Zhang, X.; Xu, X.; Wang, H.; Li, L.; Zhang, Z.; Shang, H.; Wang, Z.; Wu, J.-L. The OsABCI7 Transporter Interacts with OsHCF222 to Stabilize the Thylakoid Membrane in Rice. *Plant Physiol.* **2020**, *184*, 283–299. [CrossRef]

59. Grzesik, M.; Romanowska-Duda, Z.; Kalaji, H. Effectiveness of cyanobacteria and green algae in enhancing the photosynthetic performance and growth of willow (*Salix viminalis* L.) plants under limited synthetic fertilizers application. *Photosynthetica* **2017**, *55*, 510–521. [CrossRef]

60. Kunui, K.; Singh, S.S. Protective role of antioxidative enzymes and antioxidants against iron-induced oxidative stress in the cyanobacterium Anabaena sphaerica isolated from iron rich paddy field of Chhattisgarh, India. *Indian J. Exp. Biol. (IJEB)* **2020**, *58*.

61. Foyer, C.H.; Shigeoka, S. Understanding Oxidative Stress and Antioxidant Functions to Enhance Photosynthesis. *Plant Physiol.* **2010**, *155*, 93–100. [CrossRef] [PubMed]

62. Attia, M.S.; Younis, A.M.; Ahmed, A.F.; Elaziz, A.M.A. Comprehensive Management for Wilt Disease Caused by Fusarium Oxysporum in Tomato Plant. *Int. J. Innov. Sci. Eng. Technol.* **2016**, *4*, 2348–7968.

63. Singh, S. A review on possible elicitor molecules of cyanobacteria: Their role in improving plant growth and providing tolerance against biotic or abiotic stress. *J. Appl. Microbiol.* **2014**, *117*, 1221–1244. [CrossRef] [PubMed]

64. Attia, M.S.; El-Naggar, H.A.; Abdel-Daim, M.M.; El-Sayyad, G.S. The potential impact of *Octopus cyanea* extracts to improve eggplant resistance against Fusarium-wilt disease: In vivo and in vitro studies. *Environ. Sci. Pollut. Res.* **2021**, *28*, 35854–35869. [CrossRef] [PubMed]

65. Anaraki, Z.E.; Tafreshi, S.A.H.; Shariati, M. Transient silencing of heat shock proteins showed remarkable roles for HSP70 during adaptation to stress in plants. *Environ. Exp. Bot.* **2018**, *155*, 142–157. [CrossRef]

66. Singh, D.P.; Prabha, R.; Yandigeri, M.S.; Arora, D.K. Cyanobacteria-mediated phenylpropanoids and phytohormones in rice (*Oryza sativa*) enhance plant growth and stress tolerance. *Antonie Leeuwenhoek* **2011**, *100*, 557–568. [CrossRef]

67. Nowruzi, B.; Bouaïcha, N.; Metcalf, J.S.; Porzani, S.J.; Konur, O. Plant-cyanobacteria interactions: Beneficial and harmful effects of cyanobacterial bioactive compounds on soil-plant systems and subsequent risk to animal and human health. *Phytochemistry* **2021**, *192*, 112959. [CrossRef] [PubMed]

68. Hashem, A.; Tabassum, B.; Fathi Abd Allah, E. *Bacillus subtilis*: A plant-growth promoting rhizobacterium that also impacts biotic stress. *Saudi J. Biol. Sci.* **2019**, *26*, 1291–1297. [CrossRef] [PubMed]

69. Hashem, A.; Abdelaziz, A.; Askar, A.; Fouda, H.; Khalil, A.; Abd-Elsalam, K.; Khaleil, M. *Bacillus megaterium*-Mediated Synthesis of Selenium Nanoparticles and Their Antifungal Activity against *Rhizoctonia solani* in Faba Bean Plants. *J. Fungi* **2021**, *7*, 195. [CrossRef] [PubMed]

70. Sharma, P.; Jha, A.B.; Dubey, R.S.; Pessarakli, M. Reactive Oxygen Species, Oxidative Damage, and Antioxidative Defense Mechanism in Plants under Stressful Conditions. *J. Bot.* **2012**, *2012*, 217037. [CrossRef]

71. Jayamohan, N.S.; Patil, S.V.; Kumudini, B.S. Reactive oxygen species (ROS) and antioxidative enzyme status in *Solanum lycopersicum* on priming with fluorescent *Pseudomonas* spp. against *Fusarium oxysporum*. *Biologia* **2018**, *73*, 1073–1082. [CrossRef]

72. Bari, R.; Jones, J.D.G. Role of plant hormones in plant defence responses. *Plant Mol. Biol.* **2008**, *69*, 473–488. [CrossRef]

73. Arif, Y.; Sami, F.; Siddiqui, H.; Bajguz, A.; Hayat, S. Salicylic acid in relation to other phytohormones in plant: A study towards physiology and signal transduction under challenging environment. *Environ. Exp. Bot.* **2020**, *175*, 104040. [CrossRef]

74. Ashraf, M.; Akram, N.A.; Arteca, R.N.; Foolad, M.R. The Physiological, Biochemical and Molecular Roles of Brassinosteroids and Salicylic Acid in Plant Processes and Salt Tolerance. *Crit. Rev. Plant Sci.* **2010**, *29*, 162–190. [CrossRef]

75. Bayona-Morcillo, P.J.; Plaza, B.M.; Gómez-Serrano, C.; Rojas, E.; Jiménez-Becker, S. Effect of the foliar application of cyanobacterial hydrolysate (*Arthrospira platensis*) on the growth of *Petunia x hybrida* under salinity conditions. *J. Appl. Phycol.* **2020**, *32*, 4003–4011. [CrossRef]

76. Attia, M.S.; El-Sayyad, G.S.; Abd Elkodous, M.; El-Batal, A.I. The effective antagonistic potential of plant growth-promoting rhizobacteria against *Alternaria solani*-causing early blight disease in tomato plant. *Sci. Hortic.* **2020**, *266*, 109289. [CrossRef]

77. Kumar, G.; Teli, B.; Mukherjee, A.; Bajpai, R.; Sarma, B.K. Secondary Metabolites from Cyanobacteria: A Potential Source for Plant Growth Promotion and Disease Management. In *Secondary Metabolites of Plant Growth Promoting Rhizomicroorganisms*; Springer: Singapore, 2019; pp. 239–252.

78. Mauch-Mani, B.; Mauch, F. The role of abscisic acid in plant–pathogen interactions. *Curr. Opin. Plant Biol.* **2005**, *8*, 409–414. [CrossRef] [PubMed]

79. Schmidt, K.; Pflugmacher, M.; Klages, S.; Mäser, A.; Mock, A.; Stahl, D.J. Accumulation of the hormone abscisic acid (ABA) at the infection site of the fungus *Cercospora beticola* supports the role of ABA as a repressor of plant defence in sugar beet. *Mol. Plant Pathol.* **2008**, *9*, 661–673. [CrossRef] [PubMed]

MDPI

St. Alban-Anlage 66

4052 Basel

Switzerland

www.mdpi.com

Plants Editorial Office

E-mail: plants@mdpi.com

www.mdpi.com/journal/plants

9 783036 586908